AF342064

High Voltage Engineering in Power Systems

Khalil Denno
New Jersey Institute of Technology
Newark, New Jersey

CRC Press
Boca Raton Ann Arbor London Tokyo

Library of Congress Cataloging-in-Publication Data

Denno, K.
   High voltage engineering in power systems / author, Khalil Denno.
       p.   cm.
   Includes bibliographical references and index.
   ISBN 0-8493-4289-9
   1. Electric power distribution—High tension.   2. Electric power systems—
Protection.   3. Transients (Electricity).   I. Title.
TK3144.D38 1992
621.319′13—dc20
DNLM/DLC
                                                                                91-24037
                                                                                CIP

International Standard Book Number 0-8493-4289-9

Library of Congress Card Number 91-24037
Printed in the United States   2 3 4 5 6 7 8 9 0
Printed on acid-free paper

PREFACE

This book is the first to address in a totally unique scope the field of high voltage engineering, presenting the core of the subject matter, applications, and effects.

Subject matter presentation commences with the two major sources of high voltage surges, namely switching (accidental and intentional) and atmospheric breakdown (lightning). Next the process of field-intensified ionization will be presented and then the phenomena of inducing and induced voltages will be treated analytically in terms of partial differential equations modeling and the application of the method of magnetic moments. Comprehensive analyses for the adverse effects due to the propagation of voltage and power surges on transmission lines, transformer, and insulating systems have been presented including interaction with conducting fluids, charged clouds, corona and skin effects, as well as the concept of energy storage and extraction from lightning and the negative effects of electromagnetics on health and environments. A special chapter has been devoted to the field of protection from high voltage surges.

This book supplements the comprehensive coverage of high voltage engineering with solved examples followed by a set of problems. It blends the areas of physics, engineering analysis and applications of high voltage engineering into a unified package suitable to the reader seeking physical and engineering understanding of this field.

To
Badia, Karem, Zayd and Athra

THE AUTHOR

Khalil Denno, Ph.D., is Distinguished Professor of Electrical Engineering at Newark College of Engineering, New Jersey Institute of Technology in Newark, N.J.

Dr. Denno received his M.E.E. degree in 1959 from Rensselaer Polytechnic Institute of Troy, N.Y. and his Ph.D. degree in 1967 from Iowa State University in Ames, Iowa. He joined Newark College of Engineering of New Jersey Institute of Technology as Associate Professor in 1969, was promoted to the rank of Professor in 1973, and then became Distinguished Professor in September 1987.

Dr. Denno is a Fellow of the Institution of Electrical Engineers (IEE), a senior member of the Institute of Electrical and Electronics Engineers (IEEE), and a member of the American Nuclear Society, the American Society of Engineering Education, and Sigma Xi Society. He had the honor of winning the 1982 Harlan Perlis Award for Excellence in Research, and has numerous listings in national and international honor societies and biographies.

Dr. Denno specializes in the fields of energy conversion, renewable energy sources, conventional power system, high voltage engineering, and magnetohydrodynamics. He is a Licensed Professional Engineer in New Jersey and Chartered Engineer in the United Kingdom. He is the author of *Power System Design and Applications for Alternative Energy Sources* (Prentice Hall, Inc.) and *Engineering Economics of Alternative Energy Sources* (CRC Press). Dr. Denno has published 120 research papers in leading national and international journals covering the fields of magnetohydrodynamic power generation, fusion energy, lightning phenomenon, particle accelerators, conventional energy systems, characterization of cold plasma and various modes of renewable energy sources.

ACKNOWLEDGMENTS

With special thanks to Ms. Chiung-Wen Hueng for her help in completing the word processing of the manuscript. Also my deep thanks and gratitude to Ms. Suzanne Lassandro, Coordinating Editor at CRC Press for her great professional dedication, care and promptness throughout the process of production of this book.

The author would like to thank the following publications for release of data and diagrams for use and publication in this book.

Communication Engineering, 2nd ed., by W.L. Everitt. McGraw-Hill (1937). [Section 1.IV]

Electrical Shock Waves in Power Systems, by R. Rudenberg. Harvard University Press (1968). [Sections 1.V, 1.VII, 1.VIII, 1.XI, Figures 1.3–1.8, 1.10–1.12, 6.1, Eq. (1.135), (6.2), and (6.24)]

EPRI Journal, p. 14, July/Aug. 1984. [Section 13.I, Figures 13.1, 13.2]

EPRI Journal, p. 18, 1985. [Section 11.I]

EPRI Journal, p. 4, Oct./Nov. 1987. [Section 13.II, Figures 13.3, 13.4]

EPRI Journal, p. 4, Jan./Feb. 1990. [Section 13.III, Figures 13.5, 13.6]

Gaseous Conductors, by J.D. Cobine. Dover Publications (1941, 1958). [Sections 1.X.B, 4.I, 5.IV.B, 11.IV, Figures 1.16, 1.17, 4.1, Table 4.1, Eq. (4.73), (4.74), (8.55), (8.70), (10.24)]

IEE Proc., 127A(7), 447, 1980. [Section 5.V, Figures 5.11–5.14]

IEE Proc., 130A(3), 134, 1983. [Section 5.III, Figures 5.1–5.9]

IEE Proc., 130A(5), 264, 1983. [Sections 8.II, 8.III, Figures 8.1–8.10]

IEE Proc., 131A(2), 118, 1984. [Sections 10.IV, 10.VI.A, Figures 10.13, 10.14, Tables 10.2–10.4]

IEE Proc., 133A(8), 534, 1986. [Section 6.VII, Figures 6.6–6.10]

IEE Proc., 133A(8), 552, 1986. [Sections 7.I–7.III, Figures 7.1–7.12, Tables 7.1–7.7]

IEE Proc., 133A(8), 562, 1986. [Section 8.V, Figures 8.17, 8.18]

IEE Proc., 133A(8), 569, 1986. [Section 8.IV, Figures 8.11–8.14]

IEE Proc., 134A(9), 721, 1987. [Sections 2.X–2.XII, Figures 2.5–2.9, Tables 2.1 and 2.2]

IEE Proc., 135A(1), 22, 1988. [Sections 10.II.B, 10.II.C, 10.III, Figures 10.5–10.12, Table 10.1]

IEE Proc., 136A(2), 66, 1989. [Section 10.I, Figures 10.1–10.4]

IEEE Digest of International Electric, Electronics Conference and Exposition, Toronto, October 1983, pp. 96–98. [Section 1.X.A, Table 1.4]

IEEE Trans. Magnetics, 20(5), 1953, 1984. [Sections 3.II–3.VII, 3.X, Figure 3.1]

IEEE International Symposium on Electrical Insulation, Montreal, 1984, pp. 226–228. [Sections 5.I.B, 5.I.C, 5.I.E, Eq. (5.1) and (5.2)]

IEEE Canadian Communications and EHV Conference Record, Cat. no. 72, CHO 698-1-REG. 7, 1972, pp. 155–156. [Section 11.II, Figures 11.1–11.4]

J. Electrostat., 13, 55, 1982. [Sections 2.II–2.IX, Figures 2.1, 2.2]

J. Electrostat., 15, 43, 1984. [Sections 2.VIII, 2.IX, 4.VI, Figures 2.3, 2.4]

Proceedings of the 3rd Conference on Electrostatics, Technical University Press, University of Wroclaw, Poland, September 1985. [Sections 5.II, 6.III–6.V, Figures 6.3–6.5, Table 6.1]

Proceedings of the IEEE Canadian Communications and Power Conference, Montreal, October 1980, paper by K. Denno. [Subsections of 5.IV.A, Figure 5.10]

Proceedings of the IASTED International Symposium: POWER ENGINEERING '84, New Orleans, 1984, pp. 10–13. [Sections 6.I.A-C]

Proceedings of the International Conference on Mathematical Modelling, Berkely, CA, 1985, paper by K. Denno. [Sections 6.II.A-D, Figure 6.2]

TABLE OF CONTENTS

Chapter 1

SOURCES OF SURGE VOLTAGES

1.I INTRODUCTION

Main sources of high voltage surges are generally confined to intentional and unintentional switching operations in power systems as well as those induced by lightning phenomenon. Intentional switching may create certain sparks or discharges that may require a slow timing process to disappear due to effects caused by electrode heating and impurities accumulation across the spark gap. Unintentional switching could be caused by ground faults, sudden conductor breaks, accidental short-circuits, lightning strokes, and erroneous operation of switching devices. Current and voltage surges are usually of high amplitude and short in time duration, and of different span in frequency spectrum with a broad band in harmonies and as special distorted wave forms.

In this chapter, comprehensive presentation will be made regarding spark discharges due to switching, propagation at high voltage surges, propagation of discharge front with effects of power systems harmonizations, aspect of system oscillations, reignition, and modes of various spark arresters as well as the spectrum of natural frequencies associated with three phase systems under oscillations.

Discharges generated by lightning strokes which are much more powerful than those due to switching will be presented, including concepts of conductive and convective surges under conditions of varying pressure and temperature.

1.II PROPAGATION OF TRAVELING WAVES

Voltage or current surges initiated by switching or lightning will propagate at a velocity close to that of the velocity of light in overhead systems and at about half of that through underground installations. Numerous situations involving short distances of less than 50 miles, and instant computations for the build-up of voltage stresses on the basis of lossless representation of transmission lines could be secured with regard to short-circuit or open-circuit terminations as close equivalents for actual loading connections.

1

However, regarding long distance propagation of voltage or current surges, it is essential to carry out analytical calculations to obtain reliable information about disturbances inflicted by electromagnetic surges.

Therefore, the author felt that a systematic presentation for the theory of wave propagation is helpful to the reader to follow in the analysis for traveling voltage surges on overhead lines and cables.

Let α, μ represents the electromagnetic flux and electrostatic flux associated with the current i and voltage e waves.

$$d\alpha_m = L_i \, dx \tag{1.1}$$

and

$$d\mu_e = eC \, dx \tag{1.2}$$

where L = inductance in Henry/unit length, and C = capacitance in Farad/unit length.

Voltage drops in differential dx are expressed by

$$-\frac{\partial}{\partial t}(d\mu_m) \text{ and } -iR \, dx$$

Hence in the forward direction of X:

$$-de = -\frac{\delta e}{\delta x} \, dx$$

$$= iR \, dx + \frac{\delta}{\delta t}(d\alpha_m)$$

or

$$-de = \left(R + L\frac{\delta}{\delta t}\right)i \, dx \tag{1.3}$$

The charging current of dx is $\dfrac{\delta}{\delta t}(d\mu_e)$ in addition to the leakage current $e \, G \, dx$, so that the total change of electric current in the positive X direction becomes

$$-di = \frac{\delta i}{\delta x} \, dx$$

$$= eG \, dx + \frac{\delta}{\delta t}(d\mu_e)$$

$$= \left(G + C\frac{\delta}{\delta t}\right)e \, dx \tag{1.4}$$

Now, omitting dx from Equations (1.3) and (1.4) results in

$$-\frac{\delta e}{\delta x} = \left(R + L\frac{\delta}{\delta t}\right)i$$

$$= Z(q)i \tag{1.5}$$

$$-\frac{\delta i}{\delta x} = \left(G + C\frac{\delta}{\delta t}\right)e$$

$$= Y(q)e \tag{1.6}$$

where $q = \dfrac{\delta}{\delta t}$.

Then differentiating Equation (1.5) with respect to x and substituting into Equation (1.6),

$$\frac{\delta^2 e}{\delta x^2} = -Z(q)\left[\frac{\delta i}{\delta x}\right]$$

$$= Y(q)Z(q)e$$

$$= [RG + (RC + GL)q + LCq^2]e \tag{1.7}$$

Also, differentiating Equation (1.6) with respect to x and substituting into Equation (1.5)

$$\frac{\delta^2 i}{\delta x^2} = -Y(q)\frac{\delta e}{\delta x}$$

$$= Z(q)Y(q)i$$

$$= [RG + (RC + GL)q + LCq^2]i \tag{1.8}$$

Solving Equations (1.7) and (1.8) as ordinary differential equations in x, we obtain

$$e = e^{x\sqrt{ZY}}\, F_1(t) + e^{-x\sqrt{ZY}}\, F_2(t) \tag{1.9}$$

$$i = -Y\int e\, dx$$

$$= -\sqrt{\frac{Y}{Z}}\,[e^{x\sqrt{ZY}}\, F_1(t) - e^{-x\sqrt{ZY}}\, F_2(t)] \tag{1.10}$$

where

$$\sqrt{ZY} = \sqrt{LC\frac{RG}{LC} + \frac{R}{L} + \frac{G}{C}q + q^2} \tag{1.11}$$

Now, turning to express the energy content of a propagating pulse as follows:

$$W = W_e + W_m$$

where W_e and W_m represent electrostatic and magnetic energy content, respectively, in Joules.

$$\begin{aligned} W &= \frac{C}{2}\int e^2\, dx + \frac{L}{2}\int i^2\, dx \\ &= \frac{C}{2}\int e^2\, dx + \frac{L}{2}\int \frac{C}{L}\, e^2\, dx \\ &= C\int e^2\, dx = L\int i^2\, dx = \sqrt{LC}\int ei\, dx \end{aligned} \tag{1.12}$$

or

$$\begin{aligned} W &= \int ei\, dx = \sqrt{\frac{C}{L}}\int e^2\, dt \\ &= \sqrt{\frac{L}{C}}\int i^2\, dt \end{aligned} \tag{1.13}$$

In Equations (1.12) and (1.13), the integration process is intended to be carried out to include the scope of the entire wavelength.

Those equations show that the total energy content of a pair of traveling waves is divided equally between the potential and current waves.

$$W_e = \frac{C}{2}\int e^2\, dx = \frac{W}{2} \tag{1.14}$$

and

$$W_i = \frac{L}{2}\int i^2\, dx = \frac{W}{2} \tag{1.15}$$

However, when a pair of traveling waves reaches a transition point, or when waves traveling in opposite directions pass through each other, the energy balance is upset, and hence more energy will reside in one field than in the other.

1.III PROPAGATION CONSTANTS IN CABLES

The inductance and capacitance of cables are almost negligible, therefore,

$$Z(q) = R \text{ and } Y(q) = Cq$$

where again $q = jw$ for sinusoidal steady state situation.

The constant of propagation becomes

$$\gamma = \sqrt{ZY} = \sqrt{jR\omega C}$$
$$= \sqrt{\omega CR}\ \underline{|45°} \tag{1.16}$$

But

$$= \alpha + j\beta$$

$$\alpha = \sqrt{\frac{\omega CR}{2}} \tag{1.17}$$

$$\beta = \sqrt{\frac{\omega CR}{2}} \tag{1.18}$$

Also, it is known that

$$V = \lambda f \text{ and } \lambda = 2\pi/\beta$$
$$= \omega/\beta = \sqrt{\frac{2\omega}{CR}} \tag{1.19}$$

$$V = \sqrt{Z/Y}$$

$$Z_o = \sqrt{\frac{R}{j\omega C}}$$

$$= \sqrt{\frac{R}{\omega C}}\ \underline{|-45°} \tag{1.20}$$

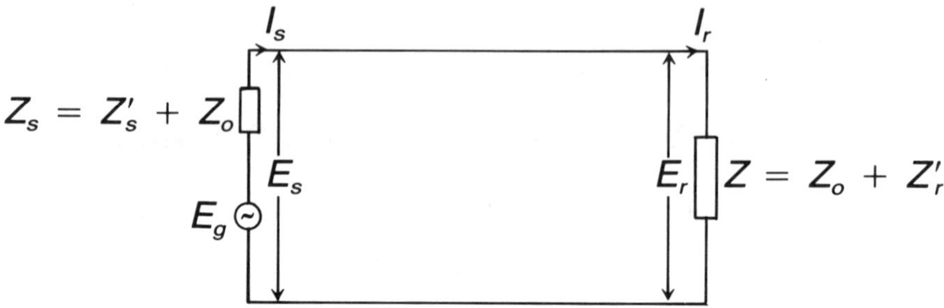

FIGURE 1.1.   Line with distributed constants.

FIGURE 1.2.   Application of compensation theorem.

1.IV EXACT SOLUTIONS FOR A LINE WITH TERMINATION $\neq Z_o$

Consider the load impedance Z_r made up of two parts, one of which is the surge impedance Z_o and the other as Z_r which may not be physically realizable. Also, the sending end Z_s can be replaced with Z_o and Z_i which may not be physically realizable.

$$Z_r = Z_o + Z_r'$$

$$Z_s = Z_o + Z_s' \tag{1.21}$$

Then, using the compensation theorem, replace any impedance by a generator of zero internal impedance and voltage equal to the total voltage drop across that impedance.

This is shown from Figures 1.1 and 1.2.

Consider the voltage at any point on the line $+ E$, where

$$\overline{E} = \overline{E}' + \overline{E}''$$

$\overline{E}'$ = voltage wave traveling toward the $+$ direction

$\overline{E}''$ = voltage wave traveling toward the $-$ direction

$\overline{E}'$ is due to $\overline{E}_g + \overline{E}_a$

$\overline{E}_p$ is due to $\overline{E}_b$ (1.22)

$$\overline{E}' = Ae^{-\gamma x} \tag{1.23}$$

$$\overline{E}'' = Be^{\gamma x} \tag{1.24}$$

$$\overline{E} = Ae^{-\gamma x} + Be^{\gamma x} \tag{1.25}$$

A could be $f_1(t)$

B could be $f_2(t)$

Also

$$I = Ce^{-\gamma x} + De^{\gamma x} \tag{1.26}$$

as the current at any point on the line.

The general solution for Figure 1.1 in terms of line, generator and the load constant can now be obtained: Let

I_s' = the component of the sending end current due to E_g and E_a
I_r' = the component of the receiving end current due to E_g and E_a
I_s'' = the component of the sending end current due to E_b (reflected wave)
I_s' and I_r' are the initial waves traveling from left to right
I_r'' and I_s'' are due to the reflected wave traveling from right to left

Then

$$\overline{I}_s = \overline{I}_s' + \overline{I}_s'' \tag{1.27}$$

$$\overline{I}_r = \overline{I}_r' + \overline{I}_r''$$

$$\overline{I}_s' = \frac{\overline{E}_g + \overline{E}_a}{2Z_o} = \frac{\overline{E}_g - \overline{I}_s Z_s'}{2Z_o} \tag{1.28}$$

$$= \frac{E_g - (Z_s - Z_o)I_s}{2Z_o}$$

$$I'_r = I'_s e^{-\gamma z} = \frac{E_g - (Z_s - Z_o)I_s}{2Z_o} e^{-\gamma x}$$

$$I''_r = \frac{E_b}{2Z_o} = -\frac{I_r Z'_r}{2Z_o} = -I\frac{(Z_r - Z_o)}{r2Z_o}$$

$$I''_s = I''_r e^{-\gamma x} \tag{1.29}$$

To determine the ratio of the currents in the initial and reflected waves at the point of reflection we can proceed to write

$$\frac{I''_r}{I'_r} = \frac{I''_r}{I'_r + I''_r} = \frac{-Z_r}{2Z_o} = -\frac{Z_o - Z_r}{2Z_o} \tag{1.30}$$

Then from the rule

$$\frac{a}{b} = \frac{b}{c}$$

and

$$\frac{a}{b - a} = \frac{c}{d - c}$$

now we can write

$$\frac{I''_r}{I'_r} = \frac{Z_r - Z_o}{2Z_o - (Z_o - Z_r)}$$

$$= \frac{Z_o - Z_r}{Z_o + Z_r} \tag{1.31}$$

And for $Z_o = Z_r$, $I''_r = 0$ no reflection.

In the same way that the current is analyzed by the principle of superposition, it is also possible to divide the voltage at any point into two components. (1) Voltage in the initial wave set up by E_g and E_a; (2) voltage in the reflected wave set up by E_b.

$$\overline{E}_s = \overline{E}'_s + \overline{E}''_s \tag{1.32}$$

$$\overline{E}_r = \overline{E}'_r + \overline{E}''_r \tag{1.33}$$

And

$$\frac{E'_s}{I'_s} = \frac{E'_r}{I'_r}$$

$$= -\frac{E''_r}{I''_r} = -\frac{E''_s}{I''_s} = Z_o \tag{1.34}$$

$$\frac{E''_r}{E'_r} = -\frac{I''_r}{I'_r}$$

$$= \frac{Z_r - Z_o}{Z_r + Z_o}$$

$$= \rho \text{ the reflection factor} \tag{1.35}$$

Again

$$I'_r = I''_s e^{-\gamma l} \qquad I'_s = I''_r e^{-\gamma l}$$

$$E_r = E_s{}^{-\gamma l} \qquad E''_s = E''_r e^{-\gamma l} \tag{1.36}$$

For solving any problem, the following steps can be followed when E_g, $Z_s - Z_r$ and the scheme of distribution of the line constants are specified.

1. Determine Z_0, α and β for the line.
2. Assume an arbitrary voltage for E'_s, say 1 volt; let the subscript 1 refer to all values obtained for this assumption.
3. Find E'_r from the relation $E'_{r1} = E'_{s1} e^{-\gamma}$.
4. Find E''_{r1} from Equation (1.35).
5. Find E''_{s1} from $E''_{s1} = E''_{r1} e^{-\gamma}{}_1$.
6. Find I'_{s1}, I'_{r1}, I''_{s1} and I''_{r1} from Equation (1.34).
7. Find E_{s1}, E_{r1}, I_{s1} and I_{r1} by adding their components according to Equations (1.27), (1.28), (1.32) and (1.33).
8. Determine the input impedance Z_{in} from the ratio of E_{s1}/I_{s1}. When Z_{in} is known, the actual sending end voltage can be found from:

$$E_s = \frac{E_g Z_{in}}{Z_s + Z_{in}} \tag{1.37}$$

9. Multiply the values of E_{s1}, E_{r1}, I_{s1} and I_{r1} obtained in #7 by the ratio of E_s/E_{s1} to obtain the actual values of current and voltage.

Or by substituting Equation (1.29) into Equations (1.28) and (1.27) the following relationships can be obtained:

$$I_s = I_s' + I_s''$$

$$= \frac{E_g - (Z_s - Z_o)I_s - I_r(Z_r - Z_o)e^{-\gamma l}}{2Z_o} \qquad (1.38)$$

$$I_r = I_r' + I_r''$$

$$= \frac{[E_g - (Z_s - Z_o)I_s]e^{-\gamma l} - I_r(Z_r - Z_o)}{2Z_o} \qquad (1.39)$$

Then solving Equations (1.38) and (1.39) for I_s and I_r:

$$I_r = \frac{2E_g Z_o}{(Z_o + Z_r)(Z_s + Z_o)e^{\gamma l} + (Z_o - Z_r)(Z_s - Z_o)e^{-\gamma l}} \qquad (1.40)$$

$$I_s = \frac{E_g - (Z_o - Z_r)e^{\gamma l} + (Z_o - Z_r)e^{-\gamma l}}{(Z_o + Z_r)(Z_s + Z_o)e^{\gamma l} + (Z_o - Z_r)(Z_s - Z_o)e^{-\gamma l}} \qquad (1.41)$$

The input impedance Z_{in} at any point x could be computed from the ratio of E_{s1}/I_{s1}. However, it is written below in completed form for convenience:

$$Z_{in} = \frac{E_r \cos h\gamma x + Z_o I_r \sin h\gamma x}{I_r \cos h\gamma x + \dfrac{E_r}{Z_o} \sin h\gamma x} \qquad (1.42)$$

Also, it is useful to write down the general form for the solution of $E(x)$ and $I(x)$ at any point x along the transmission line:

$$E(x) = E_r \cos h\gamma x + Z_o I_r \sin h\gamma x \qquad (1.43)$$

$$I(x) = I_r \cos h\gamma x + \frac{E_r}{Z_o} \sin h\gamma x \qquad (1.44)$$

1.V SWITCHING SURGES

A. SHORT-CIRCUIT TO GROUND

Figure 1.3 shows systematic sequence for the sudden initiation of a short-circuit to ground at point S. At $t < 0$,

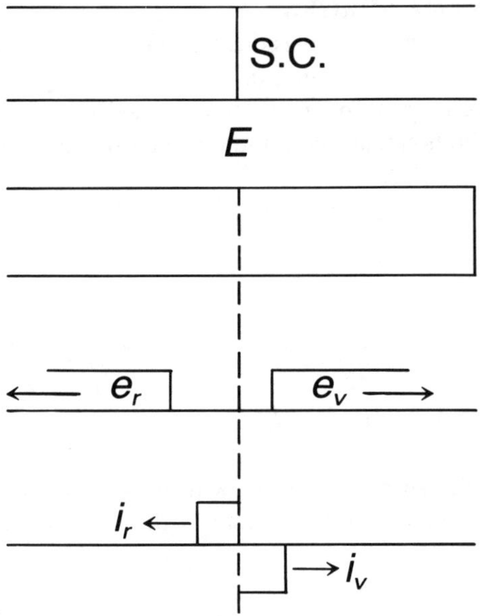

FIGURE 1.3.   Short-circuit to ground.

$$I = 0 \text{ and} \tag{1.45}$$

voltage $e = E$ volts

At $t > 10$

$$e_r = e_f = -E \tag{1.46}$$

or

$$-\frac{e_r}{Z} = \frac{e_v}{Z} = \frac{E}{Z} \tag{1.47}$$

and

$$i_r = -i_v = \frac{E}{Z} \tag{1.48}$$

where Z is the line surge impedance in ohms. The fault current i_f is expressed by:

$$i_f = 2|i_v| = 2|i_r| = 2\frac{E}{Z} \tag{1.49}$$

B. OSCILLATION OF SURGES

Surge developed at a place of sudden short-circuit will propagate instantly over a short distance to an open-end of the line whereby the voltage pulse will be reflected with a change in polarity accompanied by a reduction of current to zero according to the boundary condition at the open-end:

$$e = e_r + e_v, \quad i = 0 \tag{1.50}$$

implying,

$$e = 2e_v \tag{1.51}$$

$$i_r = -i_v \tag{1.52}$$

Consequently, the surges i_r and e_r will arrive back at the place of initial short-circuit, i.e., at point s, at which there will be another reflection subject to the boundary condition,

$$e = 0$$
$$i = I_{sc} \tag{1.53}$$

implying that

$$e_r = - e_v \tag{1.54}$$

and

$$i_r = - \frac{e_r}{Z} = \frac{e_v}{Z} = i_v \tag{1.55}$$

$$I_{sc} = 2i_v = 2i_r \tag{1.56}$$

This sequential process of reflections at the open-end and at the point of short-circuit will be repeated with a frequency expressed by

$$\pi = \frac{\pi V}{2a} \tag{1.57}$$

where a is the distance between location of short-circuit and open-end of the line and τ is the time taken by the surge to travel a distance of $2a$.

Consequently, an oscillatory wave form for the fault current i_j is shown in Figure 1.4. Of course, the author has to indicate that this analysis is applicable only to a lossless line with the distance (a) considered small, i.e., less than 50 miles.

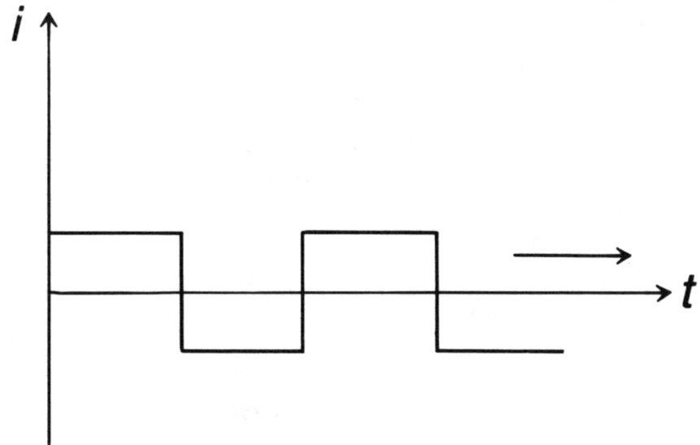

FIGURE 1.4. Oscillatory waveform.

Indeed we check the state of oscillation of i_j by returning to Equation (1.41) and substituting the following:

$$E_g = -E$$

$$Z_r = \infty$$

$$Z_s = 0$$

$$\gamma = j\beta = j\omega \tag{1.58}$$

From Equation (1.41),

$$I_s = I_f = -E\,\frac{(Z_o - Z_r)e^{\gamma a} + (Z_o - Z_r)e^{-\gamma a}}{(Z_o + Z_r)(Z_o)e^{\gamma a} + (Z_o - Z_r)(-Z_o)e^{-\gamma a}}$$

$$= -E\,\frac{\left(1 + \dfrac{Z_o}{Z_r}\right)e^{\gamma a} + \left(\dfrac{Z_o}{Z_r} - 1\right)e^{-\gamma a}}{\left(\dfrac{Z_o}{Z_r} + 1\right)Z_o\,e^{\gamma a} - \left(\dfrac{Z_o}{Z_r} - 1\right)(Z_o)e^{-\gamma a}}$$

$$I_s = I_f\Big|_{Z_r \to \infty} = -E\,\frac{e^{\gamma a} - e^{-\gamma a}}{Z_o\,e^{\gamma a} + Z_o\,e^{-\gamma a}}$$

$$= \frac{-E}{Z_o}\,\frac{e^{\gamma a} - e^{-\gamma a}}{e^{\gamma a} + e^{-\gamma a}}$$

$$= -\frac{E}{Z_o}\,\tan k\gamma a \tag{1.59}$$

For a lossless line,

$$\gamma = j\beta = j\omega$$

$$I_f = -\frac{E}{Z_o} \tan kj \frac{\omega}{V} a$$

$$= -j\frac{E}{Z_o} \tan \frac{\omega}{V} a$$

$$= -j\frac{E}{Z_o} \tan \omega t \tag{1.60}$$

where V is the velocity of propagation. For a lossless line $V = 1/\sqrt{LC}$ m/s.

Discharge through a resistance. Let the discharge take place at a termination where $Z_r = R$. Therefore,

$$E_r = I_r R \tag{1.61}$$

Voltage is felt at a line section far from the place of switching, i.e., $x \rightarrow \infty$

$$E(-x) = I_r R \cos h(\gamma x) - Z_o I_r \sin h\gamma x$$

x more than 50 miles

$$= \frac{I_r e^{\gamma x}}{2} (R - Z_o) \tag{1.62}$$

From Equation (1.62), surge voltage felt at point x from the place of switching depends upon the factor $(R - Z_o)$, whereby the surge could be intense if $R \gg Z_o$, and of course could be minimal if R is about the same order of Z_o.

C. ABRUPT OPENING OF A LINE

At $t < o$, line voltage $= E$ volt and line current $= I$ amp. For $t > o$, at point of opening $i = o$, initiating current pulses in both directions each of $(-I)$ as shown in Figure 1.5.

Voltage surge propagating to the right of the place of opening is equal to $(-IZ_o)$ and to the left is equal to (IZ_o) volts. Voltage across the opening switches $V_s = 2Z_o I$ for a current interruption of say 1 kiloamp in an overhead line where $Z_o = 500$.

$$V_s = 2 \times 1000 \times 500 = 1 \ mw$$

The development of 10^6 V across the opening switch will be accompanied by

FIGURE 1.5. Opening of a line.

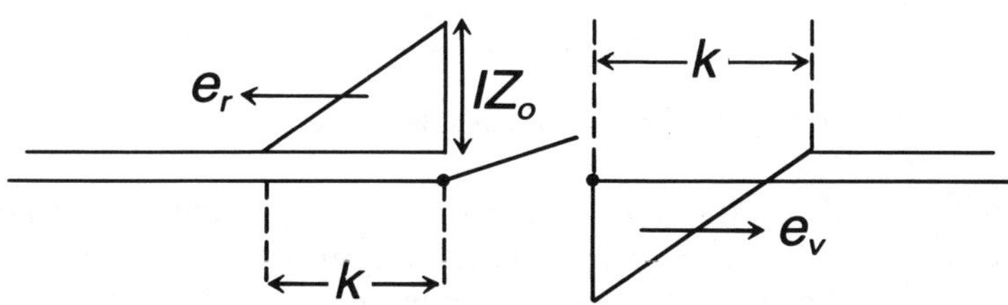

FIGURE 1.6. Spread of voltage surge across a switch.

an arc. Spread of voltage surge across the switch over a line length of K depends upon the time (τ) taken to interrupt a current of (I).

$$K = \tau v \tag{1.63}$$

when v is the velocity of propagation (m/s).

K is known as the front rise length i meters. For $\tau = 2$ ms

$$K = 2 \times 10^{-3} \times 3 \times 10^7 = 60 \text{ km}$$

However, K could be reduced by shortening τ by force of the switch mechanism such that $\tau \rightarrow 100$ ms.

$$K = 10^4 \times 3 \times 10^7 = 3 \text{ km}$$

Voltage surge propagating to the left represented by e_r which is equal to ($Z_o I + E$) may develop into a damaging surge if the line section is an open-circuit or a high impedance load. Even the forward surge of ($E - IZ_o$) may cause about the same damage build-up if the line is open to the right.

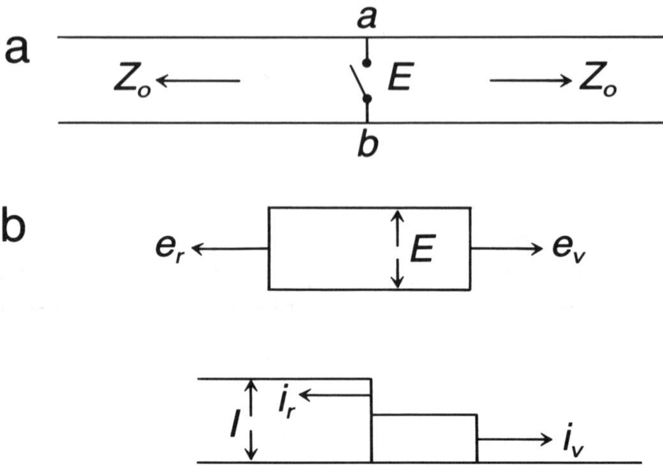

FIGURE 1.7.   Interruption of a short-circuited line.

Hence, at the first reflection to the left, the voltage build-up will reach a level of two times $(Z_o I + E)$ and to the left the level becomes $2(E - Z_o I)$. During the period of opening the switch, even in a duration of 100 ms, a process of multi-reflections between the location of line interruption and the opening end of the line may lead to development of a tremendous voltage upon which a protective over-voltage must act to stop further damaging consequences.

D. THE PROCESS OF REIGNITION

This phenomenon may generate a higher voltage surge, especially in underground cable systems where the line capacitance is a more significant portion of the surge impedance than that of overhead line.

When the cable system is disconnected relatively slowly from the source such as in the charging current i the system capacitance is interrupted at an instant of zero current which corresponds to maximum voltage, and if the time of interruption is of the order of a half cycle such that the voltage wave will attain its maximum negative level, a double nominal voltage will build across the switch such that reignition will start and hence a pulse of twice the nominal voltage will propagate through the line. At the nearest open-end the voltage will quadruple and for a sustained process of relections, levels of voltage build-up will become dangerously prohibitive.

E. INTERRUPTION OF A SHORT-CIRCUIT CURRENT

Figure 1.7a illustrates the opening of a short-circuit across (*a-b*) of a transmission line, where a current of I exists for $t < o$.

Opening of joint (*a-b*) will develop a voltage pulse of E volts as shown in Figure 1.7b as well as surge current of i_v propagating to the right and i_r traveling

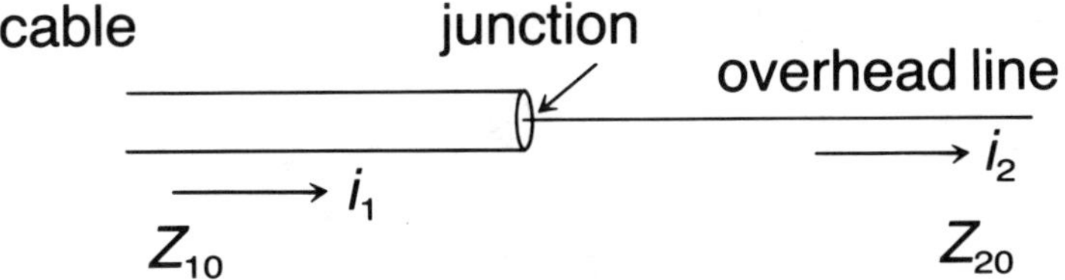

FIGURE 1.8.   Cable-overhead line junction.

to the left. However, the net pulse moving to the left is equal to $(I - i_r)$ as shown in Figure 1.7.

$$i_u = -i_r = \frac{I}{2} \text{ amp} \tag{1.64}$$

$$\text{surge voltage upon interruption} = E$$

$$= \tfrac{1}{2}IZ_o \tag{1.65}$$

The propagation of E in both directions and with incidence at a close open end or at a far loaded terminal may generate doubling and then tripling and so on of voltage build-up.

1.VI VOLTAGE BUILD-UP ALONG ASCENDING SURGE IMPEDANCE

A. CABLE-OVERHEAD LINES

At junction J, boundary conditions are:

$$i_1 = i_2$$

$$e_1 = e_2 \tag{1.66}$$

From $e_1 = e_2$, we write

$$e_{r1} + e_{t1} = e_{t2} \tag{1.67}$$

And from $i_1 = i_2$,

$$i_{t1} + i_{r1} = i_{t2} \tag{1.68}$$

Also

$$e_{t1} = i_{t1}Z_{1-0} \tag{1.69}$$

$$e_{r1} = -i_{r1}Z_{1-0} \tag{1.70}$$

From Equations (1.66) to (1.70), the results:

$$e_{t2} = \frac{2Z_{2-0}}{Z_{1-0} + Z_{2-0}} e_{t1} \tag{1.71}$$

$$e_{r1} = \frac{Z_{2-0} - Z_{1-0}}{Z_{2-0} + Z_{1-0}} e_{t1} \tag{1.72}$$

And similarly,

$$i_{t2} = \frac{2Z_{1-0}}{Z_{1-0} + Z_{2-0}} i_{t1} \tag{1.73}$$

$$i_{r1} = \frac{Z_{1-0} - Z_{2-0}}{Z_{1-0} + Z_{2-0}} i_{t1} \tag{1.74}$$

Since $Z_{2-0} > Z_{1-0}$, typical value for $Z_{1-0} \approx 150$, and typical value for $Z_{2-0} \approx$ 150. The junction J acts as an almost open-circuit leading to almost doubling of voltage and much smaller current than in region 1.

B. SINGLE LINE TO A GROUP OF LINES

Let $(n + 1)$ lines connect at a common bus-bar J to a main line of surge impedance Z_o. Each line i the group of $(n + 1)$ has similar and equal surge impedance as shown in Figure 1.9.

Resultant surge impedance for the group of $(n + 1)$ lines is expressed by:

$$Z_2 = \frac{Z_o}{n + 1 - 1} = \frac{Z_o}{n} \tag{1.75}$$

where

$$Z_o = Z_{o-1} = Z_{o-2} = Z_{o-3} = \cdots -Z_{o-n} = Z_{o-n+1}.$$

However,

$$e_{t2} = \frac{2Z_o}{Z_o + Z_2} e_{t1} \tag{1.76}$$

$$e_{t2} = \frac{2e_{t1}}{1 + \dfrac{Z}{Z}} = \frac{2}{1 + n} e_{t1} \tag{1.77}$$

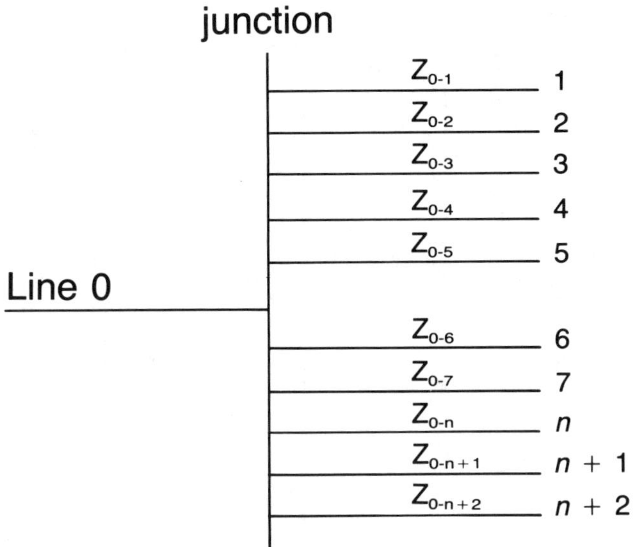

FIGURE 1.9.   Single line to a group of lines junction.

and

$$e_{r1} = \frac{1 - Z_o/Z_z}{1 + Z_o/Z_z}\, e_{t1}$$

$$= \frac{1 - n}{1 + n}\, e_{t1} \tag{1.78}$$

From Equation (1.77), we conclude the reverse, i.e.,

$$e_{t1} = e_{t2}\, \frac{1 + n}{2} \tag{1.79}$$

which implies that voltage level of an incident pulse on a line in a group of (n + 1) could be amplified several times by a factor of $(1 + n)/2$, when a single line emerges from that group. This factor of voltage surge amplification is equal to $(1 + n)/2$.

1.VII   SURGE IMPEDANCE OF TRANSFORMER

Figure 1.10 shows a front cross-section of a cylindrical transformer, illustrating the space distribution for the magnetic flux density B throughout the internal insulating region as well as the outer insulating region. Also shown is

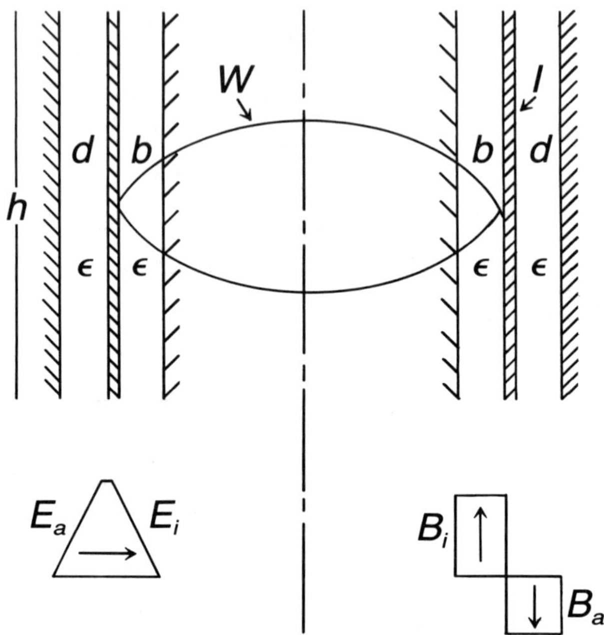

FIGURE 1.10.   Electromagnetic field inside transformer.

the space distribution for voltage across the internal and external insulating regions. This is a single phase transformer having a height of h, while w represents the length of one coil.

Now proceeding to write a simple solution for transformer surge impedance, let:

$$E = \text{winding voltage}$$

$$w = \text{length of one coil turn}$$

$$h = \text{axial length of winding}$$

Total electric charge on the winding due to E:

$$Q = \left(\frac{\epsilon E}{b} + \frac{\epsilon E}{d}\right) wh \, \frac{1}{4\pi \, V_o^2} \tag{1.80}$$

where V_o is the velocity of light.

The capacitance

$$C = \frac{Q}{E} \tag{1.81}$$

$$= \frac{E \, wh}{4\pi \, V_o^2} \left(\frac{b + d}{bd} \right) 10^9 \text{ Farad} \tag{1.82}$$

where N = total number of coil turns.

Magnetic flux in either insulating region

$$= \frac{4\pi \, \mu w \, Ni}{h} \frac{b}{1 + d/b} \tag{1.83}$$

$$= \frac{4\pi \, \mu w \, Ni}{h/b + h/d} \text{ Weber} \tag{1.84}$$

The inductance L is given by

$$L = \frac{NQ}{i} = 4\pi\mu \frac{bd}{b + d} \frac{W}{h} N^2 \text{ Henry} \tag{1.85}$$

Let also a represent the total length of wire $= Nw$; l, the self-inductance/unit length; and c, the capacitance/unit length:

$$l = L/a$$
$$c = C/a \tag{1.86}$$

and

$$V = l/\sqrt{lc} = a/\sqrt{LC}$$

Also

$$V = V_o/\sqrt{\mu\epsilon} \tag{1.87}$$

where μ is the relative permeability

$$Z_o = \sqrt{L/c} = \sqrt{\frac{L}{C}}$$

$$= 4\pi V_o \sqrt{\frac{\mu}{\epsilon} \frac{N}{h} \frac{bd}{b + d}} \, 10^{-9} \tag{1.87a}$$

For transformer ratings ranging from 5 to 50,000 kVa, the values of n/h vary from 100 to 10 to 1 turn/cm for the high voltage side. However, for the low side n/h varies from 1/5 to 1/10 of high voltage side values.

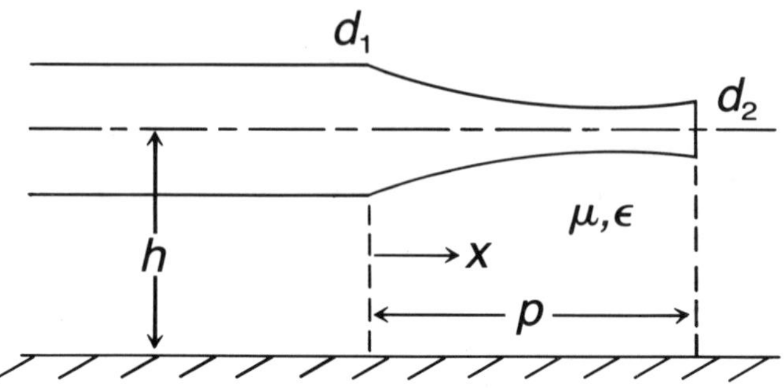

FIGURE 1.11.   Tapered line.

1.VIII   TAPERED LINES

A transmission may go through a region of linear or exponential tapering as a transition region as shown in Figure 1.11. Diameter tapering exponentially may follow the rule:

$$d(x) = d_1 e^{-\delta x}$$

where d_1 $(x = o) = D_1$ and d_2 $(x = p) = D_2$. From Equation (1.87)

$$\delta p = \ln \frac{d_1}{d_2} \tag{1.88}$$

The surge impedance for an overhead line of diameter d with its center h above ground is given by:

$$Z_o = \sqrt{\frac{L}{c}} = 2v \, \ln \frac{4h}{d} = 60 \, \ln \frac{4h}{d} \tag{1.89}$$

From Equation (1.88)

$$Z_o = 60 \, \ln \left[\frac{4h}{d_1} \, \epsilon^{\delta x} \right]$$

$$= 60 \left[\ln \frac{4h}{d_1} + \delta x \right] \tag{1.90}$$

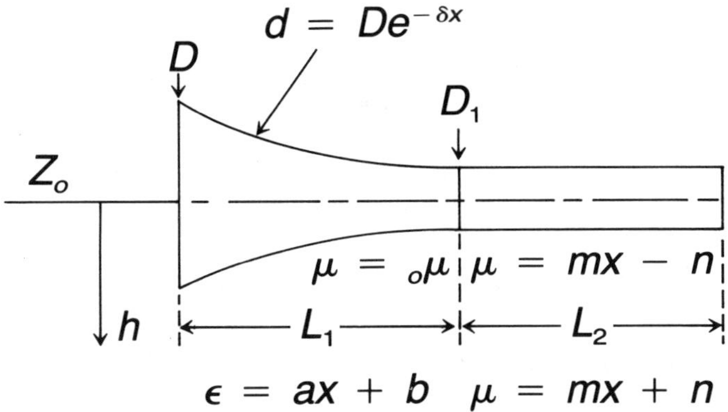

FIGURE 1.12.   Line with variable μ and ϵ.

The increment in the regular line surge impedance could be seen from Equation (1.90):

$$\Delta Z_o = 60\ \delta x$$

$$= 60\ ln\ \frac{d_1}{d_2} \tag{1.91}$$

Therefore, we can conclude from Equation (1.91) that increasing the ratio of d_1/d_2 will implicitly hike the incremental surge impedance sharply and consequently a propagating voltage surge incident on an adjacent regular untapered line will increase as the same order of increase in Z_o. It is useful to indicate that voltage amplification ratio

$$A = \sqrt{Z_1/Z_2} \tag{1.92}$$

Voltage amplification ratio is given by:

$$A = \left(ln\ \frac{4h}{d_1} + \delta x \right) \Big/ ln\ \frac{4h}{d_1} \tag{1.93}$$

A. TAPERED LINE WITH VARIABLE μ AND ϵ

Figure 1.12 represents an overhead line joined to a tapered exponential line and then to a cable untapered section. Magnetic permeability of the last two sections as well as the dielectric constants are functions of position.

We shall proceed now to express the overall voltage amplification ratio for a surge incident on line [1] and its propagation through lines [2] and [3].

Surge impedance Z_{2-o} for line #2 is:

$$Z_{2-0} = \sqrt{\frac{\mu 2}{\epsilon 2}} \left[60\ ln\ \frac{4h}{d_1} + \delta x \right] \tag{1.94}$$

or

$$Z_{2-0} = \sqrt{\frac{1}{(ax + b)\epsilon o}} \left[60 \ ln \ \frac{4h}{d_2} + \delta x \right] \qquad (1.95)$$

$$Z_{3-0} = \sqrt{\frac{mx - n}{_o\mu + kx \ _o\mu} \frac{\epsilon o}{}} \left[60 \ ln \ \frac{4h}{d_3} + \delta x \right] \qquad (1.96)$$

Hence the overall voltage amplification ratio could be expressed by:

$$\text{First } Z_{0-1} = 60 \ ln \ \frac{4h}{d_1}$$

$$A_{\text{total}} = \sqrt{\frac{Z_{0-2}}{Z_{0-1}}} \ \sqrt{\frac{Z_{0-3}}{Z_{0-2}}}$$

$$= \sqrt{\frac{Z_{0-3}}{Z_{0-1}}}$$

$$= \sqrt{\frac{mx - n}{kn + _o\mu} \frac{\epsilon o}{_o\mu}} \left[60 \ ln \ \frac{4h}{d_3} \right] \Big/ 60 \ ln \ \frac{4h}{d_1}$$

$$= \sqrt{\frac{mx - n}{kx + _o\mu}} \ ln \ \frac{4h}{d_3} \Big/ ln \ \frac{4h}{d_1} \qquad (1.97)$$

At the same time we must recognize the surge amplification factor at the second line which could be more severe due to its higher surge impedance, A_2, at any point x on line #2.

$$A_2 = \sqrt{\frac{Z_{0-2}}{Z_{0-1}}}$$

$$= \sqrt{\frac{1}{\epsilon o(ax + b)}} \left[60 \ ln \ \frac{4h}{d_2} + \delta x \right] \Big/ 60 \ ln \ \frac{4h}{d_1}$$

$$= \frac{1}{\sqrt{\epsilon o(ax + b)}} \frac{60 \ ln \ \dfrac{4h}{d_2} + \delta x}{60 \ ln \ \dfrac{4h}{d_1}} \qquad (1.98)$$

A_{total} and A_2 refer to a system of single conductor to ground. Suppose line #1 is of two conductors with spacing s, whereby in this case

$$Z_{0-1} = \frac{1}{4v^2 \ ln \ \dfrac{2s}{d}}$$

$$= 120 \, ln \, \frac{2s}{d_1} \tag{1.99}$$

Similarly for line A_2,

$$Z_{0-2} = \frac{1}{\sqrt{\epsilon o (ax + b)}} \left[120 \, ln \, \frac{2s}{d_2} + \delta x \right] \tag{1.100}$$

Then for line A_3

$$Z_{0-3} = \sqrt{\frac{mx - n}{kx + {}_o\mu}} \sqrt{\frac{\epsilon o}{{}_o\mu}} \left[120 \, ln \, \frac{2s}{d_3} \right] \tag{1.101}$$

From Equations (1.99), (1.100), and (1.101), surge amplification factor of $\sqrt{Z_{0.3}/Z_{0.1}}$ or $\sqrt{Z_{0.2}/Z_{0.1}}$ could be computed as previously done.

The other case which would be useful to consider is that if lines 1, 2 and 3 are three-phase Y systems:

$$Z_{0-1-Y} = 60 \, ln \, \frac{4h}{\Delta_1} \sqrt{\frac{u}{\epsilon}} \tag{1.102}$$

$$Z_{0-2-Y} = \sqrt{\frac{u}{\epsilon}} \left[60 \, ln \, \frac{4h}{\Delta_2} + \delta x \right] \tag{1.103}$$

and

$$Z_{0-3-Y} = \sqrt{\frac{\mu}{\epsilon}} \, 60 \, ln \, \frac{4h}{\Delta_3} \tag{1.104}$$

where Δ is known as the equivalent conductor diameter of the three conductors system

$$= \sqrt[3]{4 \, ds^2} \tag{1.105}$$

where d is the diameter for a single conductor and s is the spacing between any two conductors.

Again, the surge amplification factor could be computed by using the surge impedances of the three-phase system expressed by Equations (1.102), (1.103), and (1.104).

1.IX SWITCHING OF THREE-PHASE SYSTEMS

A. INTERRUPTION OF ONE PHASE IN A SHORT-CIRCUITED Y SYSTEM

Figure 1.13(a,b) shows the interruption of a short-circuited three-phase Y system. Let the surge impedance per phase $= Z_o$.

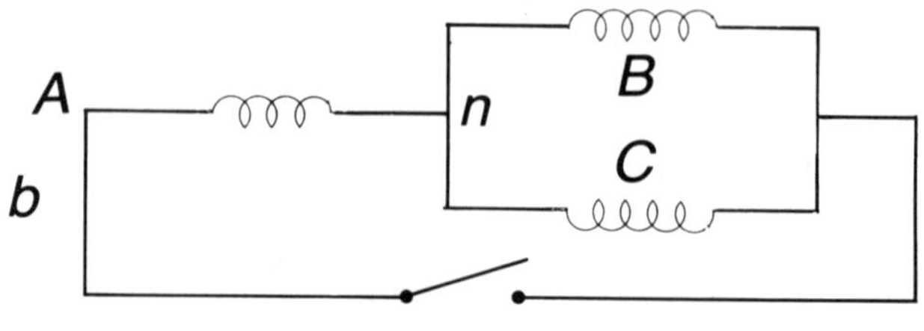

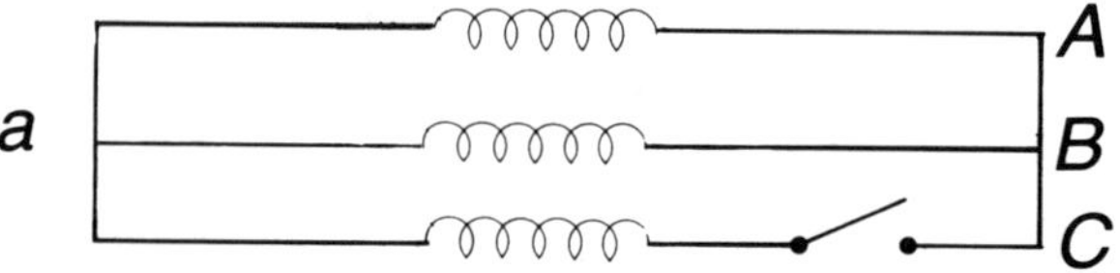

FIGURE 1.13.   Three-phase switching.

Surge impedance to the left of junction n is:

$$Z_{\text{left}} = Z_o \qquad (1.106)$$

and

$$Z_{\text{right}} = Z_o/2 \qquad (1.107)$$

Z_{left} and Z_{right} are open-circuited (o.c.) ends.

$$Z_{\text{o.c.left}} = j\, Z_o\, \text{ctn}\, \beta L_1 = -j\, Z_o\, \text{ctn}\, \frac{2\pi L_1}{\lambda} \qquad (1.108)$$

$$Z_{\text{o.c.right}} = j\, \frac{Z_o}{2}\, \text{ctn}\, \beta L_2 = -j\, \frac{Z_o}{2}\, \text{ctn}\, \frac{2\pi L_2}{\lambda} \qquad (1.109)$$

From Figure 1.13:

$$L_1 = x$$

$$L_2 = L - x$$

From Equations (1.108) and (1.109),

$$\operatorname{ctn} \frac{2\pi X}{\lambda} = \frac{1}{2} \operatorname{ctn} \frac{2\pi(1-x)}{\lambda}$$

or

$$\tan \frac{2\pi X}{L} = 2 \tan \frac{2\pi(1-x)}{\lambda} \qquad (1.110)$$

if L is an odd-multiple of $\pi/4$. $\dfrac{2\pi 1}{\lambda}$ would be an odd multiple of $\pi/2$; hence,

$$\operatorname{ctn} \frac{2\pi(1-x)}{\lambda} = \operatorname{ctn}\left[\frac{(2n-1)\pi}{2} - \frac{2\pi x}{\lambda} \right]$$

$$= \tan \frac{2\pi x}{\lambda}$$

$$2 \operatorname{ctn} \frac{2\pi(1-x)}{\lambda} = \tan \frac{2\pi x}{\lambda} \qquad (1.111)$$

or

$$\tan^2 \frac{2\pi x}{\lambda} = 2 \qquad (1.112)$$

Therefore

$$\frac{2\pi x}{\lambda} = \tan^{-1} 2 \qquad (1.113)$$

And since

$$f = \frac{V}{2\pi}, f = V\lambda$$

$$\therefore \ 2\pi x f/V = \tan^{-1} 2$$

or

$$f = \frac{(\tan^{-1} 2)v}{2\pi x}$$

$$= \frac{1.1083 \ V}{2\pi X} \qquad (1.114)$$

since $x = s$, $\therefore$

$$f = 0.1769 \, \frac{V}{s} \tag{1.115}$$

The general view to consider is that 1 may not be a sub-multiple of $\pi/4$. In this case, we can determine the frequency spectrum by using the criterion of continuity for the input impedance at the junction (n).

At point n:

$$Z_{\text{o.c.left}} = Z_{\text{o.c.right}} \text{ or}$$

$$\text{ctn } \beta 1_1 = \frac{1}{2} \text{ ctn } \beta 1_2$$

$$\tan \beta 1_2 = \frac{1}{2} \tan \beta 1_1$$

$$\text{since } \beta = w/v \tag{1.116}$$

$$\therefore \tan = \frac{w 1_2}{V} = \frac{1}{2} \tan \frac{w}{V} L \tag{1.117}$$

expanding the tangent in Equation (1.117) and stopping at the second term of the series expansion process:

$$\frac{w 1_2}{V} + \frac{1}{3} \frac{w^3 1_2^3}{V^3} = \frac{1}{2} \frac{w 1_1}{V} + \frac{1}{6} \frac{w^3 1_1^3}{v^3} \tag{1.118}$$

The expression shown in Equation (1.118) is valid under the condition:

$$\frac{w^2 1^2}{2} < \frac{\pi^2}{4}$$

Equation (1.118) could be written as:

$$1_2 + \frac{1}{3} \frac{w^2 1_2^3}{V^2} = \frac{1}{2} 1_1 + \frac{1}{6} \frac{w^2 1_1^3}{V^2}$$

$$\therefore \frac{w^2}{V^2} \left[\frac{1}{3} 1_2^3 - \frac{1}{6} 1_1^3 \right] = \frac{1}{2} 1_1 - 1_2 \tag{1.119}$$

Hence,

$$w^2 = \frac{\left[\frac{1}{2} 1_1 - 1_2\right] V^2}{\left[\frac{1}{3} 1_2^3 - \frac{1}{6} 1_1^3\right]} \tag{1.120}$$

and

$$w = +V \sqrt{\frac{\frac{1}{2} 1_1 - 1_2}{\frac{1}{3} 1_2^3 - \frac{1}{6} 1_1^3}}$$

or

$$f = \pm \frac{V}{2\pi} \sqrt{\frac{\frac{1}{2} 1_1 - 1_2}{\frac{1}{3} 1_2^3 - \frac{1}{6} 1_1^3}} \tag{1.121}$$

Equation (1.121) shows no harmonics.

A somewhat different form for the frequency (f) may result if the series expansion of tan ø could be extended to the third or higher order terms.

B. INTERRUPTION OF A GROUND FAULT

Figure 1.14 shows an interruption of short-circuit while the three-phase system is grounded at one pole. Every ground connection acts as a short-circuit of zero voltage.

Phase (a) is actually excluded from any action and remains dead because no voltage can penetrate a short-circuit.

We can proceed now to obtain expression for the frequency of surge incidence and reflection by using the criterion for the continuity of the input impedance at junction (n).

$$Z_{o.c.left} = -j \frac{Z_o}{4} \text{ ctn } \beta 1_1$$

$$Z_{o.c.right} = -j \frac{Z_o}{2} \text{ ctn } \beta 1_2 \tag{1.122}$$

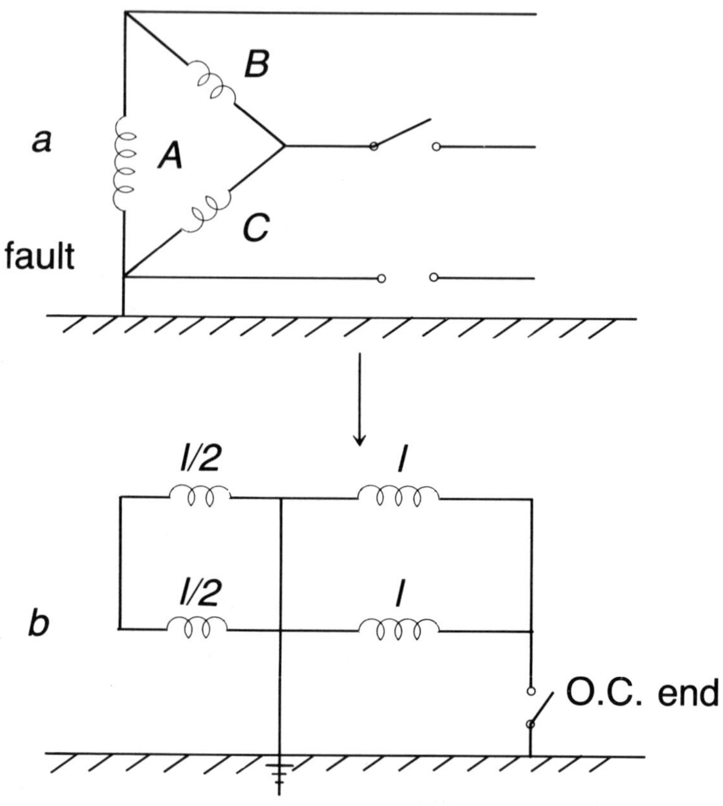

FIGURE 1.14.   Interruption of a ground fault.

where

$$l_1 = L/2$$

$$l_2 = L$$

$$\frac{1}{4} \operatorname{ctn} \beta \frac{1}{2} = \frac{1}{2} \operatorname{ctn} \beta 1$$

or

$$\operatorname{ctn} \frac{\beta 1}{2} = 2 \operatorname{ctn} \beta 1 \tag{1.123}$$

Again, expanding the (ctn) on both sides in Equation (1.123) we shall now include the third term in the series expansion process as shown below.

Equation (1.123) could be written as:

$$\tan \beta 1 = 2 \tan \beta^1/_2$$

$$\frac{2 \tan \dfrac{\beta 1}{2}}{1 - \tan^2 \dfrac{\beta 1}{2}} = 2 \tan \frac{\beta 1}{2} \qquad (1.124)$$

$$\therefore \tan^2 \frac{\beta 1}{2} = 1$$

$$\tan^2 \frac{\beta 1}{2} = \pm 1$$

and

$$\beta^1/_2 = \pm \left[2n\,\pi + \frac{\pi}{4} \right]$$

$$\beta = w/V = \frac{2\pi f}{V}$$

Hence

$$\frac{\pi f L}{V} = \pm \left(2n\,\pi + \frac{\pi}{4} \right)$$

and

$$f = \pm \frac{V}{\pi 1} \left(2n\,\pi + \frac{\pi}{4} \right) HZ \qquad (1.125)$$

Obviously, harmonics exist in the frequency spectrum.

C. INTERRUPTION OF Y CONNECTED SYSTEM WITH GROUND FAULT

Figure 1.15 illustrates the opening of a short-circuited Y system at phase (a).

$$Z_{o.c.left} = -j\,Z_o\,\operatorname{ctn} \beta 1$$

$$Z_{o.c.right} = -j\,\frac{Z_o}{2}\,\tan \beta 1_2$$

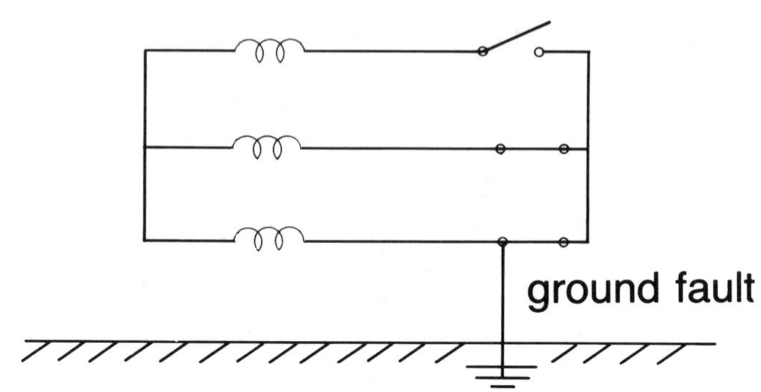

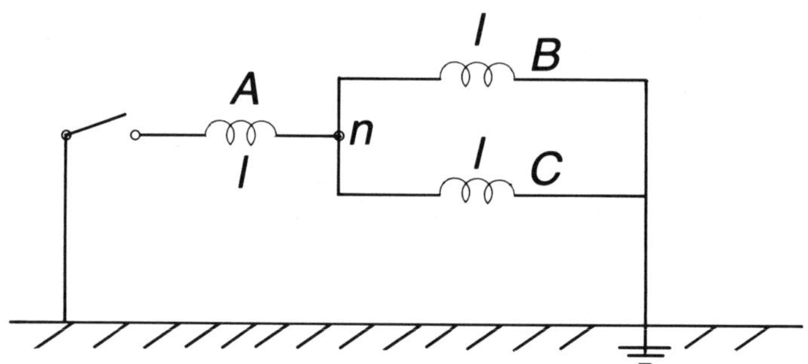

FIGURE 1.15.   Interruption of Y system.

At junction (n) $Z_{\text{o.c.left}} = Z_{\text{s.c.right}}$

$$\operatorname{ctn} \beta 1 = \frac{1}{2} \tan \beta 1$$

or

$$\tan^2 \beta 1 = 2 \tag{1.126}$$

or

$$\tan \beta 1 = \pm 2$$

$$\beta 1 = \pm(2n\pi + 0.956) \tag{1.127}$$

or

$$\frac{2\pi f 1}{V} = \pm(2n\pi \pm 0.956)$$

and

$$f = \pm \frac{v}{2\pi l} (2n\pi + 0.956) \qquad (1.128)$$

Harmonics are present for (f) according to Equation (1.128).

1.X LIGHTNING STROKES

A. PROBABILITY OF BEING STRUCK

Lightning strikes the earth 100 times in each second with varying destructive effects, depending on several parameters such as the height and area of the structure and the geometry of its peak and the nature of material structure. However, lightning strike is probabilistically expressed by empirical rule based on field measurements. For example, the probability of lightning striking a structure <600 feet is given by:

$$P_{600'} = k_1 h \qquad (1.129)$$

where h is the height in feet, and k_1 is a constant $= 2.4 \times 10^{-3}$.

For a structure height between 600 and 1300 feet the probability equation becomes:

$$P_{600-1300} = e^{4h \times 10 - 2.13} \qquad (1.130)$$

On a system of transmission of 60 to 100 feet high and on a group of 25 to 40 lines it had been reported that lightning strikes on the average of one stroke per mile of line per year.

The results have shown also that the strokes will be drawn to the line, because of its height from an effective lateral distance on each side of the line, on the average of 3.5 times the height.

Assuming an average height for the structure of 80 feet, one stroke per mile of line per year is equivalent to 9.5 strokes per year per square mile of sky area.

Other statistical data indicated that for a building of dimensions W, L and H designating width, length and height (ranging from 60 to 100 ft), the total number of strokes per year to the building is expressed approximately by:

$$P_{60-100} = K_2 \frac{(w + 2 \times 3.5H)(L + 2 \times 3.5H)}{(5280)^2} \qquad (1.131)$$

where $k_2 = 9.5$ for heights between 25 and 100 feet with isokeraunic level of 25 to 40, and $k_2 = 5.8$ for heights between 25 and 100 feet with an isokeraunic level of 20.

The isokeraunic level is defined as the number of days per year with thunderstorms occurring in a given region.

TABLE 1.1

Height in feet	Strokes to structure/year
50	0.12 (once every 8 yrs.)
100	0.24 (once every 4 yrs.)
200	0.49 (once every 2 yrs.)
400	0.97 (once every year)
600	1.45 (3 times every 2 yrs.)
800	3.00 (3 times per year)
1000	6.50 (6-7 times per year
1250	15.30 (15 times per year)

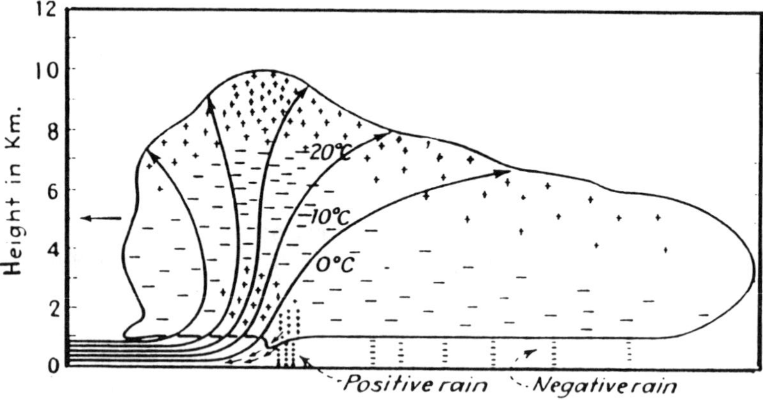

FIGURE 1.16.   Electrically charged cloud.

Table 1.1 above represents the number of strokes per year with respect to height.

B. PHYSICAL MODELING OF LIGHTNING

Electric charge distribution in thunderclouds is formed by spray electrification, whereby the powerful upward air draft at the head of the storm cloud carries with it moisture to the cool regions where it condenses into drops. The size of water drops will increase up to the limit at which the weight of any drop is in excess of the upward force produced by the surging air drafts, after which the drops will fall. Increase in drop size above the critical size will lead to a break-up associated with the release of negative ions into the air while the smaller drops will retain positive charges. This is shown in Figure 1.16. The released smaller drops now will grow in size to the critical limit, after which they will break-up again, releasing more negative charges and smaller drops with positive charge and so on. The cycle will repeat itself. It has been found that the rain at the head of the storm carries positive ions while the back of the storm is accompanied by negative ions. Voltage build-up in storms varies from 1 MV to

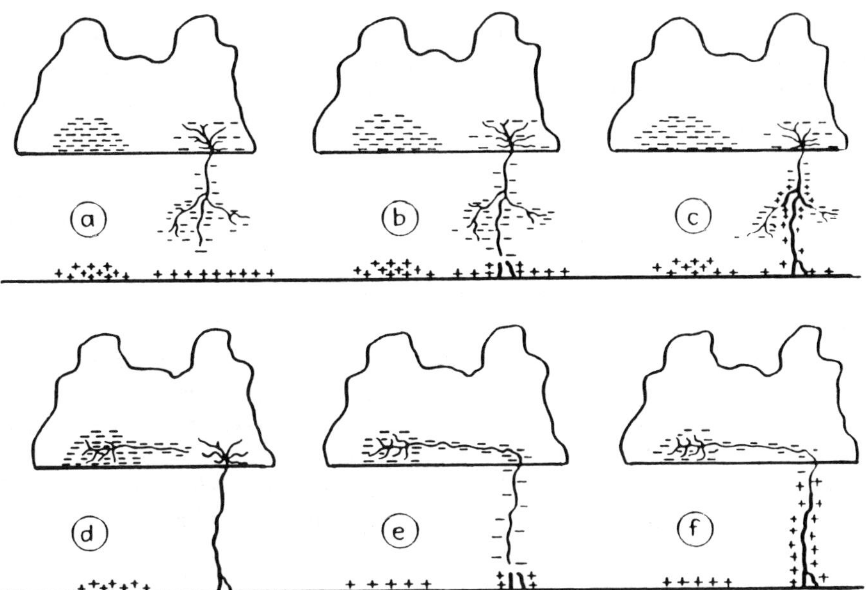

FIGURE 1.17.   Formation of lightning surge.

1 BV. Consequently, the electric field at earth's surface may range from 5 to 280 kV/meter. Potentials at transmission lines may rise to the order of 1.5 MV by the effect of storms, then may propagate at the velocity of light if the process of discharge is not instantaneous. Loeb indicated that charge build-up at transmission lines is negative due to positive ion streamers that had left the transmission structures and not due to the end of storm build-up of negative charge.

The successive growth of lightning stroke is shown in Figure 1.17 which was obtained, according to Schonland, from high speed photography of actual strokes. The first state of lightning stroke on a transmission line or any highly conductive structure starts with the formation of catholyte negatively charged layer on the line through electron avalanche called the pilot or leader stroke. The speed of propagation of the leader stroke varies from 10^7 to 2×10^8 cm/ s.

The streamer moves through an ionized bridge established by the processes of ionization due to the intensive field existing at the head of the avalanche and by photoionization of the gaseous medium surrounding streamer. At the back of the streamer, the deionization process takes action by ion recombination and diffusion resulting from the formation of distinct regions of concentration gradients. Velocity of propagation of the leader stroke is relatively much less than the velocity of light because its formation occurs in several steps of ionizations. Another process of reionization usually occurs progressively down the leader stroke by another avalanche known as the dart streamer which propagates at a speed of the order of 10^{10} cm/s. The dart streamers maintain a continuous ionizing path at the back of the leader stroke creating a highly intensive electric field.

Rudenberg indicated that the electric current at the tip of the dart streamer is given by:

$$i = \frac{Erv}{2}$$ (1.132)

where E is the field strength of the tip of radius r and v is the velocity of propagation.

Electric current at the tip of the dart streamer may be of the order of 10k amp. Usually before the leader stroke reaches the ground or line, a positive streamer springs up due to time variation of bound charges on the transmission line or the conducting object. The springing positive streamer is known as the return stroke; it can ride up to 3/4 of the separation between the leader stroke and ground. Current in the return on main stroke may approach the order of 500k amp.

Figure 1.17 illustrates formation of the leader stroke and the main or return stroke.

Modeling of the main stroke current is based on the actual visualization of a condenser system where the charged cloud having a diameter of D is separated from an equivalent ground surface by a distance of a. The self-inductance of the stroke current L is expressed by

$$L = 2h\left[ln\, \frac{D}{d} - \frac{1}{2} \right]$$ (1.133)

where d is the diameter of the stroke.

Amplitude of stroke current from Equations (1.132) and (1.133) is given by:

$$L = v_o \sqrt{\frac{C}{L}}$$

or

$$L = \frac{Dv\, 10^9}{4\sqrt{2}\, ac \sqrt{ln\, \frac{D}{d} - \frac{1}{2}}}$$ (1.134)

where c is the velocity of light, v is the potential difference between the two condenser plates, namely the cloud and ground or line, and D is in kilometers.

For an average value of V/a of 1 kV/cm, the associated current stroke could be of the order of 200k amp. Frequency of oscillation of an LC system for the

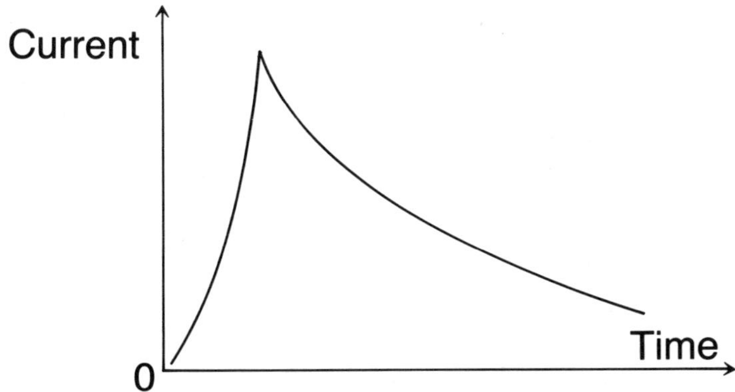

FIGURE 1.18. Waveform of lightning surge.

main stroke is of the order of 45.8 kHz. However, due to resistances contained in ground, the avalanche column and cloud, the system may not oscillate. Wave front of the main discharge current exists from 2 to 10 μs and the entire duration may approach 80 μs. Succession of return strokes may occur with high probability to as much as 40 strokes. Time interval between strokes may range from 0.0006 to 0.53 s, while the entire series of strokes may last up to 1 s. General wave form for lightning surge is shown in Figure 1.18.

We now return to the condenser system of the main stroke shown in Figure 1.19, where a cloud height (a), transmission line height of (h), electric field in the region is (E) volt/m and hence voltage build-up on the line can be expressed as below

$$V = hE \tag{1.135}$$

The surge voltage V will propagate left and right of the place of incidence.

The time-space relation in the absence of any resistance is governed by:

$$-\frac{\partial E}{\partial t} = \pm v \frac{\partial E}{\partial x} \tag{1.136}$$

Also, we have to indicate that change in voltage surge picked up from the atmosphere is composed of two components, namely the propagating forward and backward components, $E_v + E_r$.

$$\frac{\partial V}{\partial t} = \frac{\partial V_r}{\partial t} + \frac{\partial V_t}{\partial t} \tag{1.137}$$

$$= 2 \frac{\partial V}{\partial t}$$

$$= -2v \frac{\partial V}{\partial t} = 2v \frac{\partial V_r}{\partial t} \tag{1.138}$$

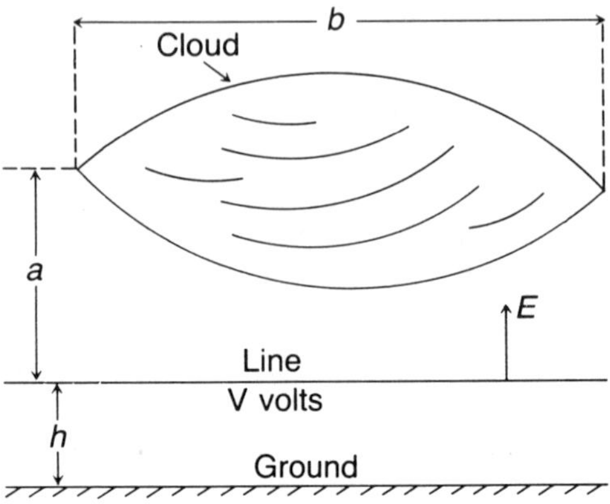

FIGURE 1.19.   Condenser system of lightning.

Voltage rise in the forward traveling surge is:

$$-\frac{\partial V_t}{\partial x} = \frac{\partial V_r}{\partial x} = \frac{1}{2v}\frac{\partial V}{\partial t} = \frac{h}{2v}\frac{\partial V}{\partial t} \tag{1.139}$$

Also, during the time interval τ of voltage rise, the surge front travels a distance given by

$$a = v\tau \tag{1.140}$$

where a is the height of cloud above ground or line.

1.XI SOLVED EXAMPLES

A. A 100-kV line has completely lumped total ductance and capacitance carrying 200 A, when suddenly the line interrupted in a time interval of 10^{-4} s.

a.   Calculate the front-rise length of the interrupted voltage pulse.
b.   Calculate magnitude of switching pulse.
c.   If switching time is increased to, say, 0.1 s, comment on the nature of front-rise length and pulse strength.
d.   Sketch the switching pulse in *b* and *c*.

Solution
    A 100 kV line
    i = 200 amp
    τ = switching time
       = 100 μs

a) k as front rise length

$$\tau = V_o \tau$$
$$= 3 \times 10^8 \times 10^2 \times 10^{-6} = 30 \text{ km}$$

b) magnitude of switching pulse

$$E_s = 2Z_i$$

For overhead line $Z = 500 \ \Omega$

For cables $Z = 50 \ \Omega$

$$E_s = 2 \times 500 \times 200 = 200 \text{ kV}$$

This is overhead line

$$E_s = 2 \times 50 \times 200 = 20 \text{ kV}$$

for underground cable

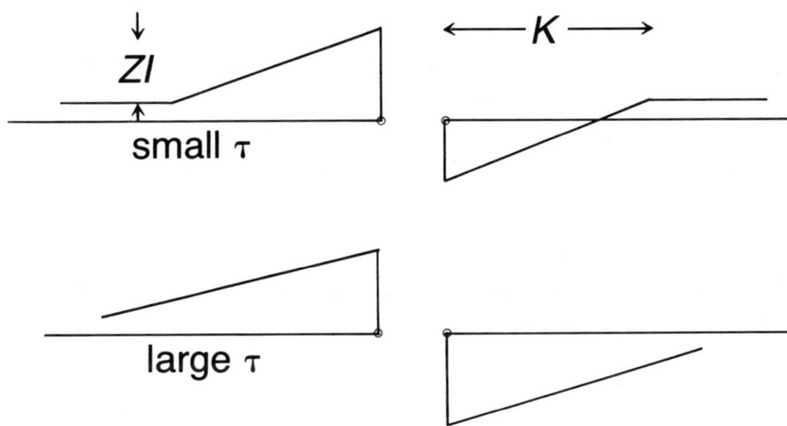

c) if τ is increased to 0.1 sec.

$$k = 3 \times 10^8 \times 0.1 = 3 \times 10^7 \ m$$
$$= 30,000 \text{ km which is a very extended front rise length.}$$

For an ordinary steady-state system operating at 60 cps, $\lambda = 5000$ km.

$\therefore$ The above value of k is of the order of $\sigma\lambda$, and hence the pulse strength will be extended over $\lambda\sigma$, resulting in substantial weakening.

τ is small

τ is large

B. In Figure B find an expression for optimum value of R where the transmission system to the right of the arrester is individual lines with increasing Z according to an arithmetical progression.

$$e_o = \text{arrester limiting voltage}$$

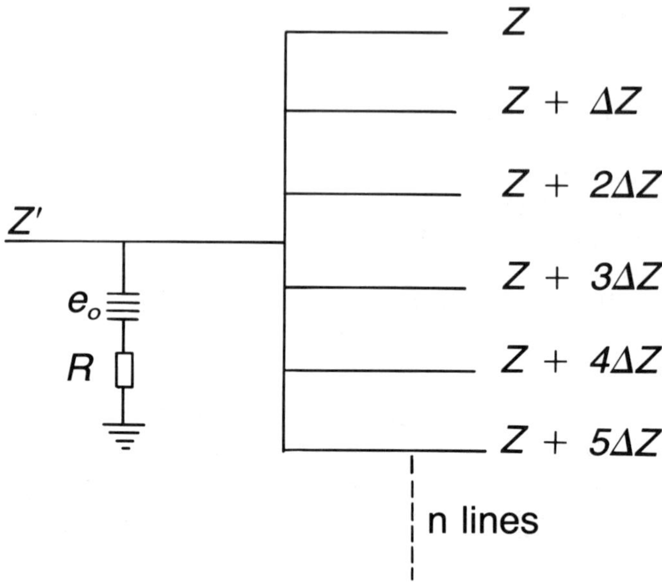

Solution

The system to the right of point 0 is n lines with gradual increase in individual surge impedance with a total parallel equivalent of Z_2.

$$Z_2 = \cfrac{1}{\dfrac{1}{Z} + \dfrac{1}{Z + \Delta Z} + \dfrac{1}{Z + 2\Delta Z} + \; - \; - \; - \; - \; + \dfrac{1}{Z + n\Delta Z}}$$

$$= \frac{1}{Y_2}$$

$$Y_2 = \frac{1}{Z} + \frac{1}{Z + \Delta Z} + \; - \; - \; - \; - \; + \frac{1}{Z + n\Delta Z}$$

$$= \frac{1}{Z}\left[1 + \frac{1}{Z + \Delta Z} + \; - \; - \; - \; - \; + \cfrac{1}{1 + n\dfrac{\Delta Z}{Z}} \right] \tag{B1}$$

Practically, ΔZ is small if all the individual Z_s are homogeneous transmission overhead or cables, but if they are due to lines, cables, transformers, etc., ΔZ could even be greater than Z.

However, as a specific case, consider all Z_s are those of homogeneous lines such that $\Delta Z \ll Z$.

$$Y_2 \approx \frac{1}{Z}\left[1 + \left(1 - \frac{\Delta Z}{Z}\right) + \left(1 - \frac{2\Delta Z}{Z}\right) + \right.$$
$$\left.\left(1 - \frac{3\Delta Z}{Z}\right) + \cdots\cdots + 1 - \frac{(n-1)\Delta Z}{Z}\right]$$

$$Z_2 = 1/Y_2 \tag{B2}$$

Then, according to Equation (1.6) of text,

$$\frac{1}{R_{opt.}} = \frac{1}{Z'} + \frac{1}{Z_2}$$

$$Z_2 \cong \frac{1}{Z}\left[n - \frac{\Delta Z}{Z}\frac{(n-1)+1}{2}\right]$$

$$\approx \frac{n}{Z}\left[1 - \frac{1}{2}\frac{\Delta Z}{Z}\right] = \frac{n}{Z}\left[\frac{2Z - \Delta Z}{2Z}\right]$$

$$R_{opt.} \approx 1 \left/ \left[\frac{1}{Z} + \frac{2Z^2}{n(2Z - \Delta Z)}\right]\right. \tag{B3}$$

For a homogenous system of small increase in Z.

C. In an inhomogeneous line where the inductance changes linearly with distance and capacitance is constant, show that at very high frequency a solution of voltage pulse can be represented in a circular function.

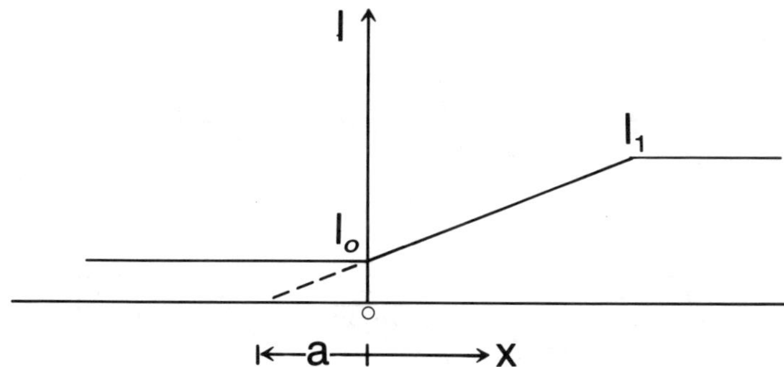

Solution

For an inhomogeneous line, with linear increase in inductance and constant lumped capacitance, d.e. of voltage pulse is:

$$\frac{\partial^2 e}{\partial x^2} - \frac{1}{a + x}\frac{\partial e}{\partial x} + wc^2 \,{}^l v\frac{a + x}{a}\,e = 0 \tag{C1}$$

$$a = \text{line length to double } 1$$

$$a + x = \xi \tag{C2}$$

$$w \sqrt{\frac{c/o}{a}} = \beta \tag{C3}$$

$$\frac{\partial^2 e}{\partial \xi 2} - \frac{1}{\xi} \frac{\partial e}{\partial \xi} + \beta^2 \xi e = 0 \tag{C4}$$

whose solution:

$$e = \xi \left[K_1 J_{2/3}\left(\frac{2}{3} \beta \xi^{3/2} \right) + K_2 N_{2/3}\left(\xi^{3/2} \beta \frac{2}{3} \right) \right] \tag{C5}$$

$$\text{For the argument } \beta \, \xi^{3/2} \gg 1 \tag{C6}$$

$$J_v(Z) \Big|_{Z \to \infty}^{\approx} \sqrt{\frac{2}{\pi Z}} \cos\left[Z - \frac{\pi V}{2} \frac{\pi}{4} \right] \tag{C7}$$

$$\therefore J_{2/3}\left(\frac{2}{3} \beta \xi^{3/2} \right) \Big|_{Z \to \infty} = \sqrt{\frac{2}{\pi}} \sqrt{\frac{1}{\frac{2}{3} \beta \xi^{3/2}}} \cos\left[\frac{2}{3} \beta \xi^{3/2} - \frac{\pi}{4} - \frac{2}{3} \frac{\pi}{2} \right]$$

$$= \sqrt{\frac{3}{\pi \beta \xi^{3/2}}} \cos\left[\frac{2}{3} \beta \xi^{3/2} - \frac{7\pi}{12} \right] \tag{C8}$$

And

$$N_v(Z) \Big|_{Z \to \infty}^{\approx} \sqrt{\frac{2}{\pi Z}} \sin\left[Z - \frac{V\pi}{2} \frac{\pi}{4} \right] \tag{C9}$$

Therefore

$$N_{2/3}\left(\frac{2}{3} \beta \xi^{3/2} \right) \Big|_{Z \to \infty} = \sqrt{\frac{2}{\pi \beta \xi^{3/2}}} \sin\left[\frac{2}{3} \beta \xi^{3/2} - \frac{7\pi}{12} \right] \tag{C10}$$

Therefore

$$e = \xi \sqrt{\frac{2}{\pi \beta \xi^{3/2}}} \, [K_1 \cos \theta + K_2 \sin \theta] \tag{C11}$$

$$= \xi \sqrt{\frac{2}{\pi \beta \xi^{3/2}}} \sin (\theta + K_4) \tag{C12}$$

D. Given:

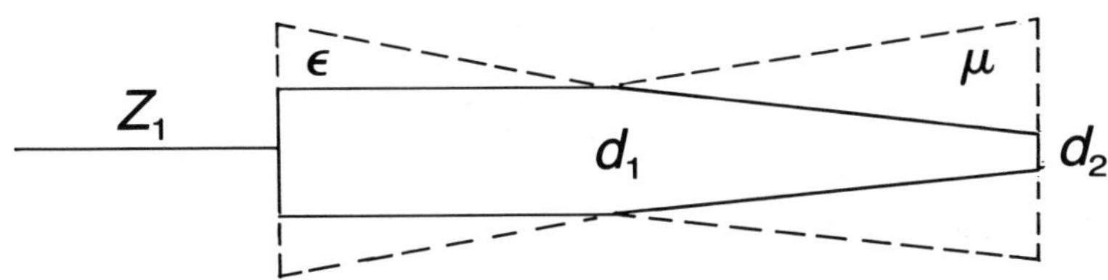

with $d_e = d_1\, e^{-dx}$, $\epsilon = 30$, and $\mu = 500$, calculate:
a)   overall voltage transformation ratio
b)   velocity of propagation in section 2
c)   total transition time in the transformation scheme.

a) $d(x) = d_1\, e^{-\delta x}$
    $\epsilon = 30,\ \mu = 500$

$$Z\alpha\ \frac{\mu}{\epsilon} \tag{D1}$$

$$Z_1\alpha\ \frac{\mu_o}{\epsilon},\quad Z_2\alpha\ \frac{\mu_2}{\epsilon_o} \tag{D2}$$

$$\therefore\ \frac{Z_2}{Z_1}\ \alpha\ \frac{\mu_2}{\epsilon_o}\,\frac{\epsilon_1}{\mu_o} \tag{D3}$$

$$\alpha\ \frac{\mu_2}{\mu_o}\,\frac{\epsilon_1}{\epsilon_o} \tag{D4}$$

$$r' = \sqrt{1 + \frac{\ln d_1/d_2}{\ln 4h/d_1}} \tag{D5}$$

with non-magnetic non-electric surrounding insulation
   And with the specified insulation

$$r' = \sqrt{\frac{\mu_2\,\epsilon_1}{\mu_o\,\epsilon_o}}\ r \tag{D6}$$

$$= 500 \times 30\ \sqrt{\frac{\ln d_1/d_2}{\ln 4h/d_1}}$$

$$= 142\ r \tag{D7}$$

b)

$$V = \frac{1}{\sqrt{\epsilon_o \, \mu}} = \frac{1}{\sqrt{\mu_r \, \mu_o \, e_o}} \tag{D8}$$

$$= \frac{1}{\sqrt{\mu_o \, \epsilon_o}} = \frac{1}{\sqrt{\mu_r}}$$

$$= \frac{V_o}{\sqrt{500}} = \frac{V_o}{22.4} \tag{D9}$$

c)   Total transition time

$$T = \frac{l_1}{V_1} + \frac{l_2}{V_2} \tag{D10}$$

$$V_1 = \frac{1}{\sqrt{\mu_o \, \epsilon}} = \frac{1}{\sqrt{\mu_o \, \epsilon_o \, \epsilon_r}} \tag{D11}$$

$$V_1 = \frac{V_o}{\sqrt{30}}$$

$$T = \left[\frac{l_1}{V_o} \sqrt{30} + \frac{l_2}{V_o} \sqrt{500} \right] 50 \tag{D12}$$

E. A transmission line of surge impedance Z is to be connected to the following sections in separate individual cases. Comment in each case with respect to the degree of reflection, velocity of pulse propagation, and voltage transformation ratio:

1.   Another line whose Z is decreasing linearly,
2.   Transformer system
3.   Transmission line whose length is of the order of 400 km.

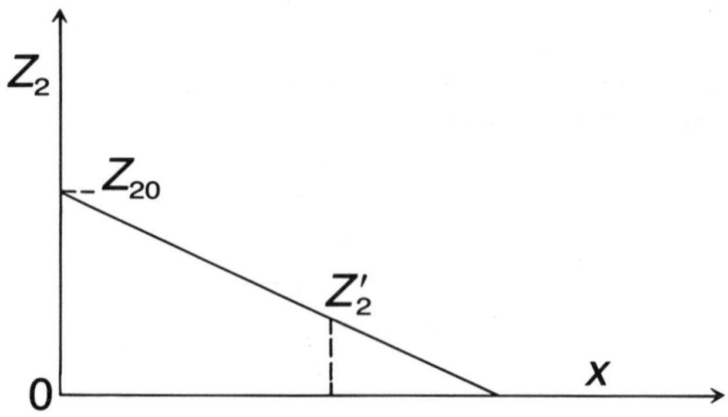

Solution

a)

$$Z_2' = Z_{20} + Kx$$

$$0 = Z_{20} + Kl \tag{E1}$$

$$\therefore K = -\frac{K_{20}}{1} \tag{E2}$$

$$Z_2' = Z_{20} - \frac{Z_{20}}{1}x = Z_{20}\left(1 - \frac{x}{1}\right) \tag{E3}$$

$$r = \sqrt{Z_2/Z_1} = \sqrt{\frac{Z_{20}(1 - x1)}{Z_1}} \tag{E4}$$

Also reflection is expected to be small if l is large

$$Z = \sqrt{L/C}$$

A decrease in Z can be attributed to either an increase in C or a decrease in L and will be affected accordingly.

b)   Transformer system of Z_2

Z_2 is expected to be of the order of 10 times that of an overhead line or 100 times that of cable. It represents a jump in ohmic value approaching an open-circuit situation.

Therefore, as r increases very largely, reflection will almost double the incident pulse without polarity change and hence velocity of propagation will diminish greatly to a value of the order of 30 m/μs.

c)   Velocity of propagation $\approx V_o$

Reflection will be very small and the ratio of transformation is of the order: $\sqrt{Z_2/Z_1}$.

F.   *A line 1000 miles long has:*

$R = 10.4$ Ω/mile
$L = 3.67$ mHenri/mile
$C = 0.00835$ μF/mile
$G = 0.8$ μmho/mile

The line is terminated in a $Z_r = 1500 \, \llcorner 45°$ and connected to a generator at the sending end having an emf of 100 volts and internal impedance of 300 Ω. Consider $w = 2000$ rad/s. Calculate voltage, current and power at Z_r using the method of separate waves along the line and the principle of superposition.

Solution

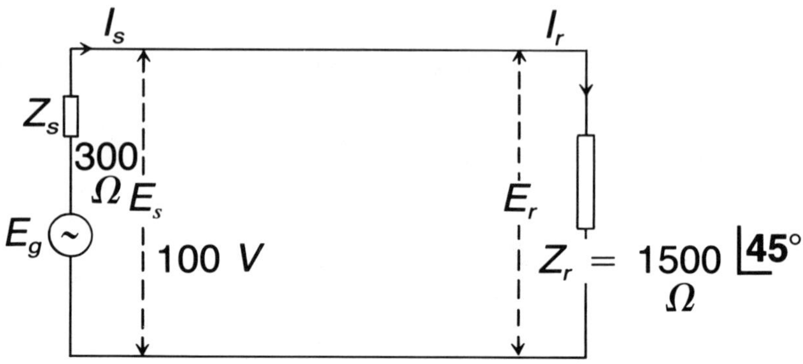

$$Z = R + jwl = (10.4 + j\,7.35) = 12.7\ \underline{|38°}$$

$$Y = G + jwc = (0.8 + j\,16.7)\ 10^{-6}\ \text{mho}$$

$$= 16.75\quad \underline{|90°}\quad 10^{-6}\ \text{mho}$$

$$Z_o = \sqrt{Z/Y} = \sqrt{\frac{12.7\ \underline{|38°}}{16.75 \times 10\ \underline{|90°}}} \tag{F1}$$

$$= 875\ \underline{|-24°}$$

$$\alpha 1 = 7.28\ \text{nepers/1000 miles} \tag{F2}$$

$$= .00728$$

$$\beta = .0126\ \text{rad/mile} \tag{F3}$$

$$\beta 1 = 12.6\ \text{radians} \approx 4\,\pi \to 0$$

$$\text{Assume } E'_{s1} = 1\ \underline{|0°}\ \text{volt}$$

$$E'_{r1} = E_{s1}\ e^{-\alpha 1}\ e^{-j\beta 1} = e^{-7.28} = 6.75\ \text{mV}$$

$$E''_{r1} = E'_{r1}\frac{Z_r - Z_o}{Z_r + Z_o} = 5.4\ \underline{|60°}\ \text{mV} \tag{F4}$$

$$E''_{s1} = E''_{r1}\ e^{-1} = E''_{r1}\ e^{-1} = 3.6\ \underline{|60°}\ \text{V} \tag{F5}$$

$$I'_{s1} = E'_{s1}/Z_o = \frac{1\ \underline{|0°}}{875\ \underline{|-24°}} \tag{F6}$$

$$= 1.15\ \underline{|24°}\ \text{mA}$$

$$I'_{r1} = \frac{E'_{r1}}{Z_o} = 7.65 \,\angle 24° \,\mu A \tag{F7}$$

$$I''_{r1} = -\frac{E''_{r1}}{Z_o} = -6.2 \,\angle 84° \,\mu A \tag{F8}$$

$$I''_{s1} = -\frac{E''_{s1}}{Z_o} = -4.1 \times 10^{-3} \,\angle 84° \,\text{amp} \tag{F9}$$

$$E_{s1} = E'_{s1} + E''_{s1}$$
$$= (1 + 3.6 \times 10^{-6} \,\angle 60°) \,\text{volt} \tag{F10}$$

$$E_{r1} = E'_{r1} + E''_{r1}$$
$$= (6.75 \times 10^{-3} \,\angle 0° + 5.4 \times 10^{-3} \,\angle 60°) \,\text{volt} \tag{F11}$$

$$I_{s1} = I'_{s1} + I''_{s1}$$
$$= (1.15 \times 10^{-3} \,\angle 24° - 6.2 \times 10^{-6} \,\angle 84°) \,\text{amp} \tag{F12}$$

$$I_{r1} = I'_{r1} + I''_{r1}$$
$$= (7.65 \times 10^{-6} \,\angle 24° - 6.2 \times 10^{-6} \,\angle 84°) \,\text{amp} \tag{F13}$$

$$\text{Actual } E_s = E_g \frac{Z_{in}}{Z_s + Z_{in}} \tag{F14}$$

$$Z_{in} = \frac{E_{s1}}{I_s} \tag{F15}$$

Then multiply E_{s1}, E_{r1}, I_{s1} and I_{r1} by the ratio of

$$\frac{E_s}{E_{s1}} \tag{F16}$$

G. To find a solution for the induced voltage when current distribution i the return stroke is an impulse $i = \delta_o(t)$.

Solution

For rectangular current in return stroke,

$$V(x,t) = [V_{11} + V_{12} + V_{21} + V_{22}]U(t - to) \tag{G1}$$

with V_{11}, V_{12}, V_{21}, and V_{22} each as a function of x,t.

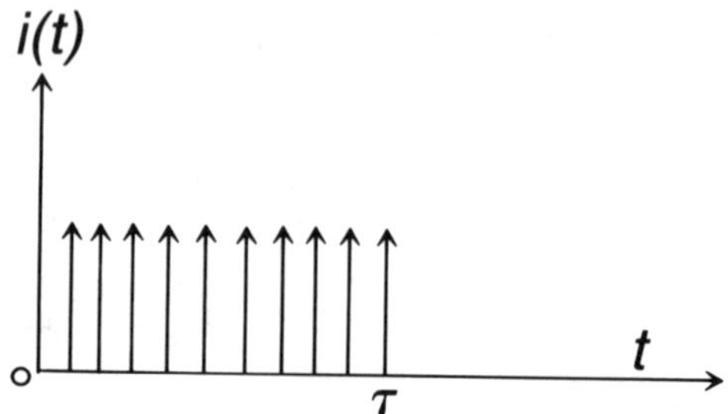

According to Duhamel Theorem, if the current i the return stroke $I'(t)$ is of exponential order and is a continuous function of t, and if its first derivative with respect to k is sectionally continuous, the induced voltage caused by this current is:

$$V(t) = I(o) V_o(t) + \int_o^t \frac{\partial}{\partial t} I(t - \tau) d\tau \tag{G2}$$

where $V_o(t)$ = induced voltage caused by unit step return current, i.e.,

$$i(t) = I_o e^{-at}$$

$$V_o(t) = \frac{d}{dt} U_{-1}(t) \tag{G3}$$

It can be concluded that the solution secured in Equation (I1) is the summation of impulses according to Equation (G1) where $V(x,t)$, due an impulse at $t = 0$, is the total derivative with respect to t of Equation (G1) since $I'(t)$ as a rectangular pulse =

$$I_o U_o(o) + I_o(t - t_1) + I_o(t - 2t_1) + \cdots + I_o(t - nt_1)$$

$$= \sum_{n=o}^{m} I_o \delta_n(t - nt) \tag{G4}$$

$$V(x,t) \text{ due to an impulse} = \frac{d}{dt} \text{ of Equation (G1)} =$$

$$(V'_{11} + V'_{12} + V'_{21} + V'_{22})\delta(t - t_1) \tag{G5}$$

$$\therefore V'_{11} = \frac{d}{dt} \frac{30 I_o h(1 - \beta)}{\beta^2(ct - x)^2 + Y_o^2} \left[\beta(ct - x) + \frac{(ct - x)x - y^2}{\sqrt{c^2t^2 + \frac{1 - \beta^2}{\beta^2}(x^2 + Y_o^2)}} \right] \tag{G6}$$

$$V'_{12} \quad \frac{d}{dt} \frac{-30\,I_o h}{\beta} \left[1 - \frac{1}{1+k^2} - \beta^2 \right] \frac{1}{ct-x} \tag{G7}$$

$$V'_{21} = \frac{d}{dt} \frac{30\,I_o h(1-\beta^2)}{(ct+x)+Y_o^2} \left[\beta(ct-x) + \frac{(ct+x)x + Y_o^2}{c^2 t^2 + \dfrac{1^2-\beta}{\beta^2}(x^2+Y^2)} \right] \tag{G8}$$

$$V'_{22} = \frac{d}{dt} \frac{-30\,I_o h}{\beta} \left[1 - \frac{1}{1+k^2} - \beta^2 \right] \frac{1}{ct-x} \tag{G9}$$

$$K \quad \frac{h_c}{\sqrt{x^2+Y_o^2}}, \; t_o = \frac{\sqrt{x^2+Y_o^2}}{c} \tag{G10}$$

All for $< t < \tau$

H. Calculate a set of natural frequencies generated when two sound phases are interrupted in a three-phase Y system with ground fault.

Solution

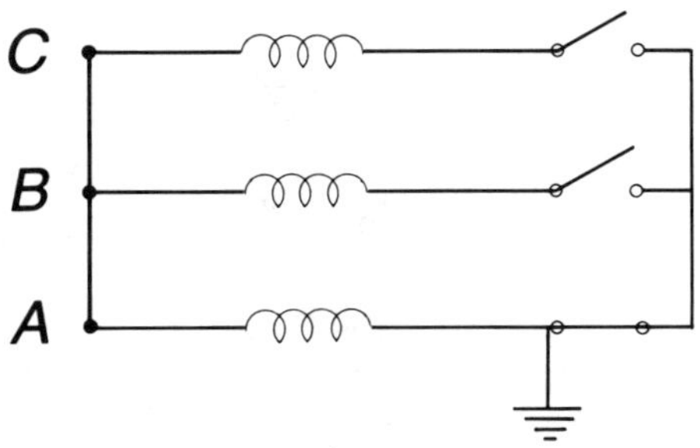

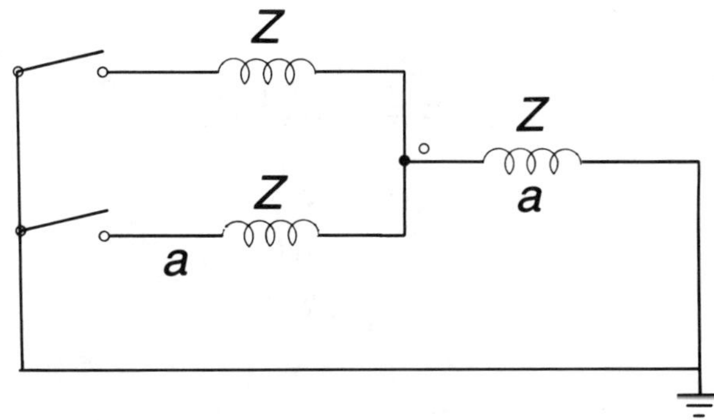

At junction (o), equating Z_{in} of the o.c. line with surge impedance $Z/2$ and that of the s.c. line with surge impedance Z,

$$Z_{in}(\text{o.c.}) = \frac{Z}{2} \tan h\gamma a \qquad (\text{H1})$$

$$Z_{in}(\text{s.c.}) = Z \tan h\gamma a \qquad (\text{H2})$$

$$Z_{in} = Z_o \frac{Z_r \cos h\gamma 1 + Z_o \sin \gamma 1}{Z_o \cos h1 + Z_r \sin \gamma 1} \qquad (\text{H3})$$

$$= Z_o \frac{Z_r \cos h\gamma 1 + \dfrac{Z_o}{Z_r} \sin \gamma 1}{Z_o \cos h\gamma 1 + \dfrac{Z_o}{Z_r} \sin \gamma 1} \qquad (\text{H4})$$

since no losses are considered $\gamma = j\beta$ and hyperbolic elements will be replaced with Trignometrics.

$$\frac{1}{2} Z \operatorname{ctn} \beta a = Z \tan \beta a \qquad (\text{H5})$$

$$\therefore \tan^2 \beta a = \frac{1}{2} \qquad (\text{H6})$$

$$\beta = \frac{v}{V} \qquad (\text{H7})$$

Therefore,

$$\tan^2 \frac{va}{V} = \frac{1}{2} \qquad (\text{H8})$$

or

$$\cos \frac{va}{V} = \pm \frac{1}{\sqrt{1 + \dfrac{1}{2}}} \qquad (\text{H9})$$

$$= \pm \sqrt{\frac{2}{3}}$$

Then,

$$\frac{va}{V} = \cos^{-1}(\pm\, 2/3)$$

$$= 0.618,\ 2.53,\ 3.70 \ldots \tag{H10}$$

And

$$f = \frac{V}{2\pi}$$

$$= \frac{.618\ V}{2\pi},\ \frac{2.53\ V}{2\pi a},\ \frac{3.76V}{2\pi a} \tag{H11}$$

1.XII PROBLEMS

1. A lossless transmission line has a length equal to 200 km. Calculate the fundamental eigen frequency and the corresponding wave number for each of the following cases:
   a.   the line is short-circuited at both ends
   b.   the line is open-circuited at both ends
   c.   the line is short-circuited at one end and open at the other.
2. Consider a bus-bar joining 6 lines, each with $Z = 500$, an incident voltage pulse of 80 kV, with a limiting arrester voltage of 32 kV, and an arrester resister $p = 120\ \Omega$. Calculate the transmitted voltage pulse and then current in the arrester.
3. A series resistance is inserted in the transmission system shown. Calculate maximum efficiency of protection provided by R and the value of R in terms of Z_1, Z_2 and n.

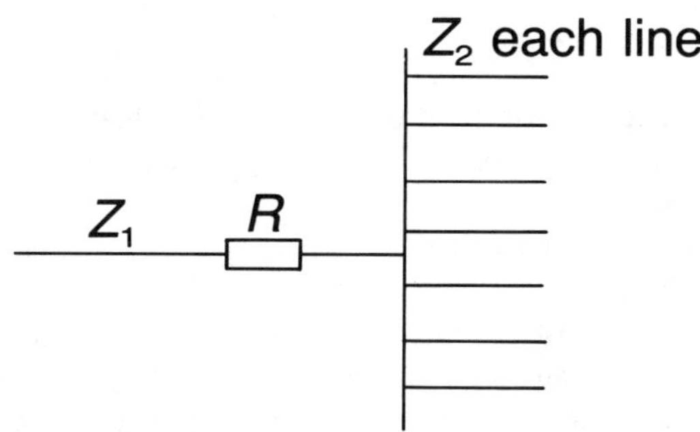

4. Repeat (3) but insert R between the junction and ground.
5. Derive an expression for amplitude V_s frequency spectrum of a step pulse applicd at a location distant from ends of transmission line.

6. Consider a dissipationless transmission line. An impulse voltage is suddenly applied at the sending end at $t = o$. The strength of the impulse is (A). Calculate the current and voltage at point (x) and at time (t) with the receiving end of the line open.

7. Repeat problem (4) when a d.c. voltage is suddenly switched at the sending end and with the receiving end short-circuited.

8. In the figure shown, find an expression for the optimum value of R. To the right of the arrester there are (n) lines with increasing surge impedance.

$$\Delta Z \ll Z$$

$$e_o = \text{arrester limiting voltage}$$

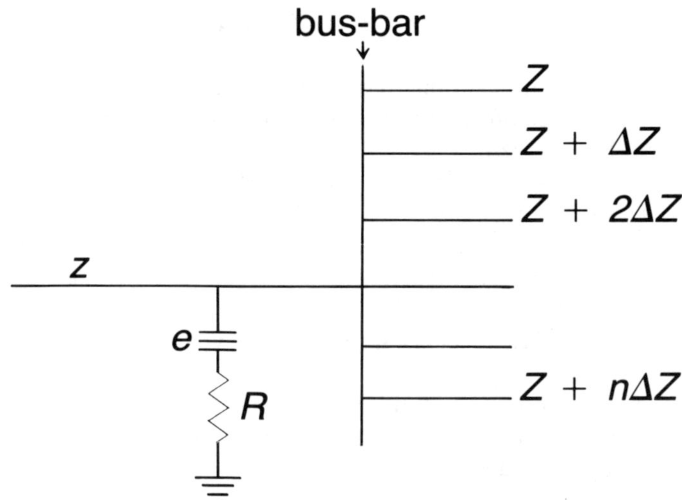

9. An atmospheric electric field of strength 150 kV/m having a time rise of 12 μs exists below a cloud of 20 km equivalent length. The line is 20 m above ground. Calculate
    a. the voltage pulse releases
    b. pulse length
    c. the time rate of change of the pulse potential gradient.

10. A dissipationless transmission $e(t)$ line is suddenly subjected to the surge-time function shown. The surge shown is applied at the sending end at $t = o$. Calculate the current and voltage at any time t along the line, for the receiving end open-circuited. Assume zero initial conditions, i.e., solve for $e(x,t)$ and $i(x,t)$.

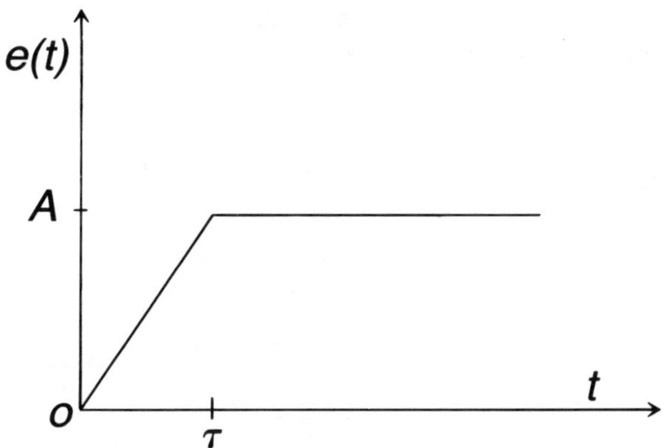

11. A line of 750 miles long has the following distributed constants:

$$R = 5 \ \Omega/\text{mile}$$
$$L = 2 \ \text{mh/mile}$$
$$C = 0.001 \ \mu f/\text{mile}$$
$$G = 1.0 \ \mu v/\text{mile}$$

The line is terminated in $Z_r = 1000 \ \angle 30°$. Generator at the sending end has an emf of 400 V and an internal impedance of 50 Ω. Consider the angular velocity $w - 800$ rad/s, compute voltage, current and power at Z_r using the step-by-step-method.

12. Consider the following figure:

1) In the figure above there are three segmented lines:

$$d = d_1 \ e^{\delta(x - l_1)}; \quad d_1 < d < d_2$$
$$\epsilon_1 = ax + b$$
$$\mu_2 = 1000 \ \mu_o$$

Calculate

a. the overall voltage transformation ratio
b. velocity of propagation in each segment
c. total transition time in l_1 and l_2.

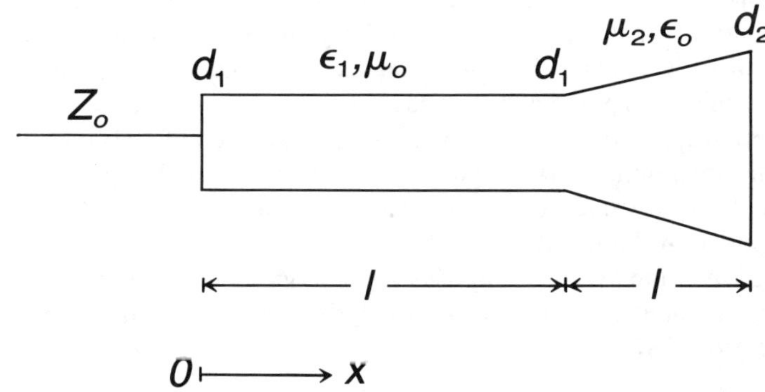

13.   In the figure below, find an expression for optimum P where the transmission system to the right of the arrester represents a pattern of decreasing surge impedance.

$$e_o = \text{arrester limiting potential}$$

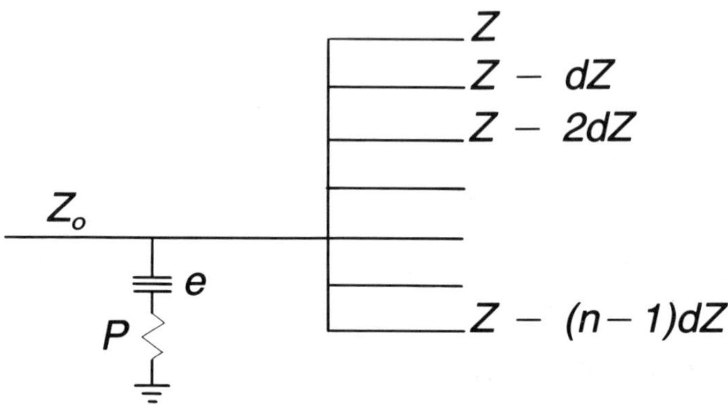

14.   a.   Explain the effect of voltage surges due to reignitions.
      b.   Explain the role of the front-rise length of a switching pulse.
      c.   Explain the effect of the equivalent resistance of the non-linear arrester with respect to the amplitude of any transmitted voltage surge.
      d.   Explain the role of skin effect in broadening a step voltage surge and hence reducing its strength.
15.   a.   Calculate the velocity of propagation of a front-discharge where breakdown field is 35 kV/cm, front dia. is 2 mm, carrying a current of 20 A.
      b.   Calculate the corresponding dielectric constant of the medium.
      c.   Calculate the average time rise and also the average front-length.
16.   A plane-parallel plate condenser has a plate separation of 0.9 cm in air under standard conditions. Determine the spark breakdown voltage between the plates. A plate of lead glass 0.3 cm thick, having a specific inductive capacity of 7, is inserted between the plates and in perfect contact with one plate. Calculate the voltage across the condenser required to break down the air-gap under this condition.
17.   Determine analytically an expression for the value of (Pd) corresponding to the minimum sparking potential for plane-parallel copper electrodes. Calculate the value of (Pd) for min. V_s for air by means of this relation comment on the result.
18.   A transmission system comprises an underground cable joined to an overhead line and then to a cylindrical type transformer with both core and tank grounded.

a.   Find an order of magnitude value of a pulse transformation ratio initially released at some point on the cable.

b.   A series resistance is placed at a junction between the line and transformer; calculate the value of that resistance for optimum protection.

c.   Repeat (*a*) if the transformer is without tank.

19.   An atmospheric electric field of 110 kV/m having a time rise of 15 μs exists below a cloud of 2 km equivalent length. The transmission line is 10 m above ground. Calculate:

a.   the amplitude of voltage pulse released

b.   the approximate length of the pulse

c.   electric field due to the released pulse

20.   Consider a 66-kV line having insulator breakdown voltage to ground of 400 kV and a line surge-impedance of $Z = 500\ \Omega$. A flashover occurred through an erected pole.

a.   Calculate the amplitude of the released current pulses.

b.   Calculate the amplitude of current pulse in the discharge breakdown.

c.   Calculate the amplitude of the released voltage pulses.

d.   Considering the value found in (*c*), write an expression of probability of flashover at any other tower in terms of frequency of occurrence. Comment on that.

21.   Analyze, in terms of order of magnitude with respect to velocity of propagation, degree of reflection and voltage transformation ratio, a transmission line of surge impedance Z to be connected to the following sections, one at a time:

a.   an underground cable three-phase system

b.   transformer with grounded tank and core

c.   tapered overhead line and terminated in open-circuit. The line is surrounded by a magnetic insulator.

22.   To protect a transformer winding whose surge impedance $= 5.5k$ against a delayed voltage pulse shown in the figure below. The surge impedance of the connecting line is 500 ohms. Peak stress in the transformer $e = 26$ kV. Design a protective series L or C, and a combination. The line lengths are $(o - a) = 2$ km, and $(a - b) = 1.2$ km.

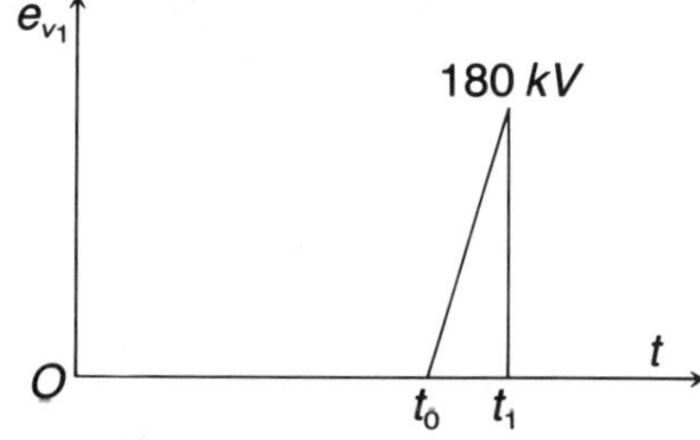

23.   In the first figure shown, a reactor with hard magnetic core is used to provide protection for region 2 against an incident pulse of strength E. The second figure illustrates R and X in terms of current passing in the reactor; obtain solution for:

$$e_{V2},\ e_{r1},\ I_{V2},\ I_{V1},\ T_2$$

Identify the time invariant voltage surge and min e_2.

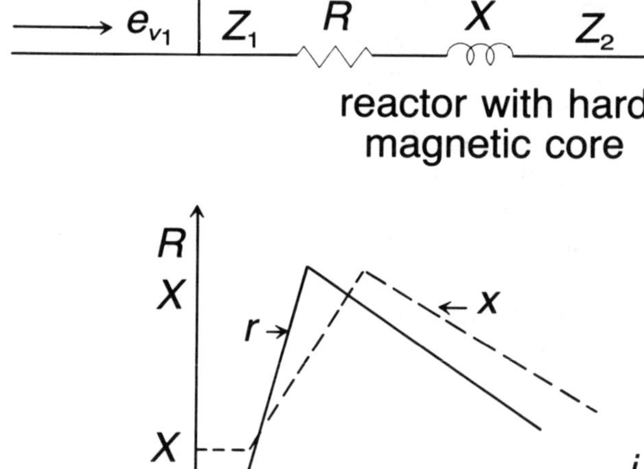

24.   Repeat problem 23 for a shunt connection of the reactor as shown in the figure below.

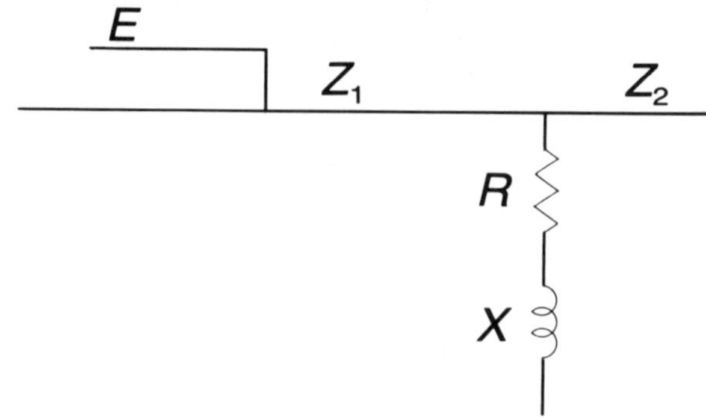

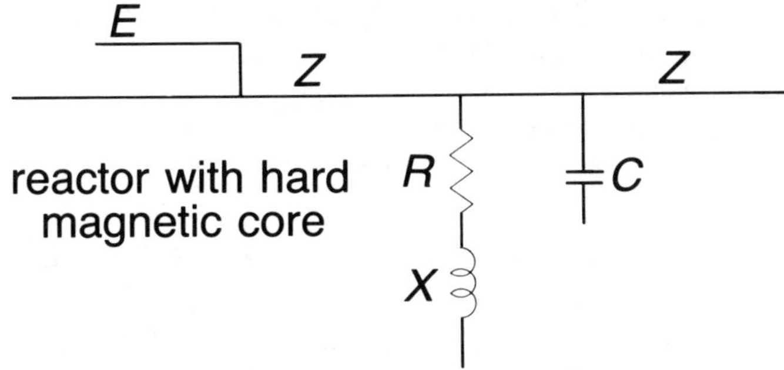

25. Repeat problem 24 for $1V_1$, a periodic pulse train. Find an optimum value for C and investigate the case of resonance.

26. a. Consider the case for the incidence of a sinusoidal train pulse in an arbitrary location on a dissipationless line. Under what condition will maximum power transfer occur on sides of pulse incidence, if one of the lines is o.c. and the other is s.c.

    b. Write down the basic criterion for zero dissipation and then for no-distortion transmission line.

27. Consider a dissipationless transmission line. A voltage surge shown is suddenly applied at the sending end at $t = o$. Calculate the current and voltage at any point (x) and time (t) for the receiving end as open-circuit.

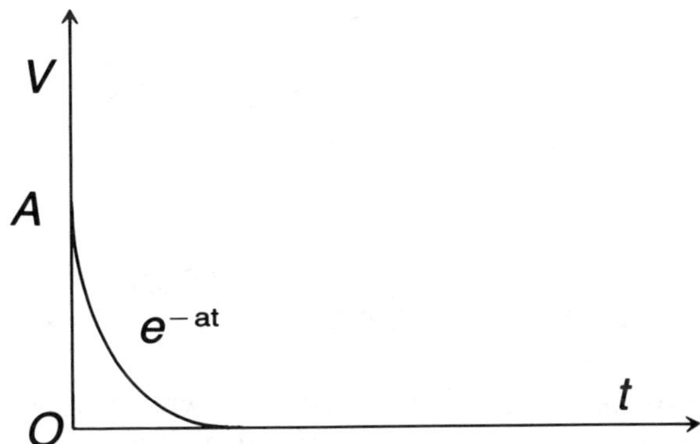

28. An atmospheric electric field of strength 222 kV/m, having a time rise of 10 μs, exists below a cloud of 13 km equivalent length. The line is 20 m above ground. Calculate:

    a. the voltage pulse released

    b. the length of the pulse formed

    c. the voltage gradient of the pulse.

29. Refer to the transmission line configuration shown where E is a linear function of position. Calculate:
    a.   total voltage transformation ratio
    b.   velocity of propagation in region 2
    c.   total transition time in the transmission system

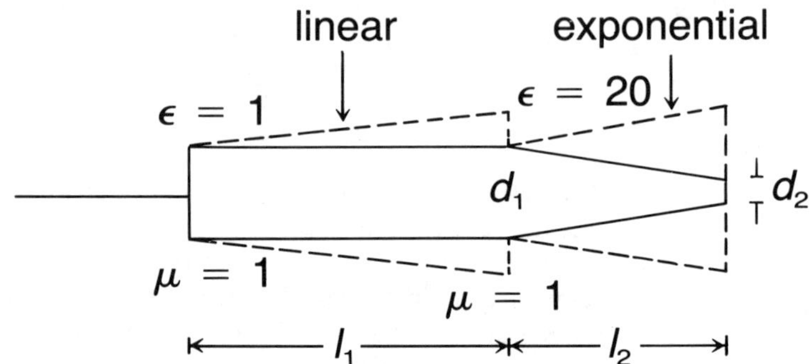

30. This figure shows a Y three-phase faulted system, where the grounded pole of the circuit breaker opens last. Calculate the spectrum of natural frequencies that may result from this operation.

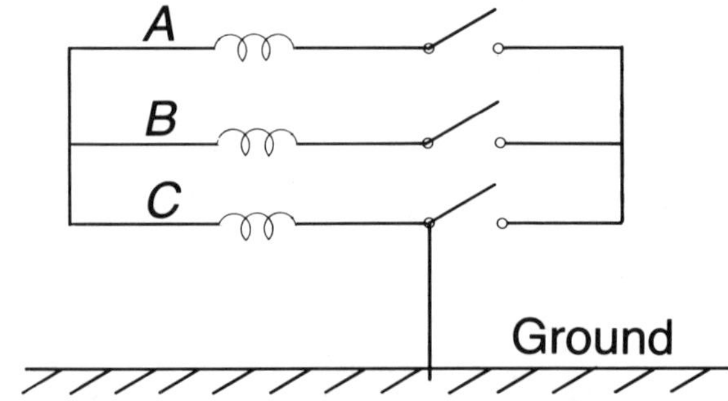

31. In the three-phase short-circuited system shown, interruption did occur at switch (s). Calculate the spectrum of natural frequencies that may result from this accidental process.

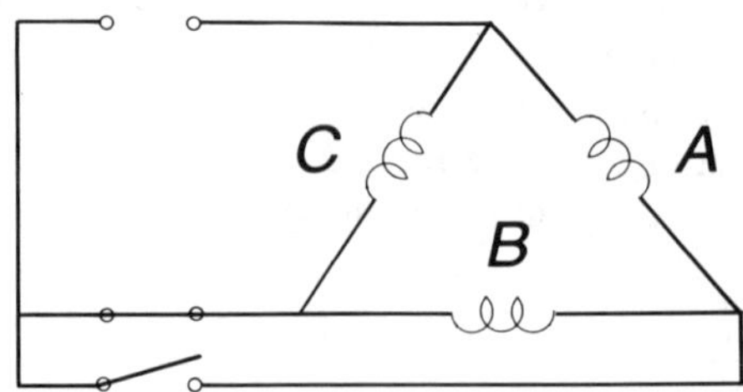

32. Repeat problem 32, but with the interruption of switch (*s*) shown in this figure.

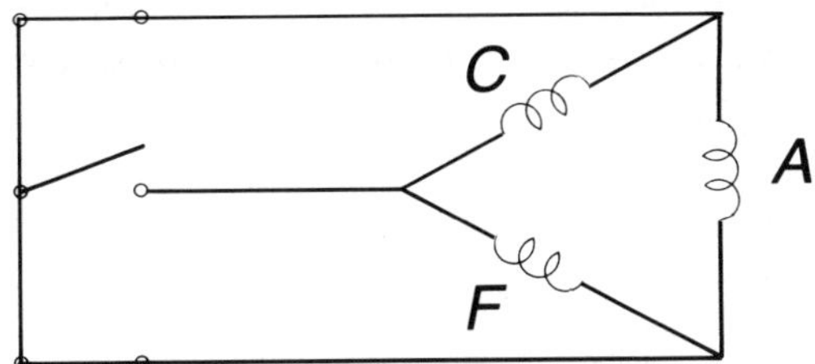

33. Repeat problem 32, but with the interruption of switch (*s*) shown in this figure.

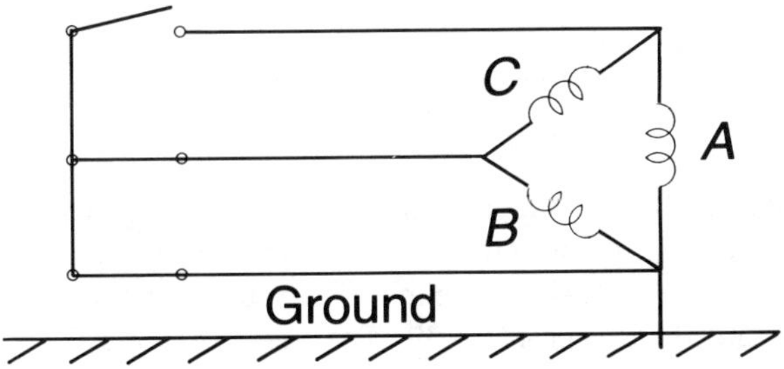

34. An atmospheric electric field generated by a time varying magnetic field density $B(t) = a_x A\ U_o(t) \cos$ cut exists below a cloud of 10 km equivalent length. There is a line of 20 m above ground. Calculate the amplitude of the voltage pulse produced. ax is a unit vector along the x direction, $V_o(t)$ is an impulse and A is a constant.

35. Repeat problem 35 but with a time varying magnetic field expressed by

$$B(t) = a_x A\ U_{-1}(t) \sin wt +$$

$$a_y K\ U_o(t) \cos wt$$

where $U_{-1}(t)$ is a step function, and A and K are constants.

36. An atmospheric breakdown surge generated a current density J vector expressed by

$$J(t) = AU_o(t)e^{-at}\ ax$$

Calculate the vectorial level of the developed magnetic induction (*B*) in the immediate vicinity.

37. From the expression obtained in problem 37 for the B vector, secure an expression for the induced electric field vector and voltage surge build-up at a transmission line 10 m above ground.

38. The actual pulse form for the magnitude of current density of lightning surge is shown in this figure. Show that this pulse form is equivalent to two linear functions.

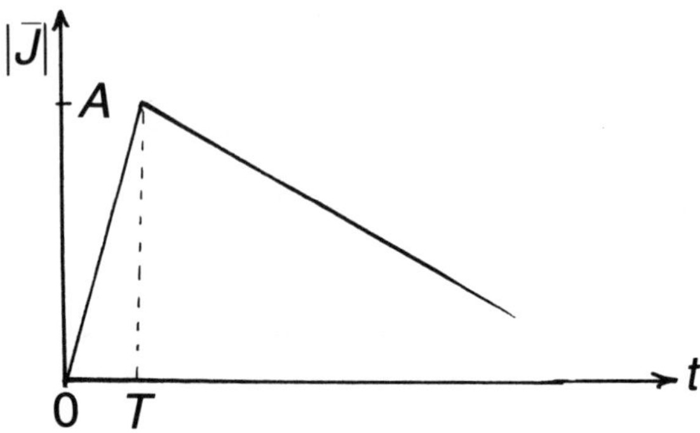

39. Using the method of superposition of wave propagation (the step-by-step method), obtain a solution for voltage and current at the terminal end of a long line whose $Z_r \neq Z_o$, due to the first function of the incident lightning surge established in problem 39. The line has α and β constants.

40. Repeat problem 39 for the other functional part of the lightning surge.

41. Show that voltage and current at Z_r obtained by adding the corresponding solutions from problems 40 and 41 is the same that could be secured by using final equations for calculating V_r and I_r.

42. Consider a situation regarding the incidence of the lightning surge form shown in problem 39. Express the terminal voltage and current in a system comprising an overhead line connected to a much large surge impedance transformer.

43. Repeat solution of problem 42 for a system comprising underground cable connected to a tapered line with exponential function of increase (i.e., proportional to e^{ax}).

44. Repeat solution of problem 43 for a system comprising an overhead line connected to a tapered line whose surge impedance increases linearly with respect to distance.

PRINCIPAL LIST OF SYMBOLS

ϕ electromagnetic flux

ψ electrostatic flux

p $\partial/\partial t$

v velocity of propagation

λ wavelength

γ propagation constant

ν frequency

a length of a line

Z_o surge impedance

k front line length

Δ equivalent diameter of a composite conductor

r radius of surge streamer

h height of a line

ρ_L reflection coefficient

REFERENCES

1. **Rudenberg, R.,** *Electrical Shock Waves in Power Systems: Traveling Waves in Lumped and Distributed Circuit Elements,* Harvard University Press, Cambridge, MA 1968.
2. **Cobine, J. D.,** *Gaseous Conductors: Theory and Engineering Applications,* Dover Publications, New York, 1941.
3. **Theriault, M. G.,** Lightning protection for substantions, buildings and towers, *The Probability of Being Struck,* Conference Digest of International Electrical, Electronics Conference and Exposition, pp. 96-98, Toronto, Canada, October 1973.
4. **Everitt, W. L.,** *Communication Engineering,* 2nd ed., McGraw-Hill, New York, 1937.

Chapter 2

INDUCING AND INDUCED EFFECTS OF LIGHTNING SURGES

2.I INTRODUCTION

This chapter undertakes as a first step the task of developing a comprehensive analytical picture centered around dynamic modeling for the phenomenon of inducing voltage, induced voltage and the corresponding associative function, with respect to the impact area of field effect.

Generalized source distribution in the return channel stroke for the conductive current and in ground surface for the convective current (which is a measure of time varying bound charges) is central to the models developed at the fields impact area. Prior to the establishment of final models, a closed-form solution for the Helmholtz function of electromagnetic radiation is secured by mathematical transformations. Mathematical models secured at the field area are broad in their dependence on any time rate of development of sources, cloud height and powerline height, velocity of propagation of the return stroke and space location of field influence area. Inducing as well as induced voltages have been calculated for the specific case of constant conductive and convective current densities at the source.

Present information regarding various aspects of the phenomenon of inducing and induced voltage surges produced by lightning points to factual effects of special distribution for the electric current and constant parametric elements in the return stroke. Those elements include velocity of propagation of the return stroke channel, certain height levels for the cloud and conducting structure, and dimensional domain of field effects with respect to the point of impact of the return stroke.

The first phase of this chapter centers on developing procedural mathematical models in three dimensional space for the inducing voltage, the corresponding associative function and the induced voltage, first when current distribution in the return stroke is in the form of step functions and then when it is of a steep linear rise with declining tail.

63

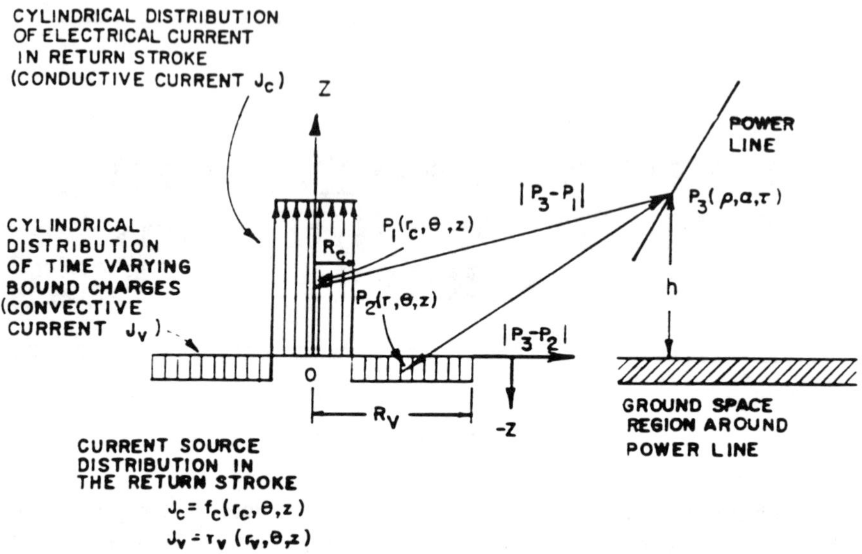

FIGURE 2.1.   Source distribution and field effects of lightning surge.

2.II MATHEMATICAL MODEL FOR INDUCING SURGES DUE TO A CONSTANT STROKE

Detailed simulation for the lightning surge, together with a representation for a tested object, will be presented in association with full analysis for the various modes of lightning surge wave form.

Given the following basic facts:

1. The inducing as well as the induced surges are the result of generalized distribution of bound charges within the local ground terrain, as shown in Figure 2.1.
2. The source function $f(r,\theta,z) = f_c(r_c,\theta_c,z_c) + f(r_v,\theta_v,z_v)$, where $f_c(r_c,\theta_c,z_c)$, $f_v(r_v,\theta_v,z_v)$ represents a step distribution function for an infinite number of concentric cylindrical conductive current sheets.
3. Any point in space (ρ,α,τ) of position vector P_3 located on a physical structure such as a power line with height h above ground.

The development of the following is required:

1. A generalized mathematical model for the inducing voltage at any point (ρ,α,τ) and the respective Legendre associative function.
2. A generalized mathematical model for the induced voltage established at a physical structure with respect to ground.

A. GENERALIZED CONDUCTIVE CURRENT DISTRIBUTION IN THE RETURN STROKE

Refer to Figure 2.1 which shows a cylindrical shell distribution for conductive electric current density in the return stroke with $J = J(r_c,\theta,z)$. Let $\hat{Z}$ and $\hat{y} =$ impedivity and admitivity of surrounding space, respectively, where $\hat{Z} = j\infty\mu$ and $\hat{y} = \infty + j\omega\epsilon$.

This problem is treated similarly to that of field radiation from the source at the coordinates (r_c,θ,z) to the point structure $P_3(\rho,\alpha,\tau)$ and uses the differential form of Maxwell's Field equations, namely,

$$\overline{\nabla} \times \overline{E} = -\hat{z}H \tag{2.1a}$$

$$\overline{\nabla} \times \overline{H} = \hat{y}E \tag{2.1b}$$

Then, by taking the curl of Equations (2.1a) and (2.1b), and with an appropriate vector identity, (2.1a) and (2.1b) become

$$V^2\overline{E} + k^2E = 0 \tag{2.2a}$$

$$V^2\overline{H} + k^2H = 0 \tag{2.2b}$$

Equations (2.2a) and (2.2b) are vector wave equations and of the same form as Helmholtz scalar wave equation,

$$\overline{\nabla}^2\phi + k^2\phi = 0 \tag{2.3}$$

where

$$k = \sqrt{-\hat{z}\hat{y}} \tag{2.4}$$

In cylindrical coordinates Equation (2.3) becomes:

$$\frac{1}{r}\frac{\partial}{\partial r}\left(r\frac{\partial\phi}{\partial r}\right) + \frac{1}{r^2}\frac{\partial^2\phi}{\partial\theta^2} + \frac{\partial^2\phi}{\partial Z^2} + k^2\phi = 0 \tag{2.5}$$

In order to secure a generalized solution for the function $\phi(|\rho - r_c|, (\alpha - \theta), |\tau - z|)$ in three dimensions, the wave function ϕ must be Fourier transformed with respect to the axial dimension z (ϕ is the transformed function),

$$\overline{\phi}_c|\rho - r_c|,(\alpha - \phi),|\tau - Z|) = \int_{-\infty}^{\infty} \phi_c(|P - r_c|,(\alpha - \phi),|\tau - z|)e^{-j\infty z}\, dz \tag{2.6}$$

and Equation (2.5) becomes a two-dimensional problem as below:

$$\frac{1}{r_c}\frac{\partial}{\partial r_c}\left(r_c\frac{\partial\phi}{\partial r_c}\right) + \frac{1}{r_c}\frac{\partial^2\phi}{\partial\theta^2} + m^2 = 0 \tag{2.7}$$

where

$$m^2 = k^2 - \omega^2 \tag{2.8}$$

Solution of ø as a two-dimensional wave function in r_c and ø at the space point (ρ, α, τ) is given in the following section.

B. SOLUTION OF THE TRANSFORMED FUNCTION ø (r_c, θ)

Details for carrying out the solution for ø is left as a problem for the student to solve. However, the final result is given below:

For $r_c < R_c$

$$\phi_c^- = \sum_{-\infty}^{\infty} A_n J_n(k|\rho - r_1|)e^{jn(\alpha - \theta)} \tag{2.9}$$

and for $r_c > R_c$

$$\phi_c^+ = \sum_{-\infty}^{\infty} B_n H_n^{(2)}(k|\rho - r_1|)e^{jn(\alpha - \theta)} \tag{2.10}$$

where J_n = Bessel function of the first kind, and $H_n^{(2)}$ = Hankel function of the second kind. The appropriate boundary conditions are

$$\bar{a}_{rc} \times (H^+ - H^-) = J_z^{(\alpha)}$$

$$(E^+ - E^-) \times \bar{a}_{rc} = 0 \tag{2.11}$$

where $\bar{a}_{rc}$ is a unit vector.

The field radiated from the multi-cylindrical current sheets is characterized as TM^z, independent of z and whose components are expressed below in terms of the wave scalar complex function ϕ_c referring to a conductive current source in the return stroke.

$$\bar{E}_r = \frac{1}{\hat{y}} \frac{\partial^2 \phi_c}{\partial r_c \partial z} \qquad \bar{H}_{r1} = \frac{1}{r_c} \frac{\partial \phi_c}{\partial \theta}$$

$$\bar{E}_\alpha = \frac{1}{\hat{y} r_c} \frac{\partial^2 \phi_c}{\partial \theta \partial z} \qquad \bar{H}_\alpha = \frac{\partial \phi_c}{\partial r_c} \tag{2.12}$$

$$\bar{E}_z = \frac{1}{\hat{y}} \left(\frac{\partial^2}{\partial z^2} + k^2 \right) \phi_c, \quad \bar{H}_z = 0$$

Then, from Equations (2.9) to (2.12), the following are obtained:

$$B_n = C_n J_n(kR_c)$$

$$A_n = C_n H_n^{(2)}(kR_c) \tag{2.12a}$$

$$\bar{\phi}_c = \sum_{-\infty}^{\infty} C_n H_n^{(2)}(kR_c) J_n(k|\rho - r_c|) e^{jn(\alpha - \theta)} \tag{2.13}$$

$$\phi_c^+ = \sum_{-\infty}^{\infty} C_n J_n(kR_c) H_n^{(2)}(k|\rho - r_c|) e^{jn(\alpha - \theta)} \tag{2.14}$$

$$H_\theta^+ = -\sum_{o}^{\infty} C_n J_n(kR_c) K\, H_n'^{(2)}(k|\rho - r_c|) e^{jn(\alpha - \theta)} \tag{2.15}$$

$$H_\theta^- = -\sum_{o}^{\infty} C_n H_n^{(2)}(kR_c) K\, J_n'(|\rho - r_c|) e^{jn(\alpha - \theta)} \tag{2.16}$$

From Equations (2.11) through (2.16), the solution of the transformed field radiation wave function ϕ_c in two dimensions is expressed as

$$\phi_c^- = \frac{\pi r_1}{2j} \sum_{-\infty}^{\infty} A_n J_n(k|\rho - r_c|) H_n^{(2)}(k|\rho - r_c|) e^{jn(\alpha - \theta)} \tag{2.17}$$

where

$$A_{nc} = \frac{1}{2\pi} \int_o^{2\pi} \bar{J}_c e^{-jn(\alpha - \theta)}\, d\theta \tag{2.18}$$

$\bar{J}_c = f_c(r_c, \theta, z)$ is the Fourier transformed function of the total generalized conduction current volume distribution in the return channel stroke. $\bar{J}_c(r_c, \theta, z)$ is the three-dimensional volume distribution function for the conductive currents.

In order to obtain the final form of the wave function ϕ_c, the inverse Fourier transform of ϕ_c must be carried out,

$$\phi_c = \frac{1}{2\pi} \int_{-\infty}^{\infty} \bar{\phi}_c(|\rho - r_1|, (\alpha, \theta), (\tau - \infty))\, e^{j\omega z}\, dz \tag{2.19}$$

The corresponding electromagnetic field components at the point $P_3\ (\rho, \alpha, \tau)$ could be secured from the set of Equation (2.12).

C. GENERALIZED CONVECTIVE CURRENT DUE TO BOUND CHARGES

Electromagnetic field components produced at point P_3 by the effect of the time variation of bound charges could follow the same procedure established in part A, with these changes: (1) $f_c(r_c,\theta,z)$ is replaced by $f_2(r_v,\theta,-z)$ which is a generalized function for the time variation of bound charges residing on the ground surface; (2) radial separation between the point of field impact and the bound charge source becomes $|P_3 - P_2|$; (3) generalized solution for the resulting transformed two-dimensional Helmholtz function ϕ_v for bound charge source becomes

$$\phi_v^- = \frac{\pi r_2}{2j} \sum_{-\infty}^{\infty} A_n J_n(k|\rho - r_v|) H_n^{(2)}(k|\rho - r_v|) e^{jn(\alpha - \theta)} \tag{2.20}$$

where

$$A_{nv} = \frac{1}{2\pi} \int_o^\infty \frac{2\pi}{\overline{J}_v} e^{-jn(\alpha - \theta)} \, d\theta \tag{2.21}$$

$J_v = f_v(r_v,\theta,-\infty)$ is the Fourier transformed function of the total generalized convection current distribution. That is θ and r_v dependent.

The final form for the wave equation ϕ_v is

$$\phi_v = \frac{1}{2\pi} \int_{-\infty}^\infty \overline{\phi}_2(|\rho - r_v|,(\alpha,\theta),(\tau - \infty) e^{j\omega z} \, dz \tag{2.22}$$

Also the corresponding electromagnetic field components generated by the convective current of the time varying bound charges could be secured from the set of Equations (2.12).

D. SIGNIFICANCE OF $\overline{J}_c$ AND $\overline{J}_v$

J_c represents the conductive current density generated by the return stroke, while J_v is the convective current density generated by the time variation of the bound electric charges.

In this section the conductive current density J_c as well as the convective component J_v are generalized functions in the three-dimensional cylindrical co-ordinate system.

The important fact is that such dependence on r, θ and z by the sources J_c and J_v implies variable status for the cloud height as well as the velocity of propagation for the return conducting stroke and the time rate for the development of bound charges in depth and width.

These source functions could be any arbitrary form provided they are analytic in order to be Fourier transformed.

2.III PROCEDURAL CALCULATION OF THE INDUCING VOLTAGE

Let the height of point P_3 above ground be h; the inducing voltage V_i is expressed as follows:

$$V_i = -\int E_P d_p \qquad (2.23)$$

where

$$\overline{E}_P = \overline{E}_c + \overline{E}_v \qquad (2.24)$$

E_c = inducing electric field due to conductive current distribution given by Equation (2.12), and E_v = inducing electric field due to convective current distribution of time varying bound charges, which could be obtained in a similar way to Equation (2.12).

A. THE ASSOCIATIVE FUNCTION

As a matter of fact, the electromagnetic effects produced at point P_3 are retarded in time with respect to their occurrence at the sources, by

$$t' = t - \frac{|\rho - R_c, R_u|}{C_o} = t - t_o \qquad (2.25)$$

where it is practical to select the time $t = 0^+$ as the instant of start for the return stroke and initial formation of bound charges. C_o is the velocity of light.

Also, because of the time retardation interval, the inducing voltage function V_i is a sectioned function, i.e.,

$$V_i = \psi[(\rho - r),(\alpha - \theta),(\tau - Z)] \, U(t - t_o) \qquad (2.26)$$

$U(t - t_o)$ is a shifted step function. $\psi(\rho,\alpha,\tau,\theta,r,Z)$ is the associative function for the entire lightning phenomenon.

2.IV THE INDUCED VOLTAGE V

Figure 2 represents the equivalent circuit of a transmission line, the induced voltage of which is expressed through the following differential equation:

$$\frac{\partial^2 V}{\partial x^2} - \frac{1}{C^2} \frac{\partial^2 V}{\partial t^2} = \frac{-1}{C^2} \frac{\partial^2 V_i}{\partial t^2} = F(x,t) \qquad (2.27)$$

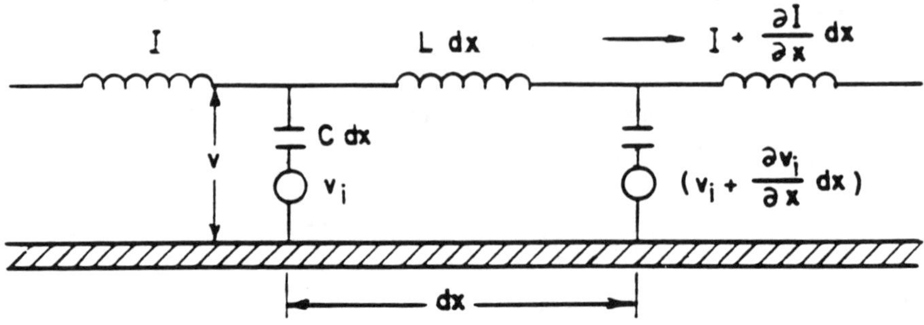

FIGURE 2.2.   Inducing voltage on power line.

where

$$C = 1/\sqrt{L'C'} \tag{2.28}$$

L' and C' = self-inductance and capacitance per unit length of the transmission line, respectively

$$V \text{ is bounded as } x \to \pm\infty \tag{2.29}$$

$$V(x,0) = V'(x,0) = 0 \tag{2.30}$$

Of course the generalized V_i modeled in this paper is in the three-dimensional cylindrical system, which must be converted in terms of cartesian coordinates upon planning to solve Equation (2.27) for $v(x,y,z,t)$.

The best approach for solving Equation (2.30) is through the introduction of Green's function $G(x,x')$.

$$V(x) = \int_a^b G(x,x')F(x')dx' \tag{2.31}$$

where $(a \cdots b)$ is the domain over which $F(x')$ is continuous, x is a point on the line, and x' is a source point.

The G function has these properties:

1. $G(x,x'^{+0}) - G(x,x'^{-0}) = 0$

2. $$\dfrac{dG(x,x'^{+0})}{dx} - \dfrac{dG(x,x'^{-0})}{dx} = 0 \tag{2.32}$$

3. $G(x,x')$ is the voltage induced at point (x) due to a unit impulse at the source point (x').

4. $G(x,x')$ satisfies the boundary condition piecewise continuous in the domain $(a \ldots b)$ and is homogeneous everywhere except at $x = x'$.

The solution for Green's function from Equation (2.31) could be secured by applying either Fourier or Laplace transformation, followed by taking the inverse transform.

Substituting the generalized solution of the G function in Equation (2.31) will establish the mathematical solution for the generalized induced voltage function.

The method of Laplace transform and its inverse will be presented through the application of Helmholtz field wave radiation function generated by constant current source distribution function. The generalized form of V_i is reflected in the arbitrary function of J_c and J_v.

Solutions of the inducing electromagnetic field components in generalized cylindrical coordinate system cover detailed parametric information as for available velocity of propagation of the conductive current in the return stroke and the time rate of developing the convective current in the layer of bound charges, height of cloud and any physical structure subjected to the field effect and how far from the source location.

2.V CALCULATION OF INDUCING FIELDS FOR CONSTANT CURRENT SOURCE DISTRIBUTION

A specific case is considered for uniform constant densities for the electric current in the return stroke as well as for the convective current produced by time variations of bound charges. Reference is made to Figure 2.1 with both J_c and J_v independent of r, θ and z, i.e.,

$J_c = A_c$, constant density

$J_v = A_v$, constant density

Fourier transform with respect to Z for $\bar{J}_c$ and $\bar{J}_v$ is:

$$\bar{J}_c(r_c,\theta,\omega) = \frac{A_c}{j\omega} \tag{2.33}$$

and

$$\bar{J}_c(r_v,0,\omega) = \frac{A_v}{j\omega} \tag{2.34}$$

Therefore,

$$A_{nc} = \frac{1}{2\pi} \int_o^{2\pi} \bar{J}_c e^{-jn(\alpha\theta)} - d\theta$$

$$= \frac{A_c e}{\pi n\omega} \quad \text{where } n \text{ is odd}$$

$$= 0 \quad \text{where } n \text{ is even} \tag{2.35}$$

Similarly,

$$A_{nv} = \frac{A_c e^{-jnx}}{\pi n \omega} \qquad \text{where } n \text{ is odd}$$

$$= 0 \qquad \text{where } n \text{ is even} \qquad (2.36)$$

The radiation field wave function transformed in the frequency domain from Equation (2.17) is expressed below for the conduction as well as for convective current sources:

$$\bar{\phi}_c = \frac{r_c A_c}{j2\omega} \sum_{n=0}^{\infty} \frac{e^{-jn\alpha}}{n} \, e^{jn\alpha} J_n(k|\rho - r_c|) H_n^{(2)}(k|\rho - r_c) \qquad (2.37)$$

where n is odd.

$$\bar{\phi}_v = \frac{r_v A_v}{jz\omega} \sum_{n=0}^{\infty} \frac{e^{-jn\alpha}}{n} \, e^{jn\alpha} J_n(k|\rho - r_v|) H_n^{(2)}(k|\rho - r_v) \qquad (2.38)$$

Solution for the radiation field wave functions ϕ_c and ϕ_v can be obtained by taking Fourier inverse transform with respect to ω:

$$\phi_c = \frac{r_c \tau A_c}{j\omega z} \sum_{n=0}^{\infty} \frac{1}{n} \, J_n(k|\rho - R_c|) H_n^{(2)}(k|\rho - R_c) \qquad (2.39a)$$

$$\phi_c = \frac{r_c \tau A_c}{2} \sum_{n=0}^{\infty} \frac{1}{n} \, J_n(k|\rho - R_c|) H_n^{(2)}(k|\rho - R_c) \qquad (2.39b)$$

$$\phi_v = \frac{r_c \tau A_v}{2} \sum_{n=0}^{\infty} \frac{1}{n} \, J_n(k|\rho - R_c|) H_n^{(2)}(k|\rho - R_c) \qquad (2.40)$$

Electromagnetic field components at a relatively far point from the source generated by the conduction constant curent density J_c are obtained through the set of Equation 2.12.

$$\bar{E}_\rho = A_c \sqrt{\frac{j}{\pi}} \, \frac{\sqrt{\rho - R_c}}{R_c \hat{y}} \sum_{n=0}^{\infty} \frac{j_n e^{-j(\rho - R_c)}}{2^n \, n!} \left[-\frac{(\rho - R_c)^n}{R_c^n} \right.$$

$$\left. - (\rho - R_c)^{n-1} - \frac{(\rho - R_c)^n \frac{3}{2}}{\sqrt{2n}} + \frac{j2^n(n-1)!}{\sqrt{2}} \right] \qquad (2.41)$$

$$\bar{E}_\alpha = \frac{jA_c}{2\hat{y}} \sqrt{\frac{2j}{\pi(\rho - R_c)}} \, e^{-j(\rho - R_c)} \sum_{n=0}^{\infty} \frac{j^n(\rho - R_c)^n}{2^n \, n!} \qquad (2.42)$$

$$\overline{E}_\tau = \frac{A_c R_c k^2 \tau}{2\hat{y}} \sqrt{\frac{j2}{\pi(\rho - R_c)}} \sum_{n=0}^{\infty} \frac{j^n(\rho - R_c)}{2^n\, n!} \tag{2.43}$$

$$\overline{H}_\rho = \frac{jA_c\tau}{2} \sqrt{\frac{2j}{\pi(\rho - R_c)}}\, e^{-j(\rho - R_c)} \sum_{n=0}^{\infty} \frac{j^n(\rho - R_c)^n}{2^n\, n!} \tag{2.44}$$

$$\overline{H}_\alpha = \frac{A_c\tau}{2} \sum_{n=0}^{\infty} \left[\frac{1}{n^2\rho}\, Jn(\rho - R_c)H_n^{(2)}(\rho - R_c) - \right.$$

$$\frac{-1}{nR_c} - H_n^{(2)}(\rho - R_c)\frac{(\rho - R_c)^{n-1}}{2^{n-1}(n-1)!} + \sqrt{\frac{2}{\pi}}\,(-j)^n + \frac{1}{2}$$

$$J_n^{(2)}(\rho - R_c)\frac{1}{nR_c}\frac{e^{-(\rho-R_c)}n-1}{(\rho - R_c)^2\, n - 1} + j^n + \frac{3}{2}\sqrt{\frac{2}{\pi(\rho - R_c)}}$$

$$\left. e^{-j(\rho - R_c)}\frac{1}{nR_c} \right] \tag{2.45}$$

where

$$J_n(\rho - R_c) \text{ at large } \rho \approx \frac{1}{n!}\frac{(\rho - R_c)^n}{2^n} \tag{2.46}$$

and

$$H_n^{(2)}(\rho - R_c) \text{ at large } \rho \approx \frac{j2}{\pi(\rho - R_c)}\, j^n e^{-j(\rho - R)} \tag{2.47}$$

$$H_\tau = 0 \tag{2.48}$$

Next the electromagnetic field components generated by the time varying bound charges and represented by the convective constant current density $J_v = A_v$ are exactly those of E_ρ, E_α, E_τ, H_ρ, H_α and H_τ expressed in the set of Equations (2.12) with replacement of A_c and A_v, and the radius of the return stroke R_c with the larger value of R_v.

2.VI CALCULATION OF INDUCING VOLTAGE

At the field point (ρ,α,τ), in cylindrical coordinates,

$$-\nabla V_i = -\left[\overline{a}_\rho \frac{\delta V_i}{\delta\rho} + \overline{a}_\alpha \frac{1}{\rho}\frac{V_i}{\delta\alpha} + \overline{a}_\tau \frac{V_i}{\delta\tau} \right]$$

$$= \hat{a}_\rho E_\rho + \hat{a}_\alpha E_\alpha + \hat{a}_\tau E_\tau \tag{2.49}$$

$$\overline{V}_i = -\int E_\rho d_\rho = -\int \frac{\delta V_i}{\delta_\rho} \, d\rho = V_{ic} + V_{iv} \tag{2.50}$$

$$V_{ic,v} = (A)\sqrt{\frac{j}{\pi}\frac{1}{R\hat{y}}} \sum_{n=0}^{\infty} \frac{j^n}{2^n n!} \quad \text{multiplied by}$$

$$\left[\frac{\rho^{n+1/2}e^{-jnp}}{jn} - \frac{n + \tfrac{1}{2}}{jn^2 R} \int \rho^{n-1/2}e^{-jnp} \, d\rho \; + \right.$$

$$\left. \frac{j2^{n+}(n-1)!}{3\sqrt{2}} \; \rho^{3/2} \right], \quad \begin{array}{l} A = A_c, \; R = R_c \text{ for conduction} \\ \text{and } A = A_v, \; R = R_v \text{ for convection} \end{array} \tag{2.51}$$

Of course the integration process could be repeated to any limit on n.

Due to retardation of field effects being felt at $\rho(\rho,\alpha,\tau)$ the inducing voltage V_i will be delayed and hence it will take the wave shape of a sectioned function in an associative form.

$$\psi = V_i(\alpha,\rho,\tau,t)U(t - t_o) \tag{2.52}$$

where

$$t_o = \frac{(\rho - R_c)}{C_o} \tag{2.53}$$

and C_o is the velocity of light.

2.VII SOLUTION OF THE INDUCED VOLTAGE ON A POINT OF A TRANSMISSION LINE

Here the method of Laplace transform is used to obtain the solution of the induced voltge function $V(x)$ through the equation

$$\frac{\partial^2 V}{\partial x^2} - \frac{1}{C^2}\frac{\partial^2 V}{\partial t^2} = \frac{1}{C^2}\frac{\partial^2 V_i}{\partial t^2} \tag{2.54}$$

where $x = \rho \cos \alpha$

$$V_i = V_i(\rho,\alpha,\tau)U(t - t_o)$$

$$\frac{\partial V_i}{\partial t} = V_i(\rho,\alpha,\tau)\delta(t - t_o)$$

$$\frac{\partial^2 V_i}{\partial t^2} = V_i(\rho,\alpha,\tau)\delta'(t - t_o) \tag{2.55}$$

or in cartesian coordinates,

$$\frac{\partial^2 V_i}{\partial t^2} = V_i(x,y,Z)\delta(t - t_o)$$

where $\delta(t - t_o)$, $\delta'(t - t_o)$ is the delayed impulse and its first derivative, respectively.

$$\frac{\partial^2 V(x,t)}{\partial x^2} - \frac{1}{C^2}\frac{\partial^2 V(x,t)}{\partial t^2} = \frac{1}{C^2} V_i(x,y,Z)\delta'(t - t_o) \tag{2.56}$$

Taking the Laplace transform of the above equation with respect to time (t), and treating V_i at constant y and Z, V_i becomes a function of x only,

$$\frac{\partial^2 V(x,s)}{x^2} - \frac{s^2}{C^2} V(x,s) = \frac{[V_i(x)s]e^{-t_o s}}{C^2} \tag{2.57}$$

Total solution of $V(x,s) = V_1(x,s)$ where $V_2(x,s)$ is the complementary solution and $V_2(x,s)$ is the particular integral solution

$$V_1(x,s) = k_1 e^{-\frac{s}{c}x} + k_2 e^{\frac{s}{c}x} \tag{2.58}$$

$$V_2(x,s) = k_3 V_i(x,s)s\,\frac{e^{-t_o s}}{C^2} = k_3 V_i(x)\,\frac{e^{-t_o s}}{C^2} \tag{2.59}$$

where $V_i(x,s) = V_i(x)/s$.

k_1, k_2 and k_3 are functions in terms of s and for convergent $V(x,s)$, $K_2 = 0$

$$V(x,s) = k_1 e^{-\frac{s}{c}x} + \frac{k_3}{C^2} V_i(x)e^{-t_o s} \tag{2.60}$$

Now taking the inverse Laplace transform of Equation

$$V(x,t) = \pounds^{-1} V(x,s) \tag{2.61}$$

$$V(x,t) = k_1 U_{-1}\left(t - \frac{x}{c}\right) + \frac{k_3}{C^2} V_i(x)U(t - t_o)t_o = \frac{x}{C_o} \tag{2.62}$$

where

$$V_i(x) = V_i(x,y,Z)| = V_i(\rho,\alpha,\tau) \text{ with}$$

$$y,X = \text{constant}$$

$$x = \rho \cos \alpha$$

$$y = \rho \sin \alpha$$

$$Z = \tau \tag{2.63}$$

The two constants could be determined from the following boundary conditions: (1) $V(0,t) = 0$, i.e., no induced voltage at the source; and (2) $V(L,t) = 2\,V_{max}$. Condition 2 is based on the fact that for a long line ($x = L$), the termination is approaching open-circuit, hence the voltage pulse is twice the maximum of the incident voltage pulse.

Boundary condition (1) above implies $k_1 = 0$ since the second term in Equation (2.62) is zero.

$$V(x,t) = \frac{k_3}{C^2} V_i(x) U_{-1}(t - t_o)$$

Then from boundary (2)

$$2\,V_{max} = \frac{k_3}{C^2} V_i(x = L) U_{-1}(t - t_o)$$

$$k_3 = 2C^2\,V_{max}/V_i(x = L) U_{-1}(t - t_o) \tag{2.64}$$

Hence, the complete and total solution of $V(x,t)$ is

$$V(x,t) = \frac{2C^2 V_{max}}{V_i(x = L) U(t - t_o)}\, V_i(x) U(t - t_o) \tag{2.65}$$

$$V_i(x,t) = V_{ic}(x,t) + V_{iv}(x,t) \tag{2.66}$$

$$V_{c,v}(x,t) = \frac{2C^2 V_{max} A}{V_i(x = L) U(t - t_o)} \sqrt{\frac{j}{\pi}\, \frac{1}{R\hat{y}}} \sum_{n=0}^{\infty} \frac{j^n}{2^n\, n!}$$

$$\left[\frac{\rho^{n+1/2} e^{-jn\rho}}{jn^2 R} - \frac{n - \frac{1}{2}}{jn^2 R} \int \rho^{r-1/2} e^{-jn\rho}\, d\rho \right.$$

$$\frac{\rho^{n-1} e^{-jn\rho}}{jn} - \frac{n-1}{jn} \int \rho^{n-2} e^{-jn\rho}\, d\rho +$$

$$\left. \frac{j2^{n+1}(n-1)}{32} \rho^{3/2} \right] \tag{2.67}$$

where

$x = \rho \cos \alpha$, α is constant

$A = A_o$, $R = R_c$ for conduction current

$A = A_v$, $R = R_v$ for time variation of bound charges.

Summarizing, the preceding presentation is confined to procedural mathematical modeling for the lightning phenomenon reflected in the arbitrary or generalized distribution functions with respect to the conductive current i, the return stroke and the convective current generated by the time varying bound charges. However, the arbitrary forms of J_c and J_v specify the variable nature with respect to cloud heights and velocity of propagation in the return channel stroke as well as the rate of build-up in the layer of bound charges of all varieties. The approach followed in solving for the generated Helmholtz wave function by each source is well established through the principles of transform theory and of radiation of electromagnetic fields from their sources to any point on an overhead power line or any other physical structure.

The basic constraints to be placed on the functional forms of J_c and J_v are that they are Fourier transformable along the Z dimension and the resulting function of the inducing voltage V_i has to be compatible with all the properties of Green's function indicated in Equation (2.32).

Now, it will be helpful to draw from the preceding presentation the following:

1. Mathematical models for the inducing voltage and the corresponding associative function due to lightning phenomenon are based on generalized functional distribution for the return stroke conductive current and the time variation of the developed bound charges.
2. Principles of obtaining solution for the Helmholtz wave function of radiated electromagnetic fields coupled to the applicability of transform theories on the arbitrary source functions are powerful tools for securing feasible mathematical models for V_i and V.
3. The well known sectioned nature in time of the inducing voltage function identifies the corresponding associative function in its most general form, namely in terms of any specific current sources established under the charged clouds.
4. Solutions of inducing voltages, induced voltages and the associative function obtained in this paper, mainly through Helmholtz radiation function, add a new approach regarding the method of scalar and vector potentials to analyzing and identifying the phenomenon of lightning.
5. Application of the principle involving the use of Helmholtz field wave radiation function has been carried out successfully for calculating the inducing as well as the induced voltages for an arbitrary point on a transmission line. Electromagnetic fields are assumed to be generated by constant current source distribution in cylindrical channel for the conductive and convective currents. Laplace transform methods with appropriate boundary conditions were used smoothly for solving the resulting differential.

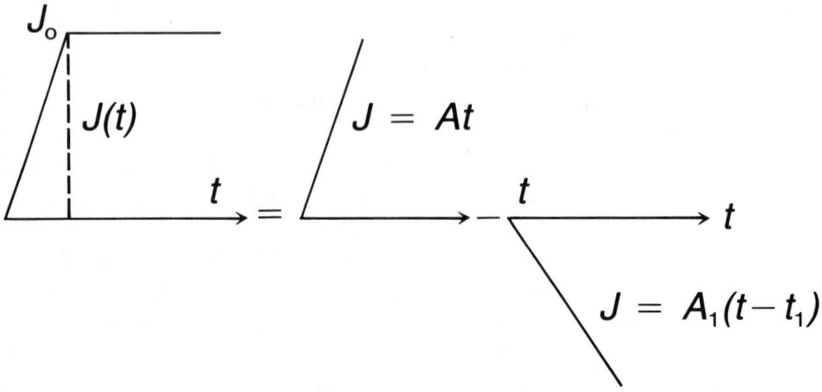

FIGURE 2.3.   Return stroke with constant tail.

2.VIII MATHEMATICAL MODELS OF PROPAGATING INDUCING AND INDUCED POWER DUE TO ACTUAL PULSE WAVE FORM OF LIGHTNING SURGE

This section presents the second phase of analytical work with respect to mathematical modeling of lightning phenomenon. The first phase identified solutions for all components of inducing electromagnetic fields as well as the voltage induced on a transmission line generated by a step pulse of the conductive and convective current densities in the return stroke.

In the following, solutions of electromagnetic inducing and induced effects as well as the induced voltage have been established due to actual pulse wave form for the current densities represented by a sharp linear rise in the pulse front and as lower linear decline in the pulse tail. Also closed form solutions are secured for the propagating real power of both the inducing and induced effects.

Phenomenological results presented here are considered the closest ever to actual calculated realization of lightning effects due to the actual pulse shape of the current densities produced by the return stroke.

Solutions for electromagnetic field components are secured at any point, developed by the inducing surge as well as the associative function and the induced voltage on a transmission line. Complete solutions will be established for the inducing current and the total propagating real power delivered to the power system by the actual lightning surge.

Refer to Figure 2.1 which shows configuration of source distribution and field effects of lightning surge, and to Figure 2.3 which identifies the conductive and convective current densities as sharp wave fronts with tails. Figure 2.2 shows a section of transmission line for calculating the induced effects.

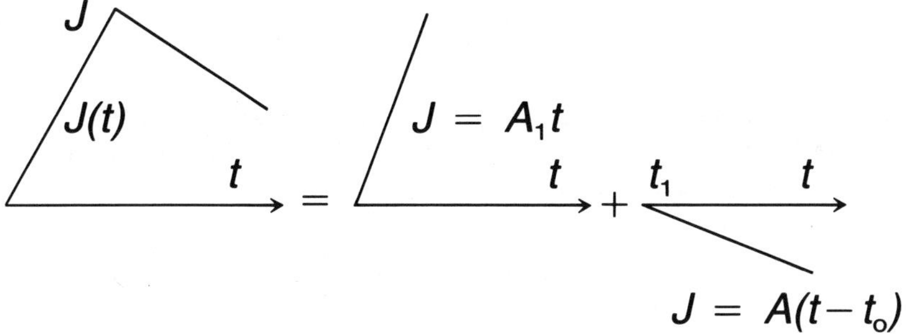

FIGURE 2.4.   Return stroke with dropping tail.

$$J = A(t - t_1) \qquad A > A_1$$

$$J_c = A_c t \qquad\qquad 0 < t < t_1 \text{ independent of } r,0,Z$$

$$\quad\; = A_c(t - t_1) \quad \text{for } t > t_1$$

$$J_v = A_v t \qquad\qquad \text{for } 0 < t < t_1$$

$$\quad\; = A_v(t - t_1) \quad \text{for } t > t_1 \tag{2.68}$$

The preceding seven sections led to the development of a mathematical model in space and time domains for the inducing as well as the induced voltages at a point on a transmission line due to a lightning surge based on the combined effects of conduction and convection, for the return stroke. That developed model was established with the aid of Helmholtz field radiation function and with the assumption of a sharp stepped function of the conductive and convective return stroke current densities.

Now, the ongoing analysis expands further on the lightning phenomenon by moving to consider the closest model of current densities for the return stroke, namely a linear rise for the front pulse, followed by dropping tail.

A. INDUCING FIELDS DUE TO ACTUAL LIGHTNING SURGE

The actual lightning surge pulse is indeed very close to a sharp linear front rise followed by a linear steep dropping as shown in Figure 2.4.

B. SOLUTION FOR THE FIELD RADIATION FUNCTION DUE TO CONDUCTIVE AND CONVECTIVE EFFECTS

These have been derived in Equations (2.39) and (2.40) and rewritten below for convenience:

$$\phi_c = \frac{c^{\tau A_c}}{2} \sum_{n=0}^{\infty} \frac{1}{n} J_n(k|\rho - R_c|) H^{(2)}(k|\rho - R_c|) \tag{2.69}$$

$$\phi_v = \frac{c^{\tau A_v}}{2} \sum_{n=0}^{\infty} \frac{1}{n} J_n(k|\rho - R_c|) H^{(2)}(k|\rho - R_c|) \tag{2.70}$$

C. SOLUTION OF ELECTROMAGNETIC FIELD COMPONENTS

These have been presented in Equations (2.41) and (2.48) and rewritten below for convenience to the reader:

$$\overline{E}_\rho = A_c \sqrt{\frac{j}{\pi}} \frac{\sqrt{\rho - R_c}}{R_c \hat{y}} \sum_{n=0}^{\infty} \frac{j_n e^{-j(\rho - R_c)}}{2^n \, n!} \left[- \frac{(\rho - R_c)^n}{R_c^n} \right.$$

$$\left. - (\rho - R_c)^{n-1} - \frac{(\rho - R_c)^n \frac{3}{2}}{\sqrt{2n}} + \frac{j2^n(n-1)!}{\sqrt{2}} \right] \tag{2.71}$$

$$\overline{E}_\alpha = \frac{jA_c}{2\hat{y}} \sqrt{\frac{2j}{\pi(\rho - R_c)}} \, e^{-j(\rho - R_c)} \sum_{n=0}^{\infty} \frac{j^n(\rho - R_c)^n}{2^n \, n!} \tag{2.72}$$

$$\overline{E}_\tau = \frac{A_c R_c k^2 \tau}{2\hat{y}} \sqrt{\frac{j2}{\pi(\rho - R_c)}} \sum_{n=0}^{\infty} \frac{j^n(\rho - R_c)}{2^n \, n!} \tag{2.73}$$

$$\overline{H}_\rho = \frac{jA_c \tau}{2} \sqrt{\frac{2j}{\pi(\rho - R_c)}} \, e^{-j(\rho - R_c)} \sum_{n=0}^{\infty} \frac{j^n(\rho - R_c)^n}{2^n \, n!} \tag{2.74}$$

$$\overline{H}_\alpha = \frac{A_c \tau}{2} \sum_{n=0}^{\infty} \left[\frac{1}{n^2 \rho} Jn(\rho - R_c) H_n^{(2)}(\rho - R_c) \right.$$

$$- \frac{-1}{nR_c} - H_n^{(2)}(\rho - R_c) \frac{(\rho - R_c)^{n-1}}{2^{n-1}(n-1)} + \sqrt{\frac{2}{\pi}} (-j)^n + \frac{1}{2}$$

$$J_n^{(2)}(\rho - R_c) \frac{1}{nR_c} \frac{e^{-(\rho - R_c)n} - 1}{(\rho - R_c)^2 \, n - 1} + j^n + \frac{3}{2} \sqrt{\frac{2}{\pi(\rho - R_c)}}$$

$$e^{-j(\rho - R_c)} \frac{1}{nR_c} \right] \tag{2.75}$$

where

$$J_n(\rho - R_c) \text{ at large } \rho \approx \frac{1}{n!} \frac{(\rho - R_c)^n}{2^n} \tag{2.76}$$

and

$$H_n^{(2)}(\rho - R_c) \text{ at large } \rho \approx \frac{j2}{\pi(\rho - R_c)} j^n e^{-j(\rho - R_c)} \tag{2.77}$$

$$H_\tau = 0 \tag{2.78}$$

Next the electromagnetic field components generated by the time varying bound charges and represented by the convective constant current density $J_v = A_v$ are exactly those of E_ρ, E_α, E_τ, H_ρ, H_α and H_τ expressed in the set of Equations (2.12) with replacement of A_c and A_v, and the radius of the return stroke R_c with the larger value of R_v.

D. SOLUTION OF THE INDUCING VOLTAGE

At the field point (ρ, α, τ), in cylindrical coordinates,

$$-\overline{\nabla}V_i = -\left[\overline{a}_\rho \frac{\partial V_i}{\partial \rho} + \overline{a}_\alpha \frac{1}{\rho}\frac{\partial V_i}{\partial \alpha} + \overline{a}_\tau \frac{\partial V_i}{\partial \tau}\right]$$

$$= \overline{a}_\rho E_\rho + \overline{a}_\alpha E_\alpha + \overline{a}_\tau E_\tau \tag{2.79}$$

$$\overline{V}_i = -\int E_\rho d_\rho = -\int \frac{\partial V_i}{\partial \rho} d_\rho = V_{ic} + V_{iv} \tag{2.80}$$

$$V_{ic,v} = (J_{c,v}) \sqrt{\frac{j}{\pi}} \frac{1}{R\hat{y}} \sum_{n=0}^{\infty} \frac{j^n}{2^n\, n!} \quad \text{multiplied by}$$

$$\left[\frac{\rho^{n+1/2}e^{-jn\rho}}{jn} - \frac{n+\frac{1}{2}}{jn^2 R}\int \rho^{n-1/2}e^{-jn\rho}\, d\rho + \right.$$

$$\frac{\rho^{n-1}e^{-jn\rho}}{jn}\frac{n-1}{jn}\int \rho^{n-2}e^{-jn\rho}\, d\rho +$$

$$\left.\frac{j^{n+1}(n-1)!}{3\sqrt{2}}\,\rho^{3/2}\right] \quad \begin{array}{l} J = J_c,\ R = R_c \text{ for conduction} \\ J = J_v,\ R = R_v \text{ for convection} \end{array} \tag{2.81}$$

Of course the integration process could be repeated to any limit on n, and that total current density $J = J_c - J_v$.

E. SOLUTION OF THE INDUCED VOLTAGE $V(x,t)$

$$V(x,t) = \frac{2C^2}{V_i(x = L)U(t - t_o)}\sqrt{\frac{j}{\pi}}\frac{1}{\hat{y}}\left[\frac{A_c}{R_c}\sum_{n=0}^{\infty}\frac{j^n}{2^n\, n!}\right.$$

$$\left[\frac{\rho^{n+1/2}e^{-jn\rho}}{jn^2 R_c} - \frac{n+\frac{1}{2}}{jn^2 R_c}\int \rho^{n-1/2}e^{-jn\rho}\, d\rho + \right.$$

$$\frac{\rho^{n-1}e^{-jn\rho}}{jn} - \frac{n-1}{jn}\int \rho^{n-2}e^{-jn\rho}\, d\rho +$$

$$\left.\frac{j2^{n+1}(n-1)!}{3\sqrt{2}}\,\rho^{3/2}\right] + \frac{A_v}{R_v}\sum_{n=0}^{\infty}\frac{j^n}{2^n\, n!}$$

$$\left[\frac{\rho^{n+1/2}e^{-jn\rho}}{jn^2R_v} - \frac{n+\tfrac{1}{2}}{jn^2R_v}\int\rho^{n-1/2}e^{-jn\rho}\,d\rho\ +\right.$$

$$\frac{\rho^{n-1}e^{-jn\rho}}{jn} - \frac{n-1}{jn}\int\rho^{n-2}e^{-jn\rho}\,d\rho\ +$$

$$\left.\frac{j2^{n+1}(n-1)}{3\sqrt{2}}\,\rho^{3/2}\right]\right] \tag{2.82}$$

and

$$J_v = A_v\,t,\ 0 < t < t_1 \text{ independent of } r, 0, Z$$

$$= A_v(t - t_1)t > t_1 \tag{2.83}$$

where J_c and J_v are conductive and convective current densities in the return stroke respectively. Restating the Helmholtz field radiation equation for the inducing electric and magnetic components,

$$\overline{\nabla}^2E k^2\overline{E} = 0$$

$$\overline{\nabla}^2H k^2\overline{H} = 0 \tag{2.84}$$

Then the scalar form is expressed as

$$\overline{\nabla}\emptyset + k^2\emptyset = 0 \tag{2.85}$$

where

$$k = \sqrt{-\hat{z}\hat{y}} \tag{2.86}$$

$\hat{z}$ and $\hat{y} = $ impedivity and admitivity of surrounding space, respectively.

By utilization of cylindrical coordinates in Equation (2.18) and identifying ρ,α,τ as the space coordinates at any point in space including a power line and using the mechanism of Fourier transform as had been done in the author's previous work (*J. Electrostat.*, **13**: 55-69, 1982) which dealt with constant lightning surge, the following solutions for $\emptyset_c$ and $\emptyset_v$ are expressed below:

$$\emptyset_c = \frac{R_c\pi_t}{2}\sum_{n=0}^{\infty}\frac{1}{2n\pi}\,[A_c(t - t_1) + jA_ctU_{-1}(t - t_1)]$$

$$[J_n(k|\rho - R_c|)H_n^{(2)}(k|\rho - R_c|)] \tag{2.87}$$

$$\emptyset_v = \frac{tR_v\pi}{2} \sum_{n=0}^{\infty} \frac{1}{2n\pi} [A_v(t - t_1) + jA_v tU_{-1}(t - t_1)]$$

$$[J_n(k|\rho - R_v|) + H_n^{(2)}(k|\rho - R_v|)] \tag{2.88}$$

E-M field components at a relatively far point from the source are generated by the truncated time function for $J_c(t)$:

$$\overline{E}_\rho = [A_c(t - t_1) + jA_c tU_{-1}(t - t_1)J_c(t)]\sqrt{j/\pi}$$

$$\sqrt{\frac{\rho - R_c}{R_{c,v}^{\hat{y}}}} \sum_{n=0}^{\infty} \frac{j^n e^{-j}(\rho - R_c)}{2^n\, n!} \left[- \frac{(\rho - R_c)}{R_c^n} \right.$$

$$\left. - (\rho - R_c)^{n-1} - \frac{(\rho - R_c)^n \frac{3}{2}}{\sqrt{2n}} + \frac{j2^n(n-1)!}{2} \right] \tag{2.89}$$

$$\overline{E}_\alpha = \frac{jJ_c}{2\hat{y}} \sqrt{\frac{2j}{\pi(\rho - R_c)}}\; e^{-j(\rho - R_c)} \sum_{n=0}^{\infty} \frac{j^n(\rho - R_c)^n}{2^n\, n!}$$

$$\overline{E}_\tau = \frac{j_c(t)R_c k^2\tau}{2\hat{y}} \sqrt{\frac{2j}{\pi(\rho - R_c)}} \sum_{n=0}^{\infty} \frac{j^n(\rho - R_c)^n}{2^n\, n!}$$

$$\overline{H}_\rho = \frac{JJ_c(t)\tau}{2\hat{y}} \sqrt{\frac{2j}{\pi(\rho - R_c)}}\; e^{-j(\rho - R_c)} \sum_{n=0}^{\infty} \frac{j^n(\rho - R_c)}{2^n\, n!}$$

$$\overline{H}_\alpha = \frac{J_c(t)\tau}{2} \sum_{n=0}^{\infty} \left[\frac{1}{n^2\rho} j_n(\rho - R_c)H_n^{(2)}(\rho - R_c) \right.$$

$$\frac{-1}{nR_c} Hn^{(2)}(\rho - R_c) \frac{(\rho - R_c)^{n-1}}{2^{n-1}(n-1)!} + \sqrt{\frac{2}{\pi}}$$

$$(-j)^{n+1/2}Jn(\rho - R_c) \frac{1}{nR_c} \frac{e^{-j(\rho - R_c)}}{(\rho - R_c)} +$$

$$\left. j^n + {}^{3/2}\sqrt{\frac{2}{\pi(\rho - R_c)}}\; e^{-j(\rho - R_c)} \frac{1}{nR_c} \right] \tag{2.90}$$

where

$$J_n(\rho - R_c) \text{ at large } P = \frac{1}{n!} \frac{(\rho - R_c)}{2^n} \tag{2.91}$$

$$H_n^{(2)}(\rho - R_c) = \frac{j2}{\pi(\rho - R_c)} j^n e^{-j(\rho - R_c)} \tag{2.92}$$

$$\overline{H}_\tau = 0 \tag{2.93}$$

Electromagnetic field components produced by time varying bound charges, and represented by the convective current density J_v, are the same as those of E_ρ, E_α, E_τ, H_ρ, H_α and H_τ expressed in Equations (2.22), (2.23) and (2.24) by replacing A_c by A_v and R_c by R_v.

F. INDUCING VOLTAGE V_i

$$V_i - \int E_\rho d_\rho$$

$$V_i = [J_c(t) - J_v(t)] \sqrt{\frac{j}{\pi}\frac{1}{R\hat{y}}} \sum \frac{j^n}{2^n\,n!} \tag{2.94}$$

$$\left[\frac{\rho^{n\,+\,1/2}e^{-jn\rho}}{jn^2R} - \frac{n\,+\,1/2}{jn^2R}\int\rho^{n-1/2}e^{-jn\rho}\,d\rho\ +\right.$$

$$\left.\frac{\rho^{n-1}e^{-jn\rho}}{j^n} - \frac{n-1}{j^n}\int\rho^{n-2}e^{-jn\rho}\,d\rho\ + \frac{j2^{n+1}(n-1)!}{3\sqrt{2}}\,\rho^{3/2}\right] \tag{2.95}$$

$$R = R_c \text{ for conductive effect}$$

$$= R_v \text{ for convective effect}$$

$$J_c = A_v t, 0 < t < t_1$$

$$= A_c(t - t_1)U - 1(t)t > t_1$$

$$J_v = A_v t, 0 < t < t_1$$

$$= A_v U_{-1}(t)[t - t_1]t > t_1 \tag{2.96}$$

due to time retardation by $t_o = \rho - R_c/C_o$ where $C_o = $ velocity of light. The associative function

$$\psi = V_i(\alpha,\rho,\tau,t)U(t - t_o) \tag{2.97}$$

The induced voltage could be obtained by starting again with the following:

$$\frac{\partial^2 V}{\partial x^2} - \frac{1}{C^2}\frac{\partial^2 V}{\partial t^2} = \frac{1}{C^2}\frac{\partial^2 V_i}{\partial t^2} \tag{2.98}$$

$$V_i = V_i(\alpha,\rho,\tau,t)U(t - t_o) \tag{2.99}$$

From Equations (2.29) and (2.32) and using the boundary conditions, namely that $V(0,t) = 0$ at the source, $V(L,t) = 2\,V_{max}$ where L is very far from the source. Solutions for $V(x,t)$ at a point on the power line are as follows:

$$V(x,t) = V_1 + V_2 + V_3 + V_4 \tag{2.100}$$

$$V_1(x,t) = \frac{2C^2 V_{max} V_i(x)}{V_i(x = L)\delta_o(t)}\, \delta_o(t)[A_c - A_v] \tag{2.101}$$

$$V_2(x,t) = \frac{2C^2 V_{max} V_i(x)}{V_i(x = L)\delta_o(t - t_1)}\, \delta_o(t - t_1)[A_c - A_v] \tag{2.102}$$

$$V_3(x,t) = \frac{-2C^2 V_{max} V_i(x)}{V_i(x = L)\delta_o(t - t_1)}\, \delta_o(t_o - t_1)[A_v - A_c](t - t_1) \tag{2.103}$$

$$V_4(x,t) = \frac{2C^2 V_{max} V_i(x)}{V_i(x = L)\delta_o(t - t_1)}\, \delta_o(t - t_1)U_{-1}(t)(A_v - A_c) \tag{2.104}$$

2.IX INDUCED AND INDUCING PROPAGATED POWER

A. INDUCED ELECTROMAGNETIC POWER

Instantaneous power density per meter squared delivered at a point on a transmission line could be expressed as

$$P_1(t) = [V(x,t)]^2\, \rho \ \text{watts/m}^2 - \text{m}$$

where ρ = resistivity of the power line

$$P_1(t) = [V_1 + V_2 + V_3 + V_4]^2\, \rho \tag{2.105}$$

Moving to calculate the average power delivered through induction (induced power), an averaging process is conducted on Equation (2.37) resulting with the following:

$$P_1 \text{ average-induced} = \frac{1}{\rho}\, \frac{2C^2 V_{max} V_i(x)}{V_i(x = L)\delta_o(t)}\, \delta_o(A_c - A_v)$$

$$+ \frac{2C^2 V_{max} V_i(x)}{V_i(x = L)}\, \delta_o(t - t_1)(A_v - A_c)$$

$$- \frac{2C^2 V_{max} V_i(x)}{V_i(x = L)}\, \delta_o(t - t_1)(t - t_1)(A_v - A_c)$$

$$+ \frac{2C^2 V_{max} V_i(x)}{V_i(x = L)}\, \delta_o(t - t_1)U_{-1}(t)(A_v - A_c)^2 \tag{2.106}$$

B. INDUCING ELECTROMAGNETIC POWER

Inducing current density vector is expressed as

$$J = J_\rho + J_\alpha + J_\tau$$

$$J_\rho = \sigma E_\rho$$

$$J_\alpha = \rho E_\alpha \text{ and}$$

$$J_\tau = \sigma E_\tau \tag{2.107}$$

Turning from cylindrical to cartesian coordinate system, J is expressed as

$$J_x = J_\rho a\rho \cos \alpha - J_\alpha a\alpha \sin \alpha$$

$$J_y = J_\rho a\rho \sin \alpha + J_\alpha a\alpha \cos \alpha$$

$$J_z = J_\tau a\tau \tag{2.108}$$

where

$$x = \rho \cos \alpha$$

$$y = \rho \sin \alpha$$

$$z = \tau \tag{2.109}$$

Solutions for E_ρ, E_α and E_τ are shown in Equation (2.22)

$$P_2(t) = \left|\frac{J}{\sigma}\right|^2$$

$$= \left[\frac{1}{\sigma} J_\rho + J_\alpha + J_\tau\right]^2 \tag{2.110}$$

To find P_2 average, an averaging process is conducted on E_ρ, E_α and E_τ in Equation (2.22). Averaging $P_2(t)$ follows

$$P_2 \text{ average} = \frac{1}{2\sigma} A_c(t - t_1)R_e jA_c tU_{-1}(t - t_1)$$

$$\frac{j}{\pi} \frac{\rho - R_c}{R\hat{y}} \sum_{n=0}^{\infty} \frac{j^n e^{-j(\rho - R_c)}}{2^n\, n!} \frac{(\rho - R_c)^n}{R_c^n}$$

$$- (\rho - R_c)^{n-1} - \frac{(\rho - R_c)^{n-3/2}}{2n} + \frac{j2^n(n - 1)!^2}{2}$$

$$+ \frac{1}{26} A_c(t - t_1)R_e \frac{j}{2\hat{y}} \frac{2j}{\pi(\rho - R_c)}$$

$$e^{-j(\rho - R_c)} \sum_{n=0}^{\infty} \frac{j(\rho - R_c)^n}{2^n \, n!} {}_2 + \frac{1}{26} A_c(t - t_1)$$

$$R_e\left[\frac{R_c k^2 \pi}{2\hat{y}} \frac{j2}{\pi(\rho - R_c)} e^{-j(\rho - R_c)} \sum_{n=0}^{\infty} \frac{j^n(\rho - R_c)}{2^n \, n!} \right]^2 \qquad (2.111)$$

where R_e = the real part of ( ).

Similarly, average convective power generated by the time varying propagating bound charges is the same as the solution expressed by Equation (2.111) with the replacement of R_c by R_v and A_c by A_v.

Work presented in Sections VIII, IX and X considered the surge lightning pulse of the return stroke to be of sharp linear rise for the pulse front and a decaying linear reduction for the pulse tail. Such a surge model is indeed very close to the actual surge which is characterized as exponential in the rise period as well as in the tail. But from field experimental measurements, the exponential pattern is almost linear, due to infinitesimal time constant for the pulse front and substantially longer time constant for the tail.

Conclusions drawn from results presented in this paper could be summarized by securing closed mathematical solutions for the following:

1. Electric magnetic inducing field components at an arbitrary location of power system physical structure
2. The inducing electric potential
3. The induced electric potential
4. The propagating real power of the inducing fields through a section of transmission line
5. The propagating induced real power through a section of transmission line

The mathematical models secured in this work as mentioned above are based on an almost actual pulse waveshape of the conductive and convective current densities in the return stroke of the lightning surge.

2.X TEST SIMULATION OF LIGHTNING SURGE

R. Sobocki[10] presented, in his paper, what he called basic impulse testing circuit, shown in Figure 2.5 which represents lightning surge simulation and the object to be tested. In the circuit: S is a direct current source for a loading capacitance C_g. R_o is a loading resistance; L_1 is an impulse generator inductance; L_2 is an external circuit inductance; R_{c1} and R_{c2} are front controlling resistances; R_r is a tail controlling resistance; C_c is a front controlling capacitance, D is the

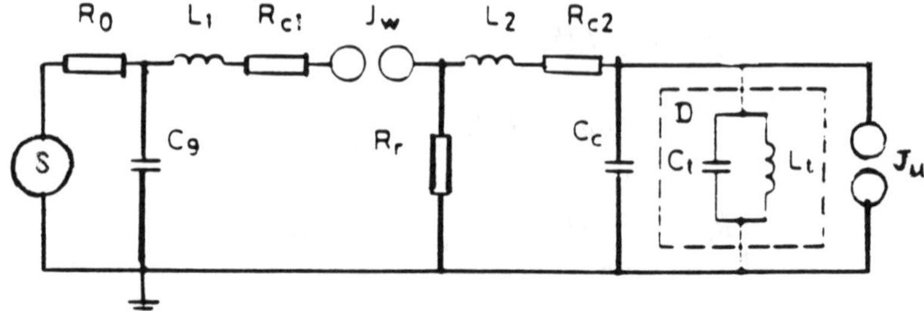

FIGURE 2.5.   Basic impulse testing circuit.

device being tested; C_t is the equivalent capacitance of the tested device; L_t is the inductance of the tested device, J_w is a switching gap; J_u is a chopping gap.

Sobocki indicated that the wave front of lightning surge could be analyzed, treating the wavetail as having infinite duration on the basis that the capacitance C_s is much larger than the sum of C_c and C_t, and that the device inductance L_t and R_r are very large. In such a circuit, the capacitance C_s does not lose its charge and after a long time interval the voltage across C_c and C_t will reach the level at which C_s was charged initially.

Then, for R_{ck} designated as a critical limit by the following,

$$R_{ck} \approx 2\sqrt{\frac{L_1 + L_2}{C_c + C_t}} \tag{2.112}$$

However, R_{ck} is accurate when $R_r \to \infty$. Different wave forms for the lightning are realized under the following: (a) $R_{c1} + R_{c2} < R_{ck}$ generating a front with superimposed oscillations; (b) $R_{c1} + R_{c2} = R_{ck}$ generating a critically damped front; (c) $R_{r1} + R_{c2} > R_{ck}$ generating an overdamped front.

Mathematical representation for cases (a), (b), and (c) mentioned above are given below according to Sobocki:

$$V(t) = \frac{V_m}{k_c}\left[1 - e^{-\alpha_{co}t}(\cos b\, \alpha_{co}t + \frac{1}{b}\sin b\, \alpha_{co}t)\right] \tag{2.113}$$

for a front with superimposed oscillations;

$$V(t) = V_m[1 - (1 + \alpha_{ck}t)e^{-\alpha_{ck}t}]$$

for the critically damped front; and

$$V(t) = V_m\left[1 - e^{\alpha_{cp}t}(\cosh b\, \alpha_{cp}t + \frac{1}{b}\sinh b\, \alpha_{cp}t)\right] \tag{2.114}$$

for an over-damped front, where V_m is the crest limit of the impulse, k_c is a constant, and α_{co}, α_{ck}, α_{cp} and b are parameters characteristic of the front wave form. The constant k_c could be obtained at a time T_m in which the surge reaches its peak.

By differentiating Equation (2.113) with respect to time and inserting $t = T_m$ and setting the result to zero, $\mathrm{sinb}\,\alpha_{co}\,T_m = 0$ is obtained. Hence,

$$T_m = \frac{n\pi}{b\alpha_{co}} \tag{2.115}$$

Then for $n = 1$, substituting T_m into Equation (2.113) will identify k_c as below:

$$k_c = 1 + e^{-\frac{\pi}{b}} \tag{2.116}$$

or

$$b = \frac{\pi}{ln(k_{c-1})} \tag{2.117}$$

For the critically damped front, instant values of (t) equal to 30 and 90% of V_m have been secured from Equation (2.93), whereby

$$0.3 = 1 - (1 + \alpha_{ck}T_{0.3})e^{-\alpha_{ck}T_{0.3}}$$

and

$$0.9 = 1 - (1 + \alpha_{ck}T_{0.9})e^{-\alpha_{ck}T_{0.9}}$$

Next, Sobocki introduced the wave front duration T_c and several characteristic parameters as shown below:

$$T_c = 1.67(T_{0.9} - T_{0.3}) \tag{2.118}$$

where $T_{0.3}$ and $T_{0.9}$ are the instants at which the impulse reaches 30% and 90% of V_m, respectively. Now, multiplying both sides of Equation (2.126) by α_{ck} and dividing by T_c, the following is obtained:

$$\alpha_{ck} = 1.67\left[\frac{\alpha_{ck}T_{0.9} - \alpha_{ck}T_{0.3}}{T_c}\right] \tag{2.119}$$

or in general

$$\alpha_{ck}\,\frac{X_{ck}}{T_c} \tag{2.120}$$

where

$$X_{ck} = 1.67(\alpha_{ck}T_{0.9} - \alpha_{ck}T_{0.3}) \qquad (2.121)$$

For the critically damped case,

$$X_{ck} = 4.653 \qquad (2.122)$$

In general, it can be written that

$$\alpha_c = X_c/T_c \qquad (2.123)$$

2.XI ANALYSIS OF THE CREST OF VOLTAGE SURGE

According to the design of the circuit shown in Figure 2.5, a flashover at J_w will generate a voltage drop across the inductance L_t that will charge the capacitance C_s, provided C_c and C_t are very small. After a long time duration, the capacitance C_s will be discharged completely whereby the voltage across L_t will fall, accompanied by a polarity change.

As for the case of wavetail, three conditions are going to be considered for the wavefront depending on the values of R_r with respect to the critical resistance R_k: (a) $R_r < R_k$ generates a crest with superimposed oscillations; (b) $R_r = R_k$ generates a critically damped crest; (c) $R_r > R_k$ generates an over-damped crest. Mathematical representation for (a), (b) and (c) are given below:

$$V(t) = V_m(1 - \alpha_{gk})e^{-\alpha_{gk}t} \qquad (2.124)$$

This corresponds to the critically damped case.

$$V(t) = V_m e^{-\alpha_{gp}t}\left(\cosh b_g\, \alpha_{gp}t - \frac{1}{b_g}\sinh b_g\, \alpha_{gp}t\right) \qquad (2.125)$$

Then, for a crest with superimposed oscillations,

$$V(t) = V_m e^{-\alpha_{go}t}\left(\cosh b_g\, \alpha_{go}t - \frac{1}{b_g}\sin b_g\, \alpha_{go}t\right) \qquad (2.126)$$

where T_g is the time interval at which the crest has been diminished to 50%, and it need be known only for the critically damped wave. For other cases, the reverse polarity amplitude will be used as the other parameter that will take time duration of T_p.

T_p could be determined by differentiating any of Equations (2.107), (2.108) and (2.109) with respect to time and equating the derivative to zero.

For the critically damped crest, T_p is given by:

$$T_{p-c} = 2/\alpha_{gk} \tag{2.127}$$

From Equations (2.127) and (2.124), the reverse polarity amplitude is obtained as the ratio to the peak V_m

$$k_g = \left| \frac{V(T_p)}{V_m} \right| = e^{-2} \approx 0.135 \tag{2.128}$$

Similarly, for the crest with superimposed oscillations, differentiating Equation (2.109) with respect to time and setting the derivative to zero results in

$$T_{p-os} = \frac{1}{b_g \alpha_{go}} \tan^{-1} \frac{2b_g}{1 - b_g^2} \tag{2.129}$$

Then substituting T_{p-os} into Equation (2.109) will lead to:

$$kg = e^{-\alpha_{go} T_{p-os}}$$

From the product of $\alpha_{go} T_{p-os}$ and rearranging:

$$\tanh b_g (1 - \ln k_g) = \frac{2b_g}{1 - b_g^2} \tag{2.130}$$

Equation (2.129) indicates that the coefficient b_g is a function of the reverse polarity amplitude.

Similarly for the over-damped case:

$$T_{p-od} \frac{1}{b_g \alpha_{go}} \tag{2.131}$$

$$k_g = e^{-\alpha_{gp} T_p} \tag{2.132}$$

$$\text{and} \tan h\, b_g (1 - \ln k_g) = \frac{2b_g}{1 + b_g^2} \tag{2.133}$$

From Equation (2.130) b_g is a function of k_g only. Each value of k_g corresponds to two values of b_g, one being >1 and the other <1. E.x. for an inductive load b_g is <1. Also where the impulse front duration T_c is infinitely short, the impulse crest duration $T_g = T_{0.5}$ as shown in Figure 2.7.

Knowing the value of X_{gk}, a value of α_{gk} can be calculated for any limit of impulse crest duration.

In general:

$$\alpha_g = \frac{X_g}{T_{0.5}} \tag{2.134}$$

where $X_g = 0.3149$ for the critically damped crest $= f(k_g)$ for both the over-damped crest and the crest with superimposed oscillations.

Figure 2.6 (a, b and c) illustrates a full wave for the lightning surge, a chopped wave and impulses with superimposed oscillations. Also for other cases, X_c is a function of the coefficient b. Numerical values for X_c and b are given in Figure 2.7 and Table 2.1

Numerical values for $X_g = f(k_g)$ are given in Figure 2.8.

2.XII CHOPPING OF THE IMPULSE

In reference to Figure 2.5, after charging capacitors C_c and C_t to a voltage V_u, a chopping gap sparks so the voltage will drop rapidly and will oscillate around zero because of the inductance effect. The highest amplitude of oscillations exceeds 5% of V_u generally.

Chopping time is characterized by:

$$\Delta T_u = 1.67(T_{0.7} - T_{0.1}) \tag{2.135}$$

Practically, in the case of the tested object is a transformer which usually presents itself as a complex impedance; the chopping time and the highest amplitude after chopping are the most important parameters.

Refer to Figure 2.7b where the moment of chopping is characterized as point (0). The variations of choppings are given below:

$$V(t) = V_u e^{-\alpha_u t}(\cos b_u \alpha_u t + \frac{1}{b_u} \sin b_u \alpha_u t) \tag{2.136}$$

Time intervals for successive oscillations after chopping are given by:

$$T_w = \frac{n\pi}{b_u \alpha_n}, \; n = 1,2,3 \; \text{-} \; \text{-} \; \text{-} \tag{2.137}$$

Because of damping, the highest maximum value will occur when $n = 1$.

Now, with T_w where $n = 1$, substitution into Equation (2.136) results in:

$$k_u = \frac{V(T_w)}{V_u} = 1 - e - \frac{\pi}{b_u} \tag{2.138}$$

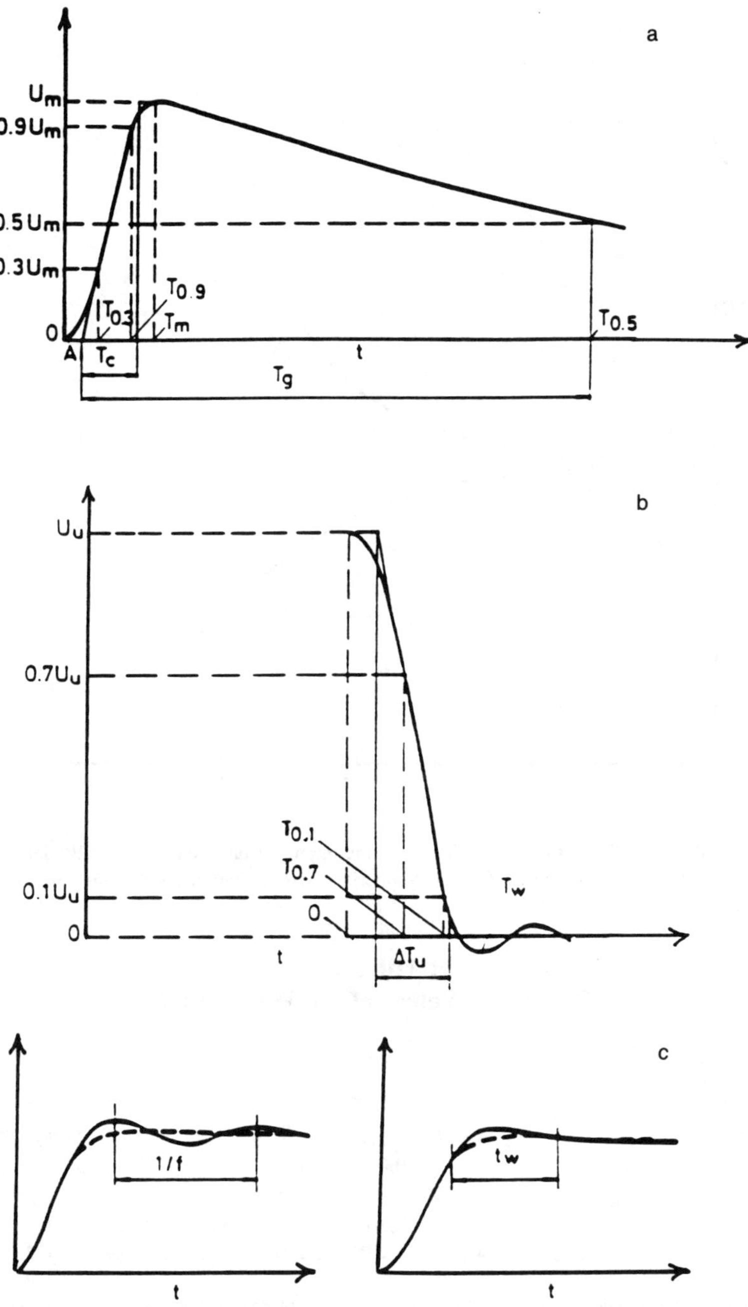

FIGURE 2.6.   Wave shape. (a) Full wave; (b) chopping of the wave; (c) impulses with superimposed oscillations.

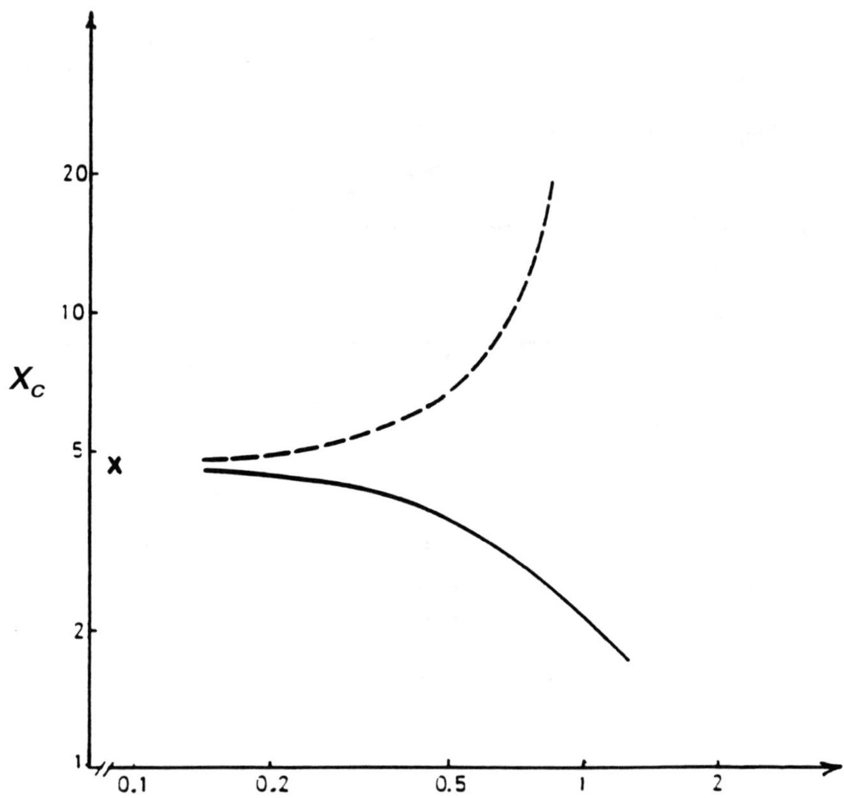

FIGURE 2.7.   Plot of the coefficient x_c of the impulse front x for the critically damped front. Solid line represents the front with superimposed oscillations; broken line represents the overdamped front.

TABLE 2.1
Parameter Values of the Wave Front

The front with superimposed oscillations × 100%								
	X_{c-1}	4.518	4.150	3.656	3.158	2.726	2.370	2.084
	$\dfrac{k_c}{V_m}$	0.0	0.0	0.1	0.5	1.5	3.0	5.0
Over-damped front	b	0.15	0.30	0.45	0.60	0.75	0.90	1.05
	X_c	4.797	5.275	6.270	8.282	13.019	32.432	—

* The highest amplitude of the oscillations near the impulse peak value, expressed as a percentage of the impulse crest value.

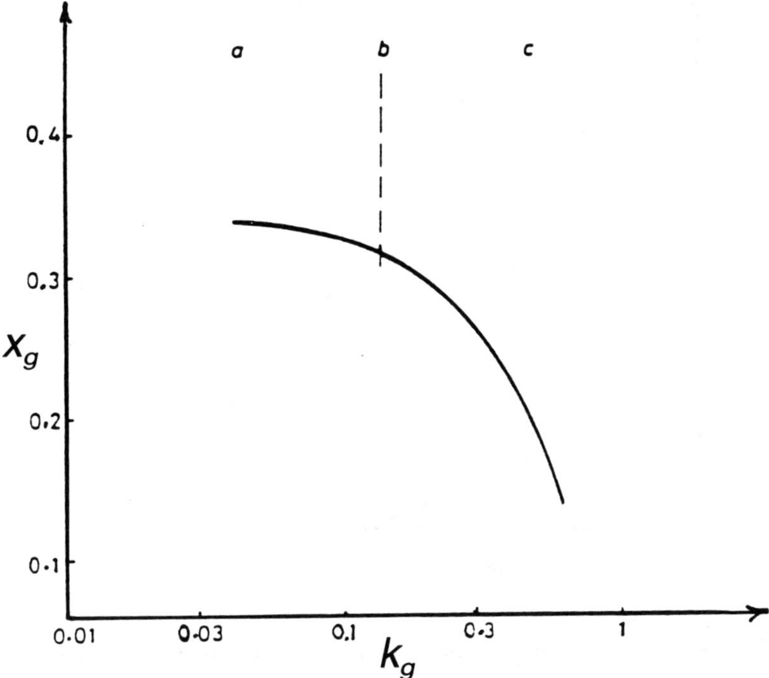

FIGURE 2.8. Plot of the coefficient x_g of the impulse tail against the reverse polarity maximum. (a) For the overdamped crest; (b) for the critically damped crest; (c) for the crest with superimposed oscillation.

k_u is a coefficient compared with a voltage value at a chopping moment. b_u in Equation (2.137) is a function of the highest amplitude of oscillation after chopping.

Similarly for the wavefront

$$\alpha_u = \frac{X_u}{T_u} \tag{2.139}$$

in which

$$X_u = 1.67[\alpha_u T_{0.7} - \alpha_u T_{0.1}] = f(k_u) \tag{2.140}$$

Parameters values of wave chopping are given in Figure 2.9 and Table 2.2. Using the oscillation frequency, $V(t)$ in Equation (2.136) is replaced by:

$$V(t) = V_u e^{-\alpha_u t}\left(\cos 2\pi f_w t + \frac{\alpha_u}{2\pi f_w}\sin 2\pi f_w t\right) \tag{2.141}$$

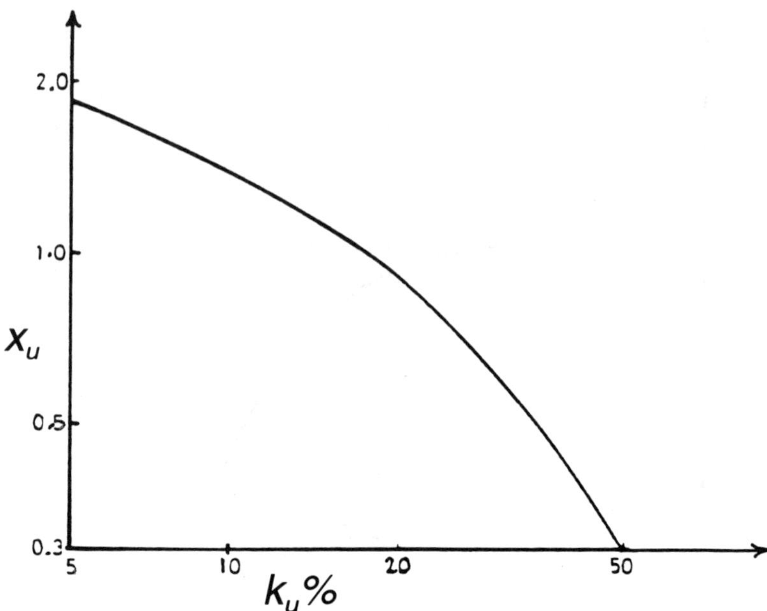

FIGURE 2.9. Plot of the coefficient x_u of the impulse chopping against the highest amplitude after chopping.

TABLE 2.2
Parameter Values of the Wave Chopping

$k_u o/o^*$	5.0	8.1	12.3	20.8	28.5	35.1	40.8	45.6
b_u	1.05	1.25	1.50	2.00	2.50	3.00	3.50	4.00
X_u	1.857	1.512	1.211	0.8325	0.6259	0.4978	0.4117	0.3503
$b_u X_u$	1.950	1.890	1.816	1.665	1.565	1.493	1.441	1.401

* The highest amplitude of the oscillation after chopping expressed as a percentage of the voltage at the chopping moment.

where f_w is the oscillation frequency. $V(t)$ from Equations (2.136) and (2.141) become identical when $b_u \alpha_u = 2\pi f_w$.

Then, using Equation (2.123), we have

$$b_u X_u = 2\pi f_w T_u$$

The coefficient X_u is a function of k_u and from Equation (2.122) is a function of b_u. For each value of k_u and consequently b_u, there will be one product of X_u b_u which is connected with the oscillation frequency and chopping time through Equation (2.141).

From Equation (2.141), Equations (2.137) and (2.138) become

$$T_u = \frac{n}{2f_w} \tag{2.142}$$

and

$$k_u = 1 = e^{-\frac{\alpha_u}{2f_w}}$$

or

$$\ln k_u = -\frac{\alpha_u}{2f_w} \tag{2.143}$$

The damping decrement is expressed by:

$$\alpha = e\frac{\alpha_u}{f_w} \tag{2.144}$$

or

$$\ln k_u = -\frac{\alpha_u}{2f_w} \tag{2.145}$$

And from Equations (2.138) and (2.143) a relationship between the damping decrement δ and the highest amplitude after chopping is expressed by:

$$\alpha = k_u 2 \tag{2.146}$$

2.XIII SOLVED EXAMPLES

A. In reference to Equation (2.4), express the constant k in a free loss medium.

Solution
   From Equation (2.4),

$$k = \sqrt{-\hat{Z}\hat{y}}$$

In a free loss medium $\sigma = 0$

$$\hat{Z} = j\omega\epsilon, \text{ also } \hat{y} = j\omega\mu$$

B. If the medium in which field effects due to lightning surge are being felt has Z and y without loss, express time of retardation for field propagation at a point in that medium.

Solution

Time of retardation for field effects from the place of its occurrence to a mean location for field impact is given by Equation (2.25):

$$\Delta t = \frac{|P - R_c, R_u|}{C_o} \tag{B1}$$

where C_o is the velocity of light.

However, if the medium has any Z and y with no loss, C_o will be by a particular velocity of propagation, where

$$V = \frac{C_o}{\sqrt{\epsilon\mu}} \tag{B2}$$

$$\Delta t = \left[\frac{|\rho - R_c, R_u|}{C_o}\right] \sqrt{\mu\epsilon} \tag{B3}$$

R_u is the radial vector of medium point of impact.

C. Refer to Equation (2.62) for the solution of $v(x,t)$. Identify the two constants if the line termination is short-circuit.

Solution

Restating Equation (2.62)

$$V_{(x,t)} = k_1 U\left(t - \frac{x}{c}\right) + \frac{k_3}{C^2} U_{-1}(t - t_o)$$

$$t = \frac{x}{C_o} \tag{C1}$$

Boundary conditions become (1) $V(0,t) = 0$, and (2) $V(L,t) = 0$. Boundary condition (1) implies $k_1 = 0$. For boundary condition (2):

$$0 = \frac{k_3}{C^3} V_i(x = L)V(t - t_o) \tag{C3}$$

Termination with a short-circuit implies $V_i\big|_{x=L} = 0$

Equation (C3) becomes

$$\frac{K_3}{C^2} V(t - t_o) = \frac{0}{0} \tag{C4}$$

Equation (C4) could be expressed in the form:

$$\frac{g(t)}{f(t)} = \frac{k_3 U_o(t - t_o)}{C^2} \rightarrow \frac{0}{0} \tag{C5}$$

Conducting a limit process on $\dfrac{g(t)}{f(t)}$

$$\frac{g'(t)}{f'(t)} \rightarrow \frac{k_3 U_o(t - t_o)}{2C} \tag{C6}$$

Proceeding to another process of differentiation:

$$\frac{g''(t)}{f''(t)} \rightarrow k_3 \frac{U_o'(t - t_o)}{2} \tag{C7}$$

$$k_3 = \frac{1}{2} U_o'(t - t_o) \tag{C8}$$

where $U_o'(t - t_o)$ is a delayed doublet singularity which gives meaningful expectation, since at the short-circuited end the transmitted and reflected voltage waves will cancel each other.

During thunderstorm, atmospheric medium is mostly considered anisotropic due to intensive ionization and strong existing electromagnetic field. In such a situation the permitivity constant is a tensor of rank 2, designated as ϵ_{ij}, $i, j = 1, 2, 3$.

D. Write solutions for the electric charge density vector D at any point in space (ρ, α, τ) due to conductive and convective lightning stroke.

Solution

A tensor of rank 2 like ϵ_{ij} is written below:

$$\epsilon_{ij} = \begin{bmatrix} \epsilon_{11} & \epsilon_{12} & \epsilon_{13} \\ \epsilon_{21} & \epsilon_{22} & \epsilon_{23} \\ \epsilon_{31} & \epsilon_{32} & \epsilon_{33} \end{bmatrix} \tag{D1}$$

Electronic field vector components established at any point in space from Equations (2.41), (2.42) and (2.43) are E_ρ, E_α and E_τ.

$$D_i = \epsilon_{ij} E_j \tag{D2}$$

$$\begin{matrix} D_\rho \\ D_\alpha \\ D_\tau \end{matrix} = \begin{bmatrix} \epsilon_{11} & \epsilon_{12} & \epsilon_{13} \\ \epsilon_{21} & \epsilon_{22} & \epsilon_{23} \\ \epsilon_{31} & \epsilon_{32} & \epsilon_{33} \end{bmatrix} \begin{bmatrix} E_\rho \\ E_\alpha \\ E_\tau \end{bmatrix} \tag{D3}$$

And then,

$$D_\rho = \epsilon_{11}\epsilon_\rho + \epsilon_{12}\epsilon_\alpha + \epsilon_{13}E_\tau$$

$$D_\alpha = \epsilon_{21}\epsilon_\rho + \epsilon_{22}\epsilon_\alpha + \epsilon_{23}E_\tau$$

$$D_\tau = \epsilon_{31}\epsilon_\rho + \epsilon_{32}\epsilon_\alpha + \epsilon_{33}E_\tau \tag{D4}$$

The unit for D_ρ, D_α and D_τ is a coloumb/m^2.

E. Obtain an expression for the total electric charge being built on a circular layer of atmospheric cloud, using the D vector established in Example D.

Solution

Electric field vectors have been given in Equations (2.41), (2.42) and (2.43).

$$Q = \int_{P_1}^{P_1} \int_0^{2\pi} A_c \sqrt{\frac{j}{\pi}} \frac{\sqrt{\rho - R_c}}{R_c y} \sum_{n=0}^{\infty} \frac{j^n e^{-j(\rho - R_c)}}{2^n\, n!} - \frac{(\rho - R_c)^n}{R_c^n}$$

$$- (\rho - R_c)^{n-1} - \frac{(\rho - R_c)^n \frac{3}{2}}{2n} + \frac{j2^n(n-1)!}{2} \rho_d \rho_{d\alpha}$$

$$+ \int_{P_1}^{P_2} \int_0^{2\pi} \frac{jA_c}{2\hat{y}} \frac{j2}{\pi(\rho - R_c)} e^{-j(-R_c)} \sum_{n=0}^{\infty} \frac{j^n(\rho - R_c)^n}{2^n\, n!} \rho_d \rho_{d\alpha} \tag{E1}$$

Let $n = 1$ to simplify computation

$$Q = \int_{P_1}^{P_2} \int_0^{2\pi} A_c \sqrt{\frac{j}{\pi}} \frac{\sqrt{\rho - R_c}}{R_c} je \frac{-j(\rho - R_c)}{2} \left[\frac{(\rho - R_c)^n}{R_c} \right.$$

$$+ \frac{(\rho - R_c)^{-1/2}}{\sqrt{2}} - \left. \frac{j2}{\sqrt{2}} \right] \rho_d \rho_{d\alpha}$$

$$- \int_{P_1}^{P_2} \int_0^{2\pi} \frac{A_c}{2\hat{y}} \sqrt{\frac{j2}{\pi(\rho - R_c)}} e^{jR_c} \frac{R_c}{2} \rho_d \rho_{d\alpha} \tag{E2}$$

$$= -\int_{P_1}^{P_2}\int_0^{2\pi} \frac{jA_c}{2R_c}\sqrt{\frac{j}{\pi}}\left[\frac{(\rho - R_c)^{3/2}}{R_c}e^{-j(\rho - R_c)}e\frac{-j(\rho - R_c)}{2\sqrt{2}}\right.$$

$$\left. + j\sqrt{2}(\rho - R_c)^{1/2}e^{-j(\rho - R_c)}\right]\cdot \rho_d\rho_{d\alpha} - \int_{P_1}^{P_2}\int_0^{2\pi}\frac{A_cR_c}{4\hat{y}}e^{-jR_c}\sqrt{\frac{j2}{\pi}}$$

$$(\rho - R_c)^{-1/2}\rho\cdot\rho_d\rho_{d\alpha} \tag{E3}$$

$$Q = \frac{jA_c\pi}{R_c}\sqrt{\frac{j}{\pi}}\left[j(\rho - R_c)^{3/2}e^{-j(\rho - R_c)}\right.$$

$$+ j(\rho - R_c)^{5/2}e^{-j(\rho - R_c)} - j\frac{3}{2}\int(\rho - R_c)^{1/2}e^{-j(\rho - R_c)}\,d\rho$$

$$- j\frac{5}{2R_c}\int(\rho - R_c)^{3/2}e^{-j(\rho - R_c)}\,d\rho$$

$$+ \frac{j}{2\sqrt{2}}\rho e^{-j(\rho - R_c)} + e^{-j(\rho - R_c)} - \sqrt{2}$$

$$\left[(\rho - R_c)^{3/2}e^{-j(\rho - R_c)} + R(\rho - R_c)^{1/2}e^{-j(\rho - R_c)}\right.$$

$$\left.\left. + \frac{3}{2}\int(\rho - R_c)^{3/2}e^{-j(\rho - R_c)}\,d\rho + \frac{1}{2}\int(\rho - R_c)^{1/2}e^{-j(\rho - R_c)}\,d\rho\right]\right]$$

$$- \frac{A_cR_c\pi e^{jR_c}}{2\hat{y}}\sqrt{\frac{j2}{\pi}}\left[\frac{2}{3}(\rho - R_c)^{3/2} - 2R_c(\rho - R_c)^{1/2}\right]\text{ coloumbs} \tag{E4}$$

F. From information secured in Example E, present a simple procedural expression for the energy stored in the medium between cloud layers, each acquiring the amount of charge represented by Equation (E4).

Solution

The dielectric constant for the medium confined between the two clouds is represented by ϵ

$$\text{capacitance C} = \frac{\epsilon A}{d}\text{ Farad} \tag{F1}$$

where A is the surface area of each cloud, and d is the distance separating the two clouds.

$$\text{Energy stored } W_e = \frac{Q^2}{2C}\text{ joules} \tag{F2}$$

$$W_e = \frac{Q^2}{2}\frac{d}{\epsilon A} \text{ joules} \tag{F3}$$

However, the (E) used is more likely to be a tensor of rank 2.

G. Identify the ratio of $U(t)$ of the front lightning pulse with superimposed oscillations when $\alpha_o t = \pi/2$ and π. Comment on the result.

Solution

From Equation (2.92)

$$V(t) = \frac{V_m}{k_c}\left[1 - e^{-\alpha_{co}t}\left(\cos_6\alpha t + \frac{1}{b}\sin b\alpha_{co}t\right)\right] \tag{G1}$$

For $\alpha_{co}t = \pi/2$

$$V_{\pi/2} = \frac{V_m}{k_c}[1 - e^{-\pi/2}/b] \tag{G2}$$

Then for $\alpha_{co}t = \pi$

$$V_\pi = \frac{V_m}{k_c}[1 - e^{-\pi}(-1)] \tag{G3}$$

$$= \frac{V_m}{k_c}(1 - e^{-\pi}) \tag{G4}$$

Therefore,

$$\frac{V_{\pi/2}}{V_\pi} = \frac{1 - e^{-\pi/2/b}}{1 + e^{-\pi}} \tag{G5}$$

$$= \frac{b - e^{-\pi/2}(.208)}{b(1 + e^{-\pi})} \tag{G6}$$

The parameter (G2) is a function of the highest amplitude of oscillations. Equation (G6) could be written as:

$$\frac{V_{\pi/2}}{V_\pi} \equiv \frac{b - 0.208}{1.043\,b}$$

$$\approx \frac{1 - \dfrac{0.208}{b}}{1.043} \tag{G7}$$

Since (G2) is a function of the highest amplitude of oscillations,

$$V_{\pi/2}/V_{\pi} \to \text{unity}$$

2.XIV PROBLEMS

1. Using proper vector identity, show the derivation of Equations (2.2a) and (2.2b).
2. Carrying out a process of Fourier transformation, show the conversion of Equation (2.5) into Equation (2.6).
3. Show that the solution of the differential Equation (2.7) is that given in Equation (2.9) for $r_c < R_c$. Use boundary conditions listed in Equation (2.11).
4. Proceed from the last stage of work for problem 3 and, using Equations (2.13) to (2.16), show the derivation of Equation (2.17).
5. Carry out the necessary research to show that electromagnetic field components for a multicylindrical current sheet which is characterized as TM^z are those expressed by Equation (2.12).
6. Convert the set of Equations (2.41) to (2.45) to represent corresponding vectors of the electric displacements. Let the medium be characterized first as having scalar dielectric constant and then when the dielectric constant is a tensor of rank 2.
7. Using the τ component in cylindrical coordinates of the electric charge density or electric displacement vector obtained in problem 6, compute the total electric charge generated by the conductive and also the convective effects on a cylindrical structure of height L.
8. Using the ρ and α components in cylindrical coordinates of the electric displacement vectors identified in problem 6, obtain a solution for the total conductive and convective total charge residing at the lower and upper bases of the cylindrical structure described in problem 7.
9. Obtain an expression for the electric energy stored in a coaxial cylindrical condenser carrying a total surface charge secured in problem 7. Consider the dielectric constant a rank 2 tensor.
10. A tensor of rank 2 like ϵ_{ij} is written below:

$$\epsilon_{ij} = \begin{bmatrix} \epsilon_{11} & \epsilon_{12} & \epsilon_{13} \\ \epsilon_{21} & \epsilon_{22} & \epsilon_{23} \\ \epsilon_{31} & \epsilon_{32} & \epsilon_{33} \end{bmatrix} \tag{a}$$

To express ϵ_{ij} in a principle coordinate system, the following procedure is to be performed by taking the determinant of the following:

$$\begin{bmatrix} \epsilon_{11}-\lambda & \epsilon_{12} & \epsilon_{13} \\ \epsilon_{21} & \epsilon_{22}-\lambda & \epsilon_{23} \\ \epsilon_{31} & \epsilon_{32} & \epsilon_{33}-\lambda \end{bmatrix} = 0 \tag{b}$$

The solved values for λ are the components of ϵ_{ij} in the new coordinate system which could be written as:

$$\epsilon_{ij} = \begin{bmatrix} \lambda_1 & & 0 \\ & \lambda_2 & \\ 0 & & \lambda_3 \end{bmatrix} \tag{c}$$

Solve λ_1, λ_2, gnd λ_3 from Equation (b).

11. Repeat the solution of problem 9 when the rank 2 tensor for ϵ_{ij} is represented in the principle coordinate system.

12. From the set of Equations (2.71) to (2.73), current density distribution in the return stroke is a sharp linear front and a dropping tail. Obtain expressions for the corresponding convective current densities.

13. Repeat problem 6 by using the set of Equations (2.71) to (2.73) for the electric displacement vectors when the dielectric constant is a tensor of rank 2.

14. Repeat problem 7 with respect to the set of Equations (2.71) to (2.73) and the lightning pulse is a sharp linear front and a linear dropping tail.

15. Repeat problem 8 with respect to the set of Equations (2.71) to (2.73) and the lightning pulse is a sharp linear front and a linear dropping tail.

16. Repeat problem 9 with respect to the set of Equations (2.71) to (2.73) and the lightning pulse is a sharp linear front and a linear dropping tail.

17. Repeat problem 11 with respect to the set of Equations (2.71) to (2.73) and the lightning pulse is a sharp linear front and a linear dropping tail.

18. Equations (2.74) to (2.78) represent the developed magnetic field components at any point in space due to a sharp linear front and a linear dropping tail for the current density in the return stroke. From those equations obtain solutions for the inducing current density vectors.

19. Using the set of Equations (2.71) to (2.73) for the inducing electric field at any point in space, obtain solutions for the time variations of the magnetic induction vectors (B field).

20. Equation (2.113) represents a solution for the induced time varying voltage wave front component with superimposed oscillations. Establish a solution for the inducing voltage on a transmission line with open end.

21. Repeat problem 20 with respect to Equation (2.114) for the critically damped wave front. Assume the line has an open end.

22. Repeat problem 20 with respect to Equation (2.115) for the over-damped wave front. Assume the line has an open end.

23. Repeat problem 20 with respect to Equation (2.124) for the critically damped wavetail. Assume the line end is a short-circuit.

24. Repeat problem 20 with respect to Equation (2.125) for the over-damped wavetail. Assume the end is a short-circuit.

25. Repeat problem 20 with respect to Equation (2.126) for the wavetail with superimposed oscillations. Assume the line end is a short-circuit.

26. Equation (2.141) represents the induced time variation chopping of the lightning surge front. Obtain a solution for the induced voltage on a transmission line with open-termination.

27. Refer to Figure 2.7 showing a plot for the lightning surge parameter X_c s b. Express X_c as a function of the parameter b for the critically damped, over-damped as well as for the oscillatory case shown in the figure.

28. Refer to Figure 2.8 showing a plot for X_g against k_g known as the reverse polarity maximum for the lightning impulse tail. Obtain empirical functions of $X_g(k_g)$ for the three cases shown in the figure.

29. Refer to Figure 2.9 showing parametric correlation between X_v and k_u for the lightning impulse chopping. Obtain a plot as well as an empirical equation for the parameter δ_u against k_u.

30. Refer to Table 2.2 showing numeration for lightning pulse chopping. Obtain the following:
    (a) empirical equation for b_u against f_w
    (b) empirical equation for $b_u X_u$ against time t
    (c) empirical equation for δ against f_w

31. In reference to Figure 2.6 (a and b), express in terms of singularity functions a chopped lightning surge as approximated by the following wave form:

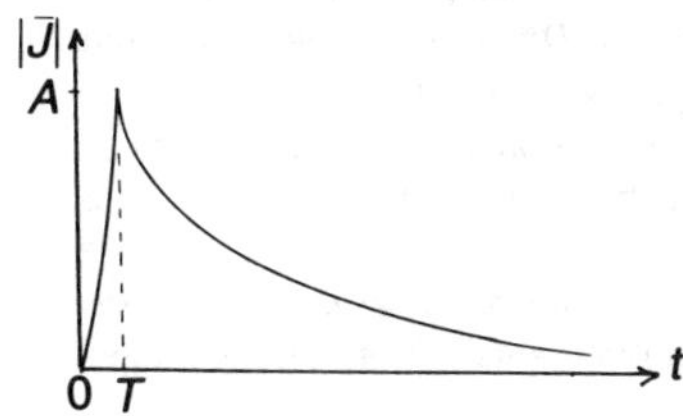

32. Using the functional distribution for lightning pulse obtained in problem 31, secure solutions for electromagnetic field components in Equations (2.71) to (2.78).

33. Using solutions of electromagnetic field components obtained in problem 32, express the corresponding solutions for the electric charge densities and electric current densities as vectors in a medium having scalar values for δ, ϵ and μ.

34. Repeat problem 3 if both δ and ϵ and μ are rank 2 tensors in a principle coordinate system.

PRINCIPAL LIST OF SYMBOLS

$\overline{E}$ electric field vector
$\overline{H}$ magnetic field vector
$\hat{z}$ impedivity
$\hat{y}$ admitivity
R_c radius of conductive stroke
R_v radius of convective stroke
t' retardation time
$\overline{A}$ magnetic vector potential
ψ associative function
c velocity of light
s Laplace transform variable
T_u chopping time
α_u wavefront length

REFERENCES

1. **Chowdhuri, P. and Gross, T. B.,** *Proc. IEE,* 114(12), 1899, 1967.
2. **Denno, K.,** *J. Electrostat.,* 13, 55, 1982.
3. **Denno, K.,** Dynamic modeling for the process of inducing and induced voltage surges due to lightning, *J. Electrostat.,* 13, 55, 1982.
4. **Denno, K.,** Mathematical modelling of propagating inducing and induced power due to actual pulse shape of lightning surge, *J. Electrostat.,* 15, 43, 1984.
5. **Garbedian, P. R.,** *Partial Differential Equations,* John Wiley & Sons, 1964.
6. **Gallet, G. and Lerogy, G.,** *IEEE Transactions on Power Apparatus and Systems,* 1973.
7. **Harrington, R. F.,** *Time-Harmonic Electromagnetic Fields,* McGraw-Hill, 1961.
8. **Papadopoulos, D. P.,** Proceedings of Midwest Power Symposium, W. Virginia University, Morgantown, 1977.
9. **Rudenberg, R.,** *Electrical Shock Waves in Power Systems,* Harvard University Press, 1968.
10. **Sobocki, R.,** Function representation of lightning impulse wave, *IEE Proc.,* 134A(9), 721, 1987.
11. **Suzuki, T. and Miyake, K.,** *IEEE Transactions on Power Apparatus and Systems,* 1973.
12. **Tripathy, S. C. and Yusuf Khan, M.,** *Power Record.,* 1572-74, 1974.

Chapter 3

LIGHTNING SURGE ANALYSIS BY MAGNETIC MOMENT

3.I INTRODUCTION

Modeling of electromagnetic phenomena resulting from released lightning voltage pulses at locations close to the return stroke as well as at any other space point (impacted area) has taken a variety of modes in theoretical research. Recently the author dealt in detail with such modeling where the conductive current of the return stroke is visualized as a solid cylindrical distribution and the convective current due to bound charges time variation is assumed to follow coaxial cylindrical distribution. Conventional methods based on the retarded propagation of inducing electromagnetic field effects at a space point had been used in the process of theoretical results modeling.

In this paper, a new mode of calculating electromagnetic field effects at a space point is presented using the concept of magnetic moments summation, to arrive successfully at a unified mathematical model for all inducing electromagnetic field components at a space point.

Principle element of moment integration at a space point is the availability of magnetic vector potential taking into consideration time retardation from the place of occurrence of the return stroke regarding conductive and convective components. Concept of magnetic vector potential will be applied to secure solutions in cylindrical three-dimensional systems of all electromagnetic field components as ground base to identify solutions for the inducing and induced voltages.

3.II MAGNETIC MOMENT AND VECTOR POTENTIAL

Figure 3.1 shows a magnetic dipole represented by a circular current loop of radius (a), carrying a constant current (I). Let ρ', ϕ', and θ' be the source coordinates with ρ, ϕ, and θ as the field point coordinates.

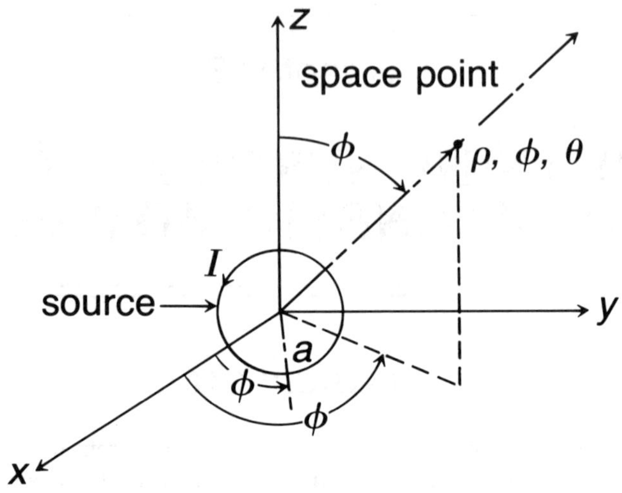

FIGURE 3.1.   Magnetic dipole.

Magnetic vector potential A from a current source of random orientation and location at a space point (X) is expressed by:

$$\overline{A}_\phi = \frac{\mu_o}{4\pi} \iiint \frac{J(x')dX'}{|X - X'|} \, e^{-JK|X - X'|} \, dX' \tag{3.1}$$

where X' is the source generalized coordinate point, and K is the intrinsic wave constant $= \sqrt{-\hat{z}\hat{y}}$

X and X' represent generalized field point and source point coordinates.

$\overline{J}(X')$ is the constant (independent of time) current density vector at a generalized source coordinate point.

$$|X - X'| = \rho^2 + \rho^2 - 2\rho\rho \cos(\phi - \phi') + (Z - Z')^{1/2} \tag{3.2}$$

$$\overline{J}(X') = I\delta(Z)\delta(\rho' - a) \tag{3.3}$$

and

$$dX' = \rho' \, d\rho' \, d\phi' \, dZ' \tag{3.4}$$

where δ is the Delta-Dirac function. To move to cylindrical coordinates $\rho = r \sin \theta$; therefore, from Equations (3.1) and (3.2) we can write

$$A_\phi(r) = \frac{\mu_o}{4\pi} \int_{-\infty}^{\infty}\int_{o}^{2\pi}\int_{o}^{\infty} \frac{a_\phi I\delta(\rho' - a)\rho' \, d\rho' \, d\phi' \, dZ' \, e^{-jk|x - x'|}}{[\rho^2 + \rho'^2 - 2\rho\rho' \cos(\phi - \phi) + (Z - Z')^2]^{1/2}} \tag{3.5}$$

$$= \frac{\mu_o I}{4\pi} \int_o^{2\pi} \frac{\hat{a}_\phi \, d\phi \; e^{-jk[\rho^2 + a^2 - 2\rho a \cos\phi + Z^2]^{1/2}}}{[\rho^2 + a'^2 - 2\rho a \cos\phi' + Z^2]^{1/2}} \qquad (3.6)$$

Now let $\rho = r \sin \phi$ and $Z = r \cos \phi$. Equation (3.6) becomes

$$\overline{A}\phi(r) = \frac{\mu_o I a}{4\pi} \, \hat{a}_\phi \int_o^{2\pi} \frac{e^{-jk(r^2\sin^2\theta + r + r^2\cos^2\theta + a^2 - 2ra\sin\theta\cos\phi')^{1/2}}}{[r^2 \sin^2\theta + r^2\cos^2\theta + a^2 - 2ra\cos^2\theta\sin\theta]^{1/2}}$$

$$(3.7)$$

where $\hat{a}_\phi$ is a unit vector. Then, since

$$\hat{a}_\phi = \hat{a}_x \sin \phi + \hat{a}_y \cos \phi \qquad (3.8)$$

$\hat{a}_x$, $\hat{a}_y$ are unit vectors in rectangular coordinates. Therefore,

$$\overline{A}_y = \frac{\mu_o I a}{4\pi} \int_o^{2\pi} \frac{e^{-jk(r^2 + a^2 - 2ra\sin\theta\cos\phi')^{1/2}}}{[r^2 + a^2 - 2ra\sin\theta']^{1/2}} \cos\phi \, d\phi' \qquad (3.9)$$

Let

$$F = \frac{e^{-jk(r^2 + a^2 - 2ra\sin\theta\cos\phi')^{1/2}}}{[r^2 + a^2 - 2ra\sin\theta'\cos\phi']^{1/2}} \qquad (3.10)$$

Hence

$$\overline{A}_y = \overline{A}_\phi\Big|_{\hat{a}_\phi \equiv 0} \frac{\mu_o I a}{4\pi} \int_o^{2\pi} F \cos\phi' \, d\phi' \qquad (3.11)$$

3.III SUMMATION OF MAGNETIC MOMENTS

A. CONDUCTIVE STROKE

Equation (3.11) identifies the y component of magnetic vector potential due to a single moment of radius a. Consider now that in a cross-section of conductive or convective lightning stroke, there are a number of finite radii represented by $a_1, a_2, a_3 \ldots a_n$.

The corresponding magnetic vector potentials are expressed by the following:

$$A_{y_1} = \mu_o \frac{I a_1}{4\pi} \int_o^{2\pi} F_1 \cos\phi' \, d\phi'$$

$$A_{y_2} = \mu_o \frac{I a_2}{4\pi} \int_o^{2\pi} F_2 \cos\phi' \, d\phi'$$

$$A_{y_3} = \mu_o \frac{I a_3}{4\pi} \int_o^{2\pi} F_2 \cos\phi' \, d\phi'$$

$$A_{y_n} = \mu_o \frac{I a_n}{4\pi} \int_o^{2\pi} F_n \cos\phi' \, d\phi' \qquad (3.12)$$

$$= \frac{Ia_n\mu_o}{4\pi} \int_o^{2\pi} \frac{e^{-jk(r_n^2 + a_n^2 - 2r_n\sin\theta\cos\phi')^{1/2}}}{[r_n^2 + a_n^2 - 2r_n \sin \theta' \cos \phi']^{1/2}} \cos \phi' \, d\phi' \qquad (3.13)$$

where

$$F_1 \text{ is a function of } a_1, r_1 = g(a_1, r_1)$$

$$F_2 = g_2(a_2, r_2)$$

$$F_3 = g_3(a_3, r_3)$$

$$\cdot$$
$$\cdot$$
$$\cdot$$

$$F_n = g_n(a_n, r_n) \qquad (3.14)$$

Therefore, total contributions from a set of magnetic moments for the magnetic vector potential at a space point is expressed by:

$$\overline{A}_{yc\text{-}H} = \hat{a}_y\mu_o \sum_{n=1}^{Mc\rightarrow\infty} \frac{I_n a_n}{4\pi} \int_o^{2\pi} F_n \cos \phi' \, d\phi' \qquad (3.15)$$

Equation (3.15) identifies the magnetic vector potential due to a conductive circular sheet containing a finite number of magnetic moments with radii ranging from $a_1 \rightarrow 0$, a_2, a_3 . . . a_n.

Now we shall express the counterpart of Equation (3.15) for the coaxial convective current stroke which commences from just the outer boundary of the conductive stroke. Let the outer radius of the conductive current stroke $= R_c$ and r_1, r_2, r_3 . . . r_n the radial vector of a space point. Therefore, from Equation (3.13):

$$A_{yv\text{-}H} = \frac{\hat{a}_y\mu_o}{4\pi} \sum_{m=1}^{M\hat{v}\rightarrow\infty} I_n(a_n - R_c)\int_o^{2\pi}$$

$$\frac{e^{-jkr_n^2 + (a_n - R_c)^2 - 2r_n\sin\theta\cos\phi}}{r_n^2 + (a_n - R_c)^2 - 2r_n \sin \theta \cos \phi} \cos \phi' \, d\phi' \qquad (3.16)$$

The next step is to compile vertically contributions for the magnetic vector potential from all magnetic moments present in the conductive and convective return strokes. This author prefers to follow the process of summation rather than integration in expressing the space vector magnetic potential as shown below.

Let the discrete number of conductive and convective dipoles or moments vertically be N_c and N_v, respectively; $\therefore$.

$$\overline{A}_{yc\text{-}H,V} = \frac{\hat{a}_y\mu_o}{4\pi} \sum_{n=1}^{M_c} \sum_{m=1}^{N_c} I_{n,m} A_{n,m} \int_o^{2\pi} F_{n,m} \cos \phi' \, d\phi' \qquad (3.17)$$

where n is the index of summation horizontally, m is the index of summation vertically, and

$$\overline{A}_{yv\text{-}H,V} = \frac{\hat{a}_y\mu_o}{4\pi} \sum_{n=1}^{M_c} \sum_{m=1}^{N_c} I_{n,m}(a_{n,m} - R_c)\left[\int_o^{2\pi}\right.$$

$$\frac{e^{-jk[r_{n,m}^2 + (a_{n,m}-R_c)^2 - 2r_n\sin\theta\cos\phi']^{1/2}}}{[r_{n,m}^2 + (a_{n,m} - R_c)^2 - 2r_{n,m} \sin \theta \cos \phi]^{1/2}} \cos \phi \, d\phi \qquad (3.18)$$

B. APPROXIMATION FOR $A_{yc\text{-}H}$ AND $A_{yv\text{-}V}$

Conductive stroke. In reference to Equations (3.15) and (3.16), let

$$S^2 = r_n^2 + a^2 - 2ar_n \sin \theta \cos \phi'$$

$$2s \, ds = (2r_n - 2a \sin \theta \cos \phi')dr_n \qquad (3.19)$$

$$r_n = a \sin \theta \cos \phi' \pm \tfrac{1}{2}\sqrt{4a^2 \sin^2 \theta \cos^2 \phi' + a^2 + S^2} \qquad (3.20)$$

$$dr_n = 2s \, ds/\sqrt{a^2 \sin^2 \theta \cos^2 \phi' - a^2 + s^2} \qquad (3.21)$$

Therefore, from Equations (3.18) through (3.21),

$$\overline{A}_y = \frac{j2I}{k} \int \frac{e^{-js} \, ds}{\sqrt{1 + \dfrac{s^2}{k^2}}} \qquad (3.22)$$

with the conditional orientation such that:

$$\sin^2 \theta \cos^2 \phi' \ll 1\ldots \qquad (3.23)$$

Equation (3.23) implies $\theta = 2m\pi$, where m is even space index and $\phi' \rightarrow m \frac{\pi}{2}$, where m is odd source index.

Conducting a process of summation horizontally, i.e., at $Z' = 0$ on all magnetic dipole moments due to conductive stroke with respect to Equation (3.22),

$$\overline{A}_{yc\text{-}H} = \frac{2I_c\hat{a}_y}{k_c} \sum_{n=0}^{\infty} \int_{o}^{s} e^{-js}\, ds\left[1 + \frac{1}{2}\frac{S^2}{k_c^2}\right]ds$$

$$= \frac{2I_c\hat{a}_y}{k_c} \sum_{n=1}^{N_c=\infty} e^{-js}\, ds + \frac{1}{k_c^3}\int e^{-js}s^2\, ds \qquad (3.24)$$

where the subscript (c) refers to conductive stroke.

Carrying out the integration process in Equation (3.24) yields

$$\overline{A}_{yc\text{-}H} = I_c \sum_{n=1}^{N_c} \left(\frac{j^2}{k_c}\, e^{-js}\frac{js^2}{k_c^3} + \frac{2se^{-js}}{k_c^3}\frac{j^2e^{-js}}{k_c^3}\right)$$

or

$$\overline{A}_{yc\text{-}H} = I_c \sum_{n=0}^{\infty} \left[\frac{2}{k_c} - \frac{s^2}{k_c^3} + \frac{2s}{k_c^3} - \frac{2}{k_c^3}\right]je^{-js} \qquad (3.25)$$

where

$$k_c^2 = a^2(\sin^2\theta\cos^2\o' - 1)$$

To account for an M_c number of magnetic moments stacked vertically along the cylindrical conductive stroke ($M_c\to\infty$), we conduct another summation process on $A_{yc\text{-}H}$ as shown below:

$$\overline{A}_{ycH\text{-}V} = jI_c \sum_{m=1}^{M_c} \sum_{n=1}^{N_c} \left[\frac{2}{k_c} - \frac{s^2}{k_c^3} + \frac{2s}{jk_c^3} - \frac{2}{k_c^3}\right]e^{-js} \qquad (3.26)$$

where

$$s = \sqrt{r_{n,m}^2 + a^2 - 2ar_{n,m}\sin\theta\cos\o'} \qquad (3.27)$$

$A_{ycH\text{-}V}$ is the magnetic vector potential produced by all magnetic moments contained in the cylindrical conductive return stroke.

To account for the magnetic vector potential generated by magnetic moments contained in the convective return stroke produced by the time variation in bound charges, we will proceed in a similar approach for that of the conductive stroke:

$$A_{ycH\text{-}V} = jI_v \sum_{n=0}^{N} \left[\frac{2}{k} - \frac{s^2}{k^3} + \frac{2s}{jk^3} - \frac{2}{k}\right]e^{-js} \qquad (3.28)$$

where $A_{ycH\text{-}V}$ is the vector magnetic potential at the plane of $Z' = 0$ generated by all magnetic moments produced by the convective return stroke. Also the subscript (v) refers to convective stroke.

The next step is to sum up all moments stacked vertically along the coaxial convective stroke whose radius at the source is $> R_c$ (R_c is the radius of the conductive stroke).

$$\overline{A}_{ycH\text{-}V} = jI \sum_{n=0}^{N} \sum_{m=0}^{M} \left[\frac{2}{k} - \frac{s^2}{k^3} + \frac{2s}{k^3} - \frac{2}{k^2} \right] e^{-js} \tag{3.29}$$

$A_{yvH\text{-}V}$ is the vector magnetic potential due to horizontal sum (H) and vertical sum (V) on all magnetic moments contained in the coaxial convective stroke.

Factual consideration in the vertical and horizontal summation process:

$$N \ll Nc$$

$$M \ll Mc \tag{3.30}$$

3.IV INDUCING VOLTAGE

From the following Maxwell field equation,

$$\overline{\nabla} V_i = -\hat{a}_y \mu \frac{\overline{A}_{H,V}}{t} \tag{3.31}$$

And with respect to our case,

$$\overline{A}_{H,V} = \overline{A}_{yc\text{-}H,V} \text{ (conductive)} + A_{y\text{-}H,V} \text{ (convective)} \tag{3.32}$$

Equation (3.31) becomes:

$$\overline{\nabla} V_i = -\mu_o \left[\frac{\partial \overline{A}_{yc\text{-}H,V}}{\partial t} + \frac{\partial \overline{A}_{y\text{-}H,V}}{\partial t} \right] \tag{3.33}$$

where V_i is the inducing electric potential.

Now, since $0 \ll Z_n^2 \ll r_n^2$ and

$$k_{c,v} = a^2 \sqrt{\frac{r_n^2 - Z_n^2}{r_n^2} \cos^2 \phi' - 1} \tag{3.34}$$

$$k_{c,v} \approx a^2 \sqrt{\cos^2 \phi' - 1} \tag{3.35}$$

Also, since

$$S = \sqrt{X_n^2 + Y_n^2 - a^2 - 2ar_n \cos \phi'} \tag{3.36}$$

Therefore,

$$S \approx \sqrt{X_n^2 + Y_n^2} \tag{3.37}$$

And as $a \to 0$

$$k_c \to \delta \tag{3.38}$$

From the above factual approximations which match the physical realities of the situation, $A_{ycH\text{-}V}$ and $A_{yvH\text{-}V}$ expressed by Equations (3.28) and (3.29) become:

$$\overline{A}_{ycH\text{-}V} \equiv 2Q_c\hat{a}_y \sum_{n=0}^{N} \sum_{m=0}^{M} e^{-j\sqrt{X_{n,m}^2 + Y_{n,m}^2}} \tag{3.39}$$

where Q is a large numerical value involving value of the conductive current in M amp. The subscript (n) refers to horizontal summation, while (m) refers to vertical summation. Also, magnetic vector potential due to convective stroke is similar to that of Equation (3.39), i.e.,

$$\overline{A}_{yvH\text{-}V} = 2Q_v\hat{a}_y \sum_{n=0}^{N_c} \sum_{m=0}^{M_c} e^{-j\sqrt{X_{n,m}^2 + Y_{n,m}^2}} \tag{3.39a}$$

Substituting Equations (3.28) and (3.29) into Equation (3.31) results in:

$$\overline{\nabla}V_{ic} = -2\mu_o Q_c, \sum_{m=1}^{\infty} \sum_{n=1}^{\infty} e^{-j\sqrt{X_{n,m}^2 + Y_{n,m}^2}} U_o(t - t_o) \tag{3.40}$$

Hence

$$V_{ic,v} = -\int_{X_n}\int_{Y_n}\int_{Z_n} - 2\mu_o Q_c, U_o(t - t_o) \sum_{m=1}^{\infty} \sum_{n=1}^{\infty}$$
$$e^{-j(X_{n,m}^2 + Y_{n,m}^2)^{1/2}}d_nX_n\, dY_n\, dZ \tag{3.41}$$

where V_{ic} is the inducing potential due to conductive stroke.

From Equation (3.37) taking into consideration that $X \ll Y$ and that $V_i = 0$ at the source of conduction and conductive strokes (i.e., $X_n, Y_n, Z_n = 0$), the solution of $V_{ic}\,(X_n, Y_n, Z_n, t)$ is:

$$V_{ic,v} = -j2\mu_o QU_o(t - t_o) \sum_{\substack{m_c=1, \\ m_v=1}}^{M_c,M_v} \sum_{\substack{n_c=1, \\ n_v=1}}^{N_c,N_v} Y_{n,m}Z_{n,m} \frac{e^{-j(X_{n,m})}}{X_{n,m}} \tag{3.42}$$

3.V   THE INDUCED POTENTIAL

At the space point X_n, Y_n, Z_n located on the transmission line having distributed inductance and capacitance, the correlation between the inducing and the induced potential is expressed by the following:

$$\frac{\partial^2 V_{c,v}}{\partial X^2} - \frac{1}{C^2} \frac{\partial^2 V_{c'}}{\partial t^2} = \frac{1}{C^2} \frac{\partial^2 V_{ic}}{\partial t^2} \tag{3.43}$$

C is the velocity of light. Equation (3.43) considers space variation in one dimension only.

Taking Laplace transform of both sides of Equation (3.43) w.r.t. (t) the following results are obtained: let

$$V_i = m' \sum_{m=1}^{\infty} \sum_{n=1}^{\infty} \frac{e^{-jX_n} Y_{n,m} Z_{n,m}}{X_{n,m}} U_o(t - t_o) \tag{3.43a}$$

where

$$m' = -j2\,\mu_o Q \tag{3.44}$$

Then differentiating w.r.t. times

$$V'_i = m' \sum_{m=1}^{\infty} \sum_{n=1}^{\infty} \frac{e^{-jX_{n,m}} Y_{n,m} Z_{n,m}}{X_{n,m}} U'_o(t - t_o) \tag{3.45}$$

$$V''_i = m' \sum_{m=1}^{\infty} \sum_{n=1}^{\infty} \frac{e^{-jX_{n,m}} Y_{n,m} Z_{n,m}}{X_{n,m}} U''_o(t - t_o) \tag{3.46}$$

Therefore,

$$\frac{\partial^2 V}{\partial t^2} = C^2 \frac{\partial^2 V}{\partial X^2} \left[\sum_{m=1}^{\infty} \sum_{n=1}^{\infty} \frac{e^{-jX_m} Y_{n,m} Z_{n,m}}{X_{n,m}} U''_o(t - t_o) \right] \tag{3.47}$$

Taking Laplace transform w.r.t. (t) of Equation (3.43), the following is obtained:

$$\frac{\partial^2 V(x,s)}{\partial X^2} - \frac{S^2}{C^2} V(X,S) = K(s) \sum_{m=1}^{\infty} \sum_{n=1}^{\infty} \frac{e^{-jX_{m,n}}}{X_{n,m}} \tag{3.48}$$

where

$$K(s) = \frac{S^2}{C^2} m Y_n Z_n e^{-t_o S} \tag{3.49}$$

Equation (3.44) is a 2nd order differential equation in X. Solution for $V(x,s)$ is left as a problem for the student to solve; however, the author is listing the final form for $V(x,s)$:

$$V_{c,v}(x,s) = \frac{-j2\mu_o QYZ_n S^2 e_{-toS}}{C^2 + S^2} \sum_{m=1}^{\infty} \sum_{n=1}^{\infty} \frac{e^{-jX_{n,m}}}{X_{n,m}} \tag{3.50}$$

$V_{c,v}(x,t)$, the induced voltage due to conductive stroke (V_c) and convective stroke (V_v), could be obtained by taking the inverse Laplace transform of $V(x,s)$. This process is left as a problem for the student to solve.

The author is listing below the final form for the induced voltage $V_{c,v}(x,t)$.

$$V_{c,v}(x,t) = -j2\mu_o Q \frac{U_1(t - t_o)}{C} \sum_{m=1}^{\infty} \sum_{n=1}^{\infty}$$

$$\left[\frac{e^{-jX_{n,m}}}{X_{n,m}} Z_{n,m} Y_{n,m}\right]\left[\frac{1}{C} \cos ct\, \delta^2(t) + \frac{3\delta(t)}{C^2} \sin ct\right.$$

$$\left. - \frac{2\delta'(t)}{C^3} \sin ct - \frac{1}{C^3} \int_o^t \sin c\tau\, \delta''(t - \tau)d\tau\right] \tag{3.51}$$

where $U_1(t - t_o)$ is a delayed step function

$$V_{c,v} = V_c(x,t) + V_v(x,t) \tag{3.52}$$

Consequently from the preceding analysis centered on the utilization of the concept of magnetic moment, the author demonstrated the useful solution for the magnetic vector potential as an effective vehicle to secure solutions for the inducing and induced voltage at any point in space rather than integration as has been used to establish expressions for the total contributions of horizontal and vertical spread of magnetic moments for the final field impact at a space point.

3.VI   THE INDUCED ELECTRIC FIELD

To calculate the induced electric field vector along the transmission line due to conductive and convective strokes, we proceed taking the gradient of $V_{c,v}(x,t)$ expressed by Equation (3.47):

$$\overline{\nabla} V_{c,v}(x,t) = \hat{a}_x \frac{V_{c,v}}{\partial X_n} + \hat{a}_y \frac{\partial V_{c,v}}{\partial Y_n} + \hat{a}_z \frac{V_{c,v}(t,t)}{\partial Z_n} \tag{3.53}$$

Let

$$P(t) = \frac{\delta^2(t)}{C^2} \cos ct + \frac{3\delta(t)}{C^2} \sin ct - \frac{2\delta'(t)}{C^3} \sin ct -$$

$$\frac{1}{C^3} \int_o^\tau \sin c\tau\, \delta''(t - \tau)dT \tag{3.54}$$

Hence,

$$V_{c,v}(x,t) = -j2Q\mu_o\rho \, \frac{U_1(t - t_o)}{} \, e \sum_{m=1}^{\infty} \sum_{n=1}^{\infty} \frac{e^{-jX_{n,m}}}{X_{n,m}} \, Y_{n,m}Z_{n,m} \qquad (3.55)$$

Therefore,

$$\frac{\partial V_{c,v}}{\partial X_n} = +j\frac{2}{C} Q\mu_o\rho U_1(t - t_o) \sum_{m=1}^{\infty} \sum_{n=1}^{\infty} Y_{n,m}Z_{n,m}\left(\frac{e^{-jX_{n,m}}}{X_{n,m}^2} + \frac{e^{-jX_{n,m}}}{X_{n,m}}\right) \qquad (3.56)$$

$$\therefore \; = \hat{a}_x E_x = \overline{E}_{2x}$$

$$\frac{\partial V_{c,v}}{\partial Y_n} = -j\frac{2}{C} Q\mu_o\rho U_1(t - t_o) \sum_{m=1}^{\infty} \sum_{n=1}^{\infty} Z_{n,m} \frac{e^{-jX_{n,m}}}{X_{n,m}^2} \qquad (3.57)$$

$$= \hat{a}_y E_y = \overline{E}_{2y}$$

$$\frac{\partial V_{c,v}}{\partial Z_n} = -j\frac{2}{C} Q\mu_o\rho U_1(t - t_o) \sum_{m=1}^{\infty} \sum_{n=1}^{\infty} Z_{n,m} \frac{e^{-jX_{n,m}}}{X_{n,m}}$$

$$= \hat{a}_z E_z = \overline{E}_{2z} \qquad (3.58)$$

It is clear from Equations (3.52) to (3.54) that the induced electric field converges at a far point in space.

Next, we can proceed to express the electric displacement of charge density vectors D from Equations (3.52), (3.53) and (3.54). If the dielectric constant is scalar in a simple medium,

$$\overline{D} = \hat{a}_x \epsilon E_x + \hat{a}_x \epsilon E_x + \hat{a}_x E_z \qquad (3.59)$$

3.VII THE INDUCED MAGNETIC FIELD

We now move on to calculate the time varying magnetic field on a transmission line whose induced electric potential has been expressed by Equation (3.47).

From the following Maxwell field equation:

$$\overline{\nabla}X\overline{E} = -\frac{\partial\overline{B}}{\partial t} \qquad (3.60)$$

The *E* field was obtained in Section VI:

$$\overline{E} = \hat{a}_x p\left[\frac{j2}{c} Q\mu_o U_1(t - t_o)\right]\sum_{m=1}^{\infty}\sum_{n=1}^{\infty} Y_{n,m}Z_{n,m}\left(\frac{e^{-jX_{n,m}}}{X_{n,m}^2} + j\frac{e^{-jX_{n,m}}}{X_{n,m}}\right)$$

$$+ \hat{a}_y p\left[\frac{-j2}{c} Q\mu_o\rho U_1(t - t_o)\right]\sum_{m=1}^{\infty}\sum_{n=1}^{\infty} Z_{n,m}\frac{e^{-jX_{n,m}}}{X_{n,m}}$$

$$+ \hat{a}_z p\left[\frac{-j2}{c} Q\mu_o\rho U_1(t - t_o)\right]\sum_{m=1}^{\infty}\sum_{n=1}^{\infty} Y_{n,m}\frac{e^{-jX_{n,m}}}{X_{n,m}} + \dots \qquad (3.60a)$$

Let

$$\frac{j2}{c} Q\mu_o\rho U_1(t - t_o) = \gamma \qquad (3.61)$$

Therefore,

$$\overline{E} = \hat{a}_x \gamma\rho \sum_{m=1}^{\infty}\sum_{n=1}^{\infty} Y_{n,m}Z_{n,m}\left(\frac{e^{-jX_{n,m}}}{X_{n,m}^2} + j\frac{e^{-jX_{n,m}}}{X_{n,m}}\right)$$

$$- \hat{a}_y\gamma p \sum_{n=1}^{\infty}\sum_{m=1}^{\infty} Z_{n,m}\frac{e^{-jX_{n,m}}}{X_n} - \hat{a}_z\gamma p \sum_{m=1}^{\infty}\sum_{n=1}^{\infty} Y_{n,m}\frac{e^{-jX_{n,m}}}{X_{n,m}} \dots \qquad (3.62)$$

$$\frac{\partial B}{\partial t} = \gamma p\sum_{m=1}^{\infty}\sum_{n=1}^{\infty}\begin{vmatrix} \hat{a}_x & \hat{a}_y & \hat{a}_z \\ \dfrac{\partial}{\partial x} & \dfrac{\partial}{\partial y} & \dfrac{\partial}{\partial z} \\ \left[Y_{n,m}Z_{n,m}\left(\dfrac{e^{-jX_{n,m}}}{X_{n,m}^2} + j\dfrac{e^{-jX_{n,m}}}{X_{n,m}}\right)\right], & \left[-Z_{n,m}\dfrac{e^{-jX_{n,m}}}{X_{n,m}}\right] & \left[-Y_{n,m}\dfrac{e^{-jX_{n,m}}}{X_{n,m}}\right] \end{vmatrix}$$

$$= \gamma_p \sum_{m=1}^{\infty}\sum_{n=1}^{\infty}\left[\hat{a}_x\left(\frac{e^{-jX_{n,m}}}{X_{n,m}} + \frac{e^{-jX_{n,m}}}{X_{n,m}}\right) + \hat{a}_y\left[Y_{n,m}\left(\frac{e^{-jX_{n,m}}}{X_{n,m}^2} \frac{e^{-jX_{n,m}}}{X_{n,m}}\right)\right.\right.$$

$$+ y_{n,m}\left(\frac{e^{-jX_{n,m}}}{X_{n,m}^2} + j\frac{e^{-jX_{n,m}}}{X_{n,m}}\right)\hat{a}_z\left[Z_{n,m}\left(\frac{e^{-jX_{n,m}}}{X_{n,m}^2} - j\frac{e^{-jX_{n,m}}}{X_{n,m}}\right)\right]$$

$$+ Z_{n,m}\left(\frac{e^{-jX_{n,m}}}{X_{n,m}^2} + j\frac{e^{-jX_{n,m}}}{X_{n,m}}\right)\right]\right]$$

or

$$\frac{\partial B}{\partial t} = \rho \sum_{m=1}^{\infty} \sum_{n=1}^{\infty} \left[\hat{a}_y 2y_{n,m}\left(\frac{e^{-jX_{n,m}}}{X_{n,m}^2} + j\frac{e^{-jX_{n,m}}}{X_{n,m}}\right) - \hat{a}_z 2Z_{n,m}\left(\frac{e^{-jX_{n,m}}}{X_{n,m}^2} + j\frac{e^{-jX_{n,m}}}{X_{n,m}}\right) \right]$$

Therefore,

$$\overline{B} = \mu_o \frac{j4}{c} Q \sum_{n,m=1}^{\infty} \sum_{m,n=1}^{\infty} \left(\frac{e^{-jX_{n,m}}}{X_{n,m}^2} + j\frac{e^{-jX_{n,m}}}{X_{n,m}}\right)(\hat{a}_y - \hat{a}_z)\int$$

$$\left[\frac{\delta^2(t)}{c} \cos ct \, dt + \frac{3\delta(t)}{C^2} \sin ct \, dt - \frac{2\delta(t)}{C^3} \sin ct \, dt\right.$$

$$\left. - \frac{1}{C^3} \delta''(t - \tau)\sin c\tau \, d\tau\right] \tag{3.63}$$

Again, the principle convergence of the B field is assured at a far space point.

Regarding the integral processes involving the singularities $\delta_o(t)$ and $\delta_o'(t)$, we can proceed to use the sifting property, namely:

$$\int_{-\infty}^{\infty} f(t)\delta(t - a)dt = f(a) \tag{3.64}$$

$$\int_{-\infty}^{\infty} \delta(x)\delta(x - u)dx = \delta(u) \tag{3.65}$$

and

$$\int_{a-\epsilon}^{a+\epsilon} f(x)\delta(x - a)dx = -f'(a) \tag{3.66}$$

Therefore,

$$\int \delta(t)\sin ct \, dt = 0$$

$$\int \sin ct \, \acute{\delta}(t)dt = -\cos ct = -c \text{ at } t = 0$$

Hence,

$$\int \delta^2(t)\cos ct \, dt = \frac{1}{C} \delta^2(t)\sin ct - \frac{2}{c}\int \sin ct \, \delta(t)dt$$

$$= \frac{1}{C} \delta^2(t)\sin ct \tag{3.67}$$

Also

$$\int\int \sin c\tau \, \delta''(t - \tau)d\tau = \int C^2 \int \cos c\tau \, d\tau = \sin ct \tag{3.68}$$

The B field from Equation (3.51) becomes:

$$\bar{B}_2 = \frac{j4}{c} Q\mu_o \sum_{m=1}^{\infty} \sum_{n=1}^{\infty} \left(\frac{e^{-jX_{n,m}}}{X_{n,m}^2} + j\frac{e^{-jX_{n,m}}}{X_{n,m}} \right)(\hat{a}_y - \hat{a}_z)$$

$$\left[\frac{1}{C2} \delta^2(t)\sin ct + \frac{2}{C2} - \frac{1}{C3} \sin ct \right] \tag{3.69}$$

or

$$\bar{H}_2 = \frac{j4}{c} Q\mu_o \sum_{m=1}^{\infty} \sum_{n=1}^{\infty} \left(\frac{e^{-jX_{n,m}}}{X_{n,m}^2} + j\frac{e^{-jX_{n,m}}}{X_{n,m}} \right)(\hat{a}_y - \hat{a}_z)$$

$$\left[\frac{2}{C2} - \frac{1}{C3} \sin ct + \frac{1}{C2} \delta^2(t)^1 \sin ct \right] \tag{3.70}$$

3.VIII INDUCED CURRENT DENSITY $\bar{J}$

We can calculate the J vector from $\bar{\nabla} \times \bar{H} = \bar{J}$. Let

$$\frac{j4}{c} Q\left[\frac{2}{C^2} - \frac{1}{C^3} \sin ct + \frac{1}{C^2} \delta^2(t)\sin ct \right] = P_1 \tag{3.71}$$

Therefore,

$$H_2 = \sum_{n,m=1}^{\infty} \sum_{m,n=1}^{\infty} \hat{a}_y P_1\left(\frac{e^{-jX_{n,m}}}{X_{n,m}^2} + j\frac{e^{-jX_{n,m}}}{X_{n,m}} \right) - \hat{a}_z P_1\left(\frac{e^{-jX_{n,m}}}{X_{n,m}^2} + j\frac{e^{-jX_{n,m}}}{X_{n,m}} \right) \tag{3.72}$$

$$\bar{\nabla}\times\bar{H}_2 = P_1 \sum_{n,m=1}^{\infty} \sum_{m,n=1}^{\infty} \begin{vmatrix} \hat{a}_x & \hat{a}_y & \hat{a}_z \\ \dfrac{\partial}{\partial x} & \dfrac{\partial}{\partial y} & \dfrac{\partial}{\partial z} \\ 0, \left(\dfrac{e^{-jX_{n,m}}}{X_{n,m}^2} + j\dfrac{e^{-jX_{n,m}}}{X_{n,m}} \right), & \left(\dfrac{e^{-jX_{n,m}}}{X_{n,m}^2} + j\dfrac{e^{-jX_{n,m}}}{X_{n,m}} \right) \end{vmatrix} \tag{3.73}$$

$$= P_1 \sum_{m=1}^{\infty} \sum_{n=1}^{\infty} \left[\hat{a}_x(0 - 0) + \hat{a}_y \left(\frac{e^{-jX_{n,m}}}{X_{n,m}^3} + \frac{e^{-jX_{n,m}}}{X_{n,m}^2} + \frac{e^{-jX_{n,m}}}{X_{n,m}} - j\frac{e^{-jX_{n,m}}}{X_{n,m}^2} \right) \right.$$

$$\left. - \hat{a}_z \left(\frac{e^{-jX_{n,m}}}{X_{n,m}^3} + \frac{e^{-jX_{n,m}}}{X_{n,m}} \right) \right] \tag{3.74}$$

Therefore,

$$\bar{J} = P_1(t) \sum_{n,m=1}^{\infty} \sum_{m,n=1}^{\infty} \left(\frac{e^{-jX_{n,m}}}{X_{n,m}^3} + \frac{e^{-jX_{n,m}}}{X_{n,m}} \right)(\hat{a}_y - \hat{a}_z) \tag{3.75}$$

where P_1 is expressed by Equation (3.74).

Consider the condition of field breakdown which will occur when $J \to \infty$. From Equation (3.63) $J \to \infty$ will take place where

$$1) \quad e^{jX_{n,m}}X_{n,m}^3 \to 0$$

$$2) \quad e^{jX_{n,m}}X_{n,m} \to 0$$

The observation from conditions 1 and 2 is that $J \to \infty$ will take place where $X_{n,m} \to 0$ and $X_{n,m} \to -\infty$.

3.IX EXPRESSION FOR ELECTRICAL CONDUCTIVITY σ

The electrical conductivity σ characteristic of the medium between conductive plus convective lightning strokes could now be expressed under two prevailing conditions: (i) σ is a scalar, and (ii) σ is a tensor of rank 2, i.e., σ_{ij}, i, $j = 1, 2, 3$.

From the set of Equations (3.56), (3.57) and (3.58) σ is a scalar in a medium characterized simply as

$$\bar{J} = \sigma\bar{E} \tag{3.76}$$

Of course, $\bar{J}_x$ in Equation (3.56) $= 0$ because $\sigma_{11} = 0$. Then

$$\bar{J} = \hat{a}_y \left[-j\frac{2Q}{c} PU_1(t - t_o) \sum_{n,m=1}^{\infty} \sum_{m,n=1}^{\infty} Z_{n,m} \frac{e^{-jX_{n,m}}}{X_{n,m}} \right] \sigma$$

$$+ \hat{a}_z \left[-j\frac{2Q}{c} PU_1(t - t_o) \sum_{n,m=1}^{\infty} \sum_{m,n=1}^{\infty} Y_n \frac{e^{-jX_{n,m}}}{X_{n,m}} \right] \tag{3.77}$$

where ρ was given by Equation (3.54).

From Equations (3.75) and (3.77), and for a scalar σ, it can be expressed by:

$$\sigma = \sum_{n,m=1}^{\infty} \sum_{n,m=1}^{\infty} P_1 \left(\frac{e^{-jX_{n,m}}}{X_{n,m}^3} + \frac{e^{-jX_{n,m}}}{X_{n,m}} \right) \Big/$$

$$\frac{2Q}{c} P_1 U(t - t_o) \sum_{n,m=1}^{\infty} \sum_{m,n=1}^{\infty} \frac{e^{-jX_{n,m}}}{X_{n,m}} \sqrt{Y_{n,m}^2 + Z_{n,m}^2} \qquad (3.78)$$

$$= \frac{P_1(t)c}{2Q\rho} \frac{1}{U_1(t - t_o)} \sum_{n=1}^{\infty} \sum_{m,n=1}^{\infty} \frac{X_{n,m}^{-2} + 1}{(Y_{n,m}^2 + Z_{n,m}^2)^{1/2}} \qquad (3.78a)$$

We observe from Equation (3.65) that the scalar σ_{ij} is a function of X_n where $n = 1, 2, 3, 4 \ldots$ and time with the presence of singularity functions, their derivatives and sinusoids.

The second case to consider is when σ is a tensor of rank 2 in a medium characterized as anisotropic when σ_{ij} is written below:

$$\sigma_{ij} = \begin{bmatrix} \sigma_{11} & \sigma_{12} & \sigma_{13} \\ \sigma_{21} & \sigma_{22} & \sigma_{23} \\ \sigma_{31} & \sigma_{32} & \sigma_{33} \end{bmatrix} \qquad (3.79)$$

σ_{ij} can be expressed as a tensor in a much simplified form when referred to the principle system of coordinates, where

$$\sigma_{ij} = \begin{bmatrix} \sigma_{11}' & & 0 \\ & \sigma_{22}' & \\ 0 & & \sigma_{33}' \end{bmatrix} \qquad (3.80)$$

The diagonal elements in Equation (3.70) can be solved for ω from the following equation:

$$\begin{bmatrix} \sigma_{11}-\lambda & \sigma_{12} & \sigma_{13} \\ \sigma_{21} & \sigma_{22}-\lambda & \sigma_{23} \\ \sigma_{31} & \sigma_{32} & \sigma_{33}-\lambda \end{bmatrix} = 0 \qquad (3.81)$$

The three values of λ from Equation (3.81) are σ_{11}', σ_{22}' and σ_{33}'. Therefore, following the case of σ a tensor of rank 2 in a principle coordinate system, again referring to Equations (3.63) and (3.66), now we can write

$$\bar{J}_i = \sigma_{ij}E_j \tag{3.82}$$

Therefore,

$$= (\hat{a}_y - \hat{a}_z)P_1 \sum_{m=1}^{\infty} \sum_{n=1}^{\infty} \left(\frac{e^{-jX_{n,m}}}{X_{n,m}^2} + \frac{e^{-jX_{n,m}}}{X_{n,m}} \right) = \begin{bmatrix} \sigma_{11} & 0 & 0 \\ 0 & \sigma_{22} & 0 \\ 0 & 0 & \sigma_{33} \end{bmatrix}$$

$$(\hat{a}_y + \hat{a}_z)\left[-\frac{j}{c} 2QPU_1(t - t_o) \right] \sum_{m=1}^{\infty} \sum_{n=1}^{\infty} (Y_{n,m} + Z_{n,m}) \frac{e^{-jX_{n,m}}}{X_{n,m}} \tag{3.83}$$

$\therefore \sigma_{11}' = 0$, and

$$P_1 \sum_{n,m=1}^{\infty} \sum_{m,n=1}^{\infty} \left(\frac{e^{-jX_{n,m}}}{X_{n,m}^3} + \frac{e^{-jX_{n,m}}}{X_{n,m}} \right)$$

$$= \sigma_{22} \left[\frac{-j}{c} 2QPU_1(t - t_o) \sum_{n=1}^{\infty} \sum_{m=1}^{\infty} (Y_{n,m} + Z_{n,m}) \frac{e^{-jX_{n,m}}}{X_{n,m}} \right] \tag{3.84}$$

Also

$$- P_1 \left[\frac{e^{-jX_{n,m}}}{X_{n,m}^3} + \frac{e^{-jX_{n,m}}}{X_{n,m}} \right]$$

$$= \sigma_{33}' \left[\frac{-j}{c} 2QPU_1(t - t_o) \sum_{n=1}^{\infty} \sum_{m=1}^{\infty} Y_{n,m}Z_{n,m} \frac{e^{-jX_{n,m}}}{X_{n,m}} \right] \tag{3.85}$$

or from Equations (3.83), (3.84) and (3.85):

$$\sigma_{11}' = 0; \text{ then,}$$

$$\sigma_{22}' = \frac{cP_1 \left[\dfrac{e^{-jX_{n,m}}}{X_n^3} + \dfrac{e^{-jX_{n,m}}}{X_n} \right]}{-j2QPU_1(t - t_o) \displaystyle\sum_{n,m=1}^{\infty} \sum_{m,n=1}^{\infty} (y_{n,m} + Z_{m,n}) \dfrac{e^{-jX_{n,m}}}{X_{n,m}}}$$

$$= j \frac{cP_1}{2QPU_1(t - t_o)} \sum_{n,m=1}^{\infty} \sum_{m,n=1}^{\infty} \frac{X_{n,m}^{-2}}{Y_{n,m} + Z_{n,m}} + 1 \tag{3.86}$$

Then

$$\sigma_{33}' = -j \frac{cP_1}{2QPU_o(t - t_o)} \sum_{n,m=1}^{\infty} \sum_{m,n=1}^{\infty} \frac{X_{n,m}^{-2}}{Y_{n,m} + Z_{n,m}} + 1 \tag{3.87}$$

A similar procedure is used to calculate all nine elements of σ_{ij} in the regular coordinate system. This is left as a problem for the student to solve.

3.X PROPAGATION OF SURGE POWER DUE TO INDUCED FIELD

The next step in our analysis of inducing and induced field effects due to lightning surge is the mode of power propagation along a transmission line of any length. This mode of power transfer is in the form of the Poynting vector propagation as shown below.

First refer to the set of Equations (3.56) to (3.58) expressing the induced electric field vectors in rectangular coordinates, where

$$\bar{E} = \hat{a}_x E_x + \hat{a}_y E_y + \hat{a}_z E_z \tag{3.88}$$

The corresponding set of the induced magnetic field equations is that given by Equation (3.58), where

$$\bar{H} = \hat{a}_y H_y + \hat{a}_z H_z \tag{3.89}$$

Power propagation in watts/m² is expressed by the Poynting vector P:

$$\bar{P} = \bar{E} \times \bar{H} \ \text{watts/m}^2$$

$$\bar{P} = \begin{bmatrix} \hat{a}_x & \hat{a}_y & \hat{a}_z \\ E_x & E_y & E_z \\ 0 & H_y & H_z \end{bmatrix} \tag{3.90}$$

$$= \hat{a}_x(E_y H_z - E_z H_y) - \hat{a}_y(E_x H_z) + \hat{a}_z(E_x H_z) \tag{3.91}$$

Therefore,

$$\hat{a}_x P_x = \left[-\frac{8}{C^2} Q^2 \mu_o P U_1(t - t_o) \sum_{n,m=1}^{\infty} \sum_{m,n=1}^{\infty} Z_n \frac{e^{-jX_{n,m}}}{X_{n,m}} \right]$$

$$\left[\sum_{n,m=1}^{\infty} \sum_{m,n=1}^{\infty} \left(\frac{e^{-jX_{n,m}}}{X_{n,m}^2} + j\frac{e^{-jX_{n,m}}}{X_{n,m}} \right) \right]\left[\frac{2}{C^2} - \frac{1}{C^3} \sin ct \right.$$

$$\left. + \frac{1}{C^2} \delta^2(t) \sin ct \right] - \left[\frac{8}{C^2} Q^2 \mu_o P U_1(t - t_o) \right.$$

$$\left. \sum_{n,m=1}^{\infty} \sum_{m,n=1}^{\infty} Y_n \frac{e^{-jX_{n,m}}}{X_{n,m}} \right] \sum_{n,m=1}^{\infty} \sum_{m,n=1}^{\infty} \left(\frac{e^{-jX_{n,m}}}{X_{n,m}^2} + \frac{e^{-jX_{n,m}}}{X_{n,m}} \right) \right]$$

$$\left[\frac{2}{C^2} - \frac{1}{C^3} \sin ct + \frac{1}{C^2} \delta^2(t)\sin ct \right] \text{W/m}^2 \tag{3.92}$$

Writing Equation (3.92) in simplified form:

$$\hat{a}_x P_x = \hat{a}_x \left[\frac{8}{C^2} Q^2 \mu_o P U_1(t - t_o) \right] \left[\frac{2}{C^2} - \frac{1}{C^3} \sin ct + \frac{1}{C^2} \delta^2(t)\sin ct \right]$$

$$\left[\sum_{n,m=1}^{\infty} \sum_{m,n=1}^{\infty} \frac{e^{-jX_{n,m}}}{X_{n,m}} \left(\frac{e^{-jX_{n,m}}}{X_{n,m}^2} + j\frac{e^{-jX_{n,m}}}{X_{n,m}} \right)(Y_n - Z_n) \right] \text{W/m}^2 \qquad (3.93)$$

Next $\overline{P}_y$:

$$\hat{a}_y P_y = - \hat{a}_y E_x H_z \qquad (3.94)$$

Therefore,

$$\hat{a}_y P_y = \left[- \frac{8}{C^2} Q^2 \mu_o P U_1(t - t_o) \right] \left[\sum_{n,m=1}^{\infty} \sum_{m,n=1}^{\infty} Y_{n,m} Z_{n,m} \left[\frac{e^{-jX_{n,m}}}{X_{n,m}^2} + j\frac{e^{-jX^2}}{X_{n,m}} \right] \right]$$

$$\left[\frac{2}{C^2} - \frac{1}{C^3} \sin ct + \frac{1}{C^2} \delta^2(t)\sin ct \right] \text{W/m}^2 \qquad (3.95)$$

And then $\overline{P}_z$:

$$\hat{a}_z P_z = \hat{a}_z E_x H_y \qquad (3.96)$$

$$= \left[- \frac{8}{C^2} Q^2 \mu_o P U_1(t - t_o) \right] \left[\sum_{m=1}^{\infty} \sum_{n=1}^{\infty} Y_{m,n} Z_{n,m} \left(\frac{e^{-jX_{n,m}}}{X_{n,m}^2} + j\frac{e^{-jX_{n,m}^2}}{X_{n,m}} \right) \right.$$

$$\left(\frac{2}{C^2} - \frac{1}{C^3} \sin ct \frac{1}{C^2} \delta^2(t)\sin ct \right) \text{W/m}^2 \qquad (3.97)$$

It is clear that $\overline{P}_y = \overline{P}_z$.

Also it is useful to remind the reader that:

$$X_n = X_1, X_2, X_3 \ldots$$
$$Y_n = Y_1, Y_2, Y_3 \ldots \qquad \text{and}$$
$$Z_n = Z_1, Z_2, Z_3 \ldots$$

C is the velocity of light

Q is an arbitrary large number

P is a time dependent function given by Equation (3.54).

3.XI INTRINSIC WAVE SURGE IMPEDANCES

It is time to move on to express representation in three dimensions for the intrinsic surge impedances along the transmission line due to conductive and convective strokes for the lightning surge.

Let

$$Z_x = \frac{\overline{E}_x}{\overline{H}_x}$$

$$Z_z = \frac{\overline{E}_y}{\overline{H}_y}$$

and

$$Z_z = \frac{\overline{E}_z}{\overline{H}_z}$$

starting first with Z_x:

$$Z_x \rightarrow \infty \text{ since } H_x = 0$$

$$Z_y = \cfrac{\dfrac{-j2}{c} Q\mu_o P U_1(t - t_o) \sum\limits_{n,m=1}^{\infty} \sum\limits_{m,n=1}^{\infty} Z_{n,m} e^{-jX_{n,m}/X_{n,m}}}{\dfrac{j4}{c} Q \sum\limits_{n=1}^{\infty} \sum\limits_{m=1}^{\infty} \left(\dfrac{e^{-jX_{n,m}}}{X_{n,m}^2} + j\dfrac{e^{-jX_{n,m}}}{X_{n,m}}\right)\left[\dfrac{2}{C^2} - \dfrac{1}{C^3}\sin ct + \dfrac{1}{C^2}\delta^2(t)\sin ct\right]} \tag{3.98}$$

or

$$Z_y = -\frac{1}{2} P\mu_o U_1(t - t_o) \sum\limits_{n,m=1}^{\infty} \sum\limits_{m,n=1}^{\infty} Z_{n,m} \frac{e^{-jX_{n,m}}}{X_{n,m}} \Bigg/$$

$$\sum\limits_{n,m=1}^{\infty} \sum\limits_{m,n=1}^{\infty} \left(\frac{e^{-jX_{n,m}}}{X_{n,m}} \frac{e^{-jX_{n,m}}}{X_{n,m}^2}\right)\left[\frac{2}{C^2} - \frac{1}{C^2}\sin ct + \frac{1}{C^2}\delta(t)^2\sin ct\right] \tag{3.99}$$

Next:

$$Z_z = -\frac{1}{2} \cfrac{\mu_o P U_1(t - t_o) \sum\limits_{n,m=1}^{\infty} \sum\limits_{m,n=1}^{\infty} Y_{n,m} e^{-jX_{n,m}/X_{n,m}}}{\sum\limits_{n,m=1}^{\infty} \sum\limits_{m,n=1}^{\infty} \left(\dfrac{e^{-jX_{n,m}}}{X_{n,m}^2} + j\dfrac{e^{-jX_{n,m}}}{X_{n,m}^2}\right)\left[\dfrac{2}{C^2} - \dfrac{1}{C^3}\sin ct + \dfrac{1}{C^2}\delta^2(t)\sin ct\right]}$$

$$\tag{3.100}$$

where P is expressed by Equation (3.54) and repeated below for convenience
to the reader:

$$P(t) = \frac{1}{C}\delta^2(t)\cos ct + \frac{1}{3}\frac{\delta(t)}{C^2}\sin ct - 2\frac{\delta'(t)}{C^2}\sin ct$$

$$- \frac{1}{C^3}\int \sin C\tau\, \delta''(t - \tau)d\tau \tag{3.101}$$

$\delta(t)$ could be given the symbol $U_o(t)$ as the impulse singularity function.

3.XII SOLVED EXAMPLES

A. Express the wave number k and wave surge impedance for the most general form.

Solution

$$\text{The wave number } k = \sqrt{-\hat{z}\hat{y}} \tag{A1}$$

$$\hat{y} = a + wE' + jwE'' \tag{A2}$$

$$\hat{z} = W\mu' + jW\mu'' \tag{A3}$$

Wave surge impedance Z_w

$$Z_w = \sqrt{\frac{\hat{z}}{\hat{y}}} = \sqrt{\frac{W\mu' + jW\mu''}{a + W\epsilon'' + jW\epsilon''}} \tag{A4}$$

Also the wave number k is a complex number $= k' + jk''$

$$k' + jk'' = \sqrt{-(a + W\epsilon' + jW\epsilon'')(W\mu' + jW\mu'')} \tag{A5}$$

where a, ϵ', and μ' are loss Y elements.

B. Express the electric displacement vector induced along the transmission line medium characterized as linear to the first order of differentiation with respect to E.

Solution

In a linear medium, the electric displacement vector, also known as the electric charge density vector, is expressed to the first order in time differentiation of the E vector.

$$\overline{D} = \epsilon\overline{E} + \epsilon_1 \frac{\partial \overline{E}}{\partial t} \tag{B1}$$

$$\overline{E} = \hat{a}_x E_x + \hat{a}_y E_y + \hat{a}_z E_z \tag{B2}$$

E_x, E_y, and E_z are given in Equations (3.56), (3.57) and (3.58).
Time dependent parameters in E_x, E_y and E_z are $V_1(t - t_o)$ and $P(t)$ where

$$P(t) = \frac{1}{C}\,\delta^2(t)\cos ct + \frac{3}{2}\,\delta(t)\sin ct$$

$$- \frac{2}{C^3}\,\delta'(t)\sin ct + \frac{1}{C^3}\,\sin ct\,\delta'(t) + \frac{1}{C}\,\sin ct \tag{B3}$$

Now, let the time independent parameters in Equations (3.52), (3.53) and (3.54) be k_1, k_2 and k_3, respectively

$$\overline{D}_x = \epsilon\overline{E} + \epsilon_1 \frac{\partial E_x}{\partial t}$$

$$\overline{D}_y = \epsilon\overline{E} + \epsilon_1 \frac{\partial E_y}{\partial t}$$

$$\overline{D}_z = \epsilon\overline{E} + \epsilon_1 \frac{\partial E_z}{\partial t}$$

The terms $E\,E_x$, $E\,E_y$ and $E\,E_z$ are those in Equations (3.56), (3.57) and (3.58) multiplied by ϵ. The term involving time differentiation:

$$\frac{\partial}{\partial t} U_1(t - t_o)P(t)$$

$$= U_o(t - t_o)P(t) + U_1(t - t_o)\left[\frac{2\delta(t)}{C} \cos ct\right.$$

$$- \delta^2(t)\sin ct + \frac{3}{C^2} \delta'(t)\sin ct + \frac{3}{C} \delta(t)\cos ct$$

$$- \frac{1}{C^3} \delta''(t)\sin ct - \frac{2}{C^2} \delta'(t)\cos ct + \frac{1}{C^2} \cos ct\, \delta'(t)$$

$$\left. + \cos ct\right] \tag{B4}$$

$$\overline{D}_x = \epsilon E_x + \epsilon_1 k_1 \frac{\partial U_1(t - t_o)P(t)}{\partial t}$$

$$\overline{D}_y = \epsilon E_y + \epsilon_1 k_2 \frac{\partial U_1(t - t_o)P(t)}{\partial t}$$

$$\overline{D}_z = \epsilon E_z + \epsilon_1 k_3 \frac{\partial U_1(t - t_o)P(t)}{\partial t} \tag{B5}$$

C. Express the induced current density vector J in a medium characterized as linear to the first differentiation in the E field.

Solution

For a medium characterized as linear:

$$\overline{J} = \sigma\overline{E} + \sigma_1 \frac{\partial\overline{E}}{\partial t} \tag{C1}$$

In Equations (3.56), (3.57) and (3.58)

$$\overline{E} = \hat{a}_x E_x + \hat{a}_x E_y + \hat{a}_z E_z \qquad \text{(C2)}$$

Again, time dependent parameters in each E_x, E_y and E_z are $V_1(t - t_o)$, $P(t)$ and those time independents could be assigned respectively as k_1, k_2 and k_3 as in Example B. Therefore,

$$\overline{J}_x = \sigma E_x + \sigma_1 \frac{\partial E_x}{\partial t}$$

$$= \sigma E_x + \sigma_1 \frac{\partial}{\partial t} U_1(t - t_o)P(t)k_1 \qquad \text{(C3)}$$

$$\overline{J}_y = \sigma E_y + \sigma_1 \frac{\partial E_y}{\partial t}$$

$$= \sigma E_y + \sigma_1 \frac{\partial}{\partial t} U_1(t - t_o)P(t)k_2 \qquad \text{(C4)}$$

$$\overline{J}_z = \sigma E_z + \sigma_1 \frac{\partial E_z}{\partial t}$$

$$= \sigma E_z + \sigma_1 \frac{\partial}{\partial t} U_1(t - t_o)P(t)k_3 \qquad \text{(C5)}$$

where $\dfrac{\partial}{\partial t} U_1(t - t_o)P(t)$ was obtained in Example B.

D. Revise the expression for the induced voltage on a transmission line due to conductive and convective effects of lightning stroke, when $t = 2_o$ and $y = Z - 10x$.

Solution

Refer to Equation (3.47) and inserting $t = 2t_o$, $y = Z = 10x$. First convert the term $\int_o^t \sin c\tau \, \delta''(\tau - t)d\tau$ in terms of t. Using the sifting property:

$$\int_{a-\epsilon}^{a+\epsilon} f(x)\delta'(x - a)d_x = -f'(a) \qquad \text{(D1)}$$

Equation (D1) could represent the following case:

$$\int_{a-E}^{a+\epsilon} f(x)\delta''(x - a)d_x - ?$$

Let $u = f(x)$, $\delta''(x - a) = dv$, $du = f'(x)$, and $= \delta'(x - a)$. Therefore,

$$\int f(x)\delta''(x - a)d_x = f(x)\delta'(x - a) - \int f'(x)\delta'(x - a)dx$$

Using Equation (D1),

$$= f(x)\delta'(x - a) + f''(a) \tag{D2}$$

Now

$$\int \sin c\tau\, \delta''(t - \tau)d = \sin c\tau\, \delta'(\tau - t)\Big|_o^t - c^2 \sin ct$$

$$= -\sin ct\, \delta'(t) - C^2 \sin ct \tag{D3}$$

Now turning to Equation (3.47):

$$V_{c,v}(x,t) = -j2\mu_o Q\, \frac{U_1(t_o)}{C} \sum_{m,n=1}^{\infty} \sum_{m,n=1}^{\infty}$$

$$\left[\frac{e^{-jX_n}}{X_n}\, 100\, X_n\right]\left[\frac{1}{C}\cos(2Ct_o)\delta^2(2Ct_o) +\right.$$

$$+ \frac{3\delta(2t_o)}{C^2}\sin(2Ct_o) - \frac{2\delta'(2t_o)}{C^3}\sin(2Ct_o) +$$

$$\left.\frac{1}{C^3}\left[\sin(2Ct_o)\delta'(2t_o) - C^2\sin(2Ct_o)\right]\right] \tag{D4}$$

$$= -j2\mu_o Q\, \frac{U_1(t_o)}{C}\sum_{n=1}^{\infty}\sum_{n=1}^{\infty} 100e^{-jX_{n,m}}\left[\frac{1}{C}\cos(2Ct_o)\delta^2(2Ct_o)\right.$$

$$\frac{3}{C^2}\delta(2Ct_o)\sin(2Ct_o) - \frac{1}{C^3}\delta'(2t_o)\sin(2Ct_o)$$

$$\left.- \frac{1}{C}\sin(2Ct_o)\right] \tag{D5}$$

E. Express representation for σ as a tensor of rank 2 in a principle coordinate system at $t = 0$ and $X_n = Y_n = Z_n = 0$. Comment on the result.

Solution

$$\sigma'_{11} = 0 \tag{E1}$$

From Equation (3.76):

$$\sigma'_{22} = j \frac{CP_1}{2QPU_1(t - t_o)} \left[\sum_{\substack{n=1 \\ m=1}}^{\infty} \sum_{\substack{n=1 \\ m=1}}^{\infty} \frac{1 + \dfrac{1}{X^2_{n,m}}}{Y_{n,m} + Z_{n,m}} \right] \tag{E2}$$

$$X_n = Y_n = Z_n = 0 = \infty$$

$$\text{since } U_1(t - t_o) = 0 \tag{E3}$$

Similarly from Equation (3.77)

$$\sigma'_{33} = j \frac{CP_1}{2QPU_1(t - t_o)} \sum_{\substack{n=1 \\ m=1}}^{\infty} \sum_{\substack{n=1 \\ m=1}}^{\infty} \frac{1 + \dfrac{1}{X^2_{n,m}}}{Y_{n,m} + Z_{n,m}} \tag{E4}$$

$$= \infty$$

The physical meaning from the result that $\sigma'_{11} = 0$, $\sigma'_{22} = \sigma'_{33} = \infty$ is that at a location on the transmission line just under the lightning stroke. The surge is facing a perfectly conductive plane perpendicular to the line itself, while conductivity at an axis normal to that plane is zero, which means no conduction.

F. Equations (3.25) and (3.28) represent the y component for the horizontal sum of all conductive and convective magnetic moments under prescribed conditions of approximations, where the current flowing is stationary independent of time. Show the necessary revision required when electric current in the dipole is a chopped lightning impulse shown below.

Solution

Consider $J(x')$ as a chopped pulse as indicated below:

$$J(x',t) = \frac{A'}{\tau'} t - A'U_1(t - \tau') \tag{F1}$$

Equation (3.25) states:

$$A_{yC-H} \approx \hat{a}_y \sum_{m,n=1}^{\infty} \frac{j2}{k_c} e^{-js} - j \frac{S^2}{k_c^3} + \frac{2S}{k_c^3} e^{-js} - \frac{j2}{k_c^3} e^{-js} \tag{F2}$$

To have $I(x',t)$, the only thing to do is replacing it by the function:

$$I(t) = \frac{A'}{\tau'} t - A'U_1(t - \tau') \qquad \text{(F3)}$$

Equation (F2) becomes:

$$\overline{A}_{yc\text{-}H} \approx \hat{a}_y \left[\frac{A'}{\tau'} t - A'U_1(t - \tau') \right] \sum_{n,m=1}^{\infty} \left[\frac{j2}{k} e^{-js} c - j \frac{S_c^2}{k_c^2} + \right.$$

$$\left. \frac{2S_c}{k_c^3} e^{-js_c} - \frac{j2}{k_c^3} e^{-jS_c} \right] \qquad \text{(F4)}$$

with respect to Equation (3.28)

$$\overline{A}_{yv\text{-}V} = \hat{a}_y \left[\frac{A'}{\tau'} t - A'U(t - \tau') \right] \sum_{n,m=1}^{\infty} \left[\frac{2}{k} - \frac{S_v^2}{k_v^3} + \frac{2S_v}{jk^3} - \frac{2}{k_v^3} \right] je^{-jS_v} \qquad \text{(F5)}$$

where

$$S = [(r_{n,m} - R_c^2) + a^2 - 2a(r_{n,m} - R_c)\sin\theta\cos\phi']^{1/2} \qquad \text{(F6)}$$

where R_c is the radius of the conductive stroke.

3.XIII PROBLEMS

1. Show systematic derivation of $\overline{A}_y$ in Equation (3.9) starting from $\overline{A}_\phi$ in Equation (3.5).
2. Refer to Equation (3.44) which represents a second order differential equation in $V_c(x,s)$. Show that the solution for $V(x,s)$ is that expressed in Equation (3.46).
3. Show systematic steps that the inverse Laplace transform of $V_{c,v}(x,s)$ is that expressed by Equation (3.47).
4. Modify $P(t)$ given by Equation (3.50) if the velocity of surge propagation is b which is less than c (the velocity of light). Of course b is a function of μ and ϵ of the surrounding medium.
5. Based on the modified solution for $P(t)$ secured in problem (4), revise the induced electric field vector expressed by Equation (3.54).
6. Based on the new electric field vector obtained in problem 5, present a corresponding solution for the induced current density vector J on the transmission line when the electrical conductivity (σ) is a rank 2 tensor in a principle coordinate system where

$$\sigma_{ij} = \begin{bmatrix} \sigma_{11} & & 0 \\ & \sigma_{22} & \\ 0 & & \sigma_{33} \end{bmatrix}$$

7. Repeat the solution of problem 6 when σ_{ij} is a tensor of rank 2 in the regular Cartesian system, where:

$$\sigma_{ij} = \begin{bmatrix} \sigma_{11} & \sigma_{12} & \sigma_{13} \\ \sigma_{21} & \sigma_{22} & \sigma_{23} \\ \sigma_{31} & \sigma_{32} & \sigma_{33} \end{bmatrix}$$

8. Modify Equation (3.55) representing the induced magnetic field H on a transmission line when the velocity of surge propagation b is $< c$.
9. Recalculate the induced $\overline{H}_y$ and $\overline{H}_z$ induced on a transmission line under breakdown conditions which implies that the current density vector $\overline{J} \rightarrow \infty$.
10. Regarding the solution of $\overline{H}_y$ and $\overline{H}_z$ obtained in problem 9 under breakdown, indicate quantitatively the effect on σ, E and as scalars.
11. In a linear medium, the current density vector J is expressed by:

$$\overline{J} = \sigma\overline{E} + \sigma_a \frac{\partial \overline{E}}{\partial t} + \sigma_b \frac{\partial^2 E}{\partial t^2} + \cdots$$

Express the vector J up to and including the second order differential term shown above; σ, σ_a and σ_b all are scalars.
12. Based on the new solution for J obtained in problem 11, solve for the condition of voltage breakdown.
13. Repeat the solution for J in problem 11 if σ, σ_a and σ_b are tensors of rank 2 in a principle coordinate system where, say,

$$\sigma = \begin{bmatrix} \sigma_{11}' & & 0 \\ & \sigma_{22}' & \\ 0 & & \sigma_{33}' \end{bmatrix}$$

$$\sigma_a = \begin{bmatrix} \sigma_{a-11} & & 0 \\ & \sigma_{22} & \\ 0 & & \sigma_{a-33} \end{bmatrix}$$

and

$$\alpha_b = \begin{bmatrix} \sigma_{b-11} & & 0 \\ & \sigma_{b-22} & \\ 0 & & \sigma_{b-33} \end{bmatrix}$$

14. Following the new solution for J obtained in problem 13, obtain conditions for voltage breakdown.

15. Based on the solution for J obtained in problem 11, obtain a new solution for the Poynting vector P.

16. Based on the solution for J obtained in problem 12, obtain a new solution for the Poynting vector $\overline{P}$.

17. Express the mathematical and physical implications if the intrinsic wave surge impedances $Z_y = Z_z$ according to Equations (3.99) and (3.101).

18. Modify the expression for Z_y given by Equation (3.89) when the velocity of surge propagation $v \ll c$.

19. Repeat solution of Example B when the D vector is linear to the second order of differentiation with respect to $\overline{E}$.

20. Based on the new solution for the $\overline{D}$ vector obtained in problem 19, solve for the total conductive and convective electric charge residing on A square miles in the $Y - Z$ plane.

21. Based on the solution for the total electric charge residing on A square miles obtained in problem 20, express the total electric energy stored in a condenser system formed between the cloud's and earth's surfaces with a separation of d meters. Assume the medium in between as simple implying $\overline{J} = \sigma\overline{E}, \overline{D} = \epsilon\overline{E}$.

22. Repeat problem 21 for the total magnetic energy stored in an equivalent inductive system. Also obtain a solution for the equivalent self-inductive of the system.

23. Repeat the solution of Example D when $t = 1000\, t_o$ and $Y = Z = 10,000\, X$.

24. Repeat the solution of example F when the lightning surge is generated by two chopped consecutive strokes as shown below:

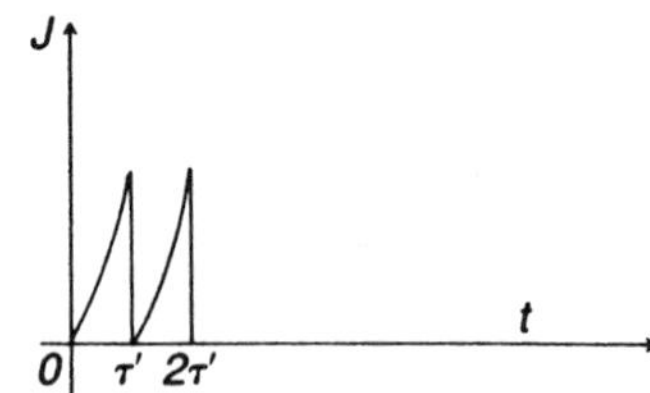

25. Repeat problem 24 when the lightning surge is generated by three consecutive strokes as shown below:

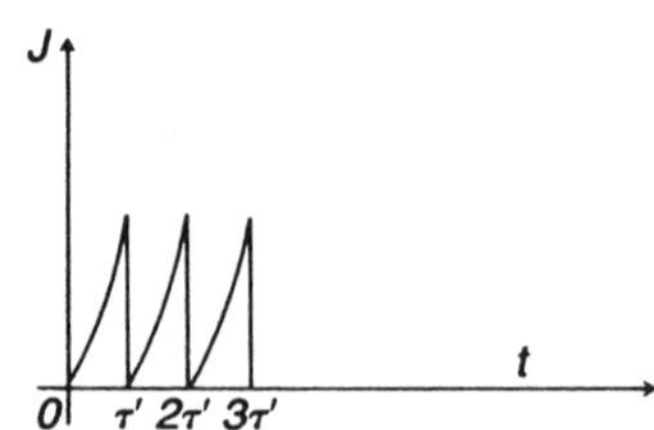

26. From the solution of the magnetic vector potential obtained in problem 24, solve for the inducing voltage on a transmission line.

27. From the solution of the magnetic vector potential in problem 24, obtain a solution for the induced voltage on the transmission line.

28. From problem 26, obtain a solution for the inducing electric field vectors at any point in space.

29. From problem 27, obtain a solution for the induced electric field vector on a transmission line.

30. From the solution of the inducing electric field vector obtained in problem 25, obtain expression for the surface charge density vector $(\overline{D})$ if the medium is characterized as simple.

31. Repeat problem 30, if the medium is characterized as linear to the first order in the differentiation of $\overline{E}$ with respect to time.

32. Repeat problem 30, if the medium is characterized as linear to the second order of differentiation of $\overline{E}$ with respect to time.

PRINCIPAL LIST OF SYMBOLS

$\overline{A}_{\phi}$	magnetic vector potential in the ϕ direction
s	radial distance space point
$\overline{B}$	magnetic induction vector
$\delta(t)$	delta dirac function $= U_o(t)$
λ	element of a matrix in the principle coordinate system
$\overline{P}$	Poynting vector
C	velocity of light
ω	angular velocity in rad/s
Z_x	the x oriented surface impedance
ϕ	a large numerical containing strength of surge current

REFERENCES

1. **Denno, K.,** Computation of electromagnetic lightning response using moments method, *IEEE Trans. Magnetics,* 20(5), 1953, 1984.
2. **Denno, K.,** Dynamic modeling for the process of inducing and induced voltage surges due to lightning, *J. Electrostat.,* 13, 55, 1982.
3. **Denno, K.,** Mathematical modelling of propagating inducing and induced power due to actual pulse shape of lightning surge, *J. Electrostat.,* 15, 43, 1984.
4. **Harrington, R. F.,** *Field Computation By Moment Methods,* Macmillan New York, London, 1968.
5. **Harrington, R. F.,** *Time Harmonic Electromagnetic Fields,* McGraw-Hill, New York, 1961.

Chapter 4

RELEASE OF IONS DUE TO INDUCED AND INDUCING FIELDS BY LIGHTNING DISCHARGE

4.1 INTRODUCTION

The process of ionization preceding the occurrence of lightning surge is basically intensive cycles accompanied by a powerful electric field, where both electrons and positive ions and their clusters take cumulative roles in ionization that will eventually release the electron and positive ion avalanche.

J.S. Townsend had indicated through established empirical and experimental work that when the ratio of the number of new ions produced per unit length by an accelerated electron under the influence of intensive electric field to the prevailing pressure is a smooth and continuous function of the ratio of the prevailing electric field to the same pressure, i.e.,

$$\frac{\gamma}{P} = g\left(\frac{E}{P}\right) \tag{4.1}$$

Actually, the expression of γ/P had been given by the following equation, published in a book authored by J.D. Cobine.[1]

$$\frac{\gamma}{P} = Ae^{-\frac{B}{E/P}} \tag{4.2}$$

where γ is the number of ionizing collisions produced or charge carriers released by field accelerated electron per centimeter.

In Equation (4.2), P is the pressure, in millimeters, of Hg. A and B are constants of the gaseous continuum. Figure 4.1 illustrates the functional relationship of γ/P v_x E/P for a number of gases.

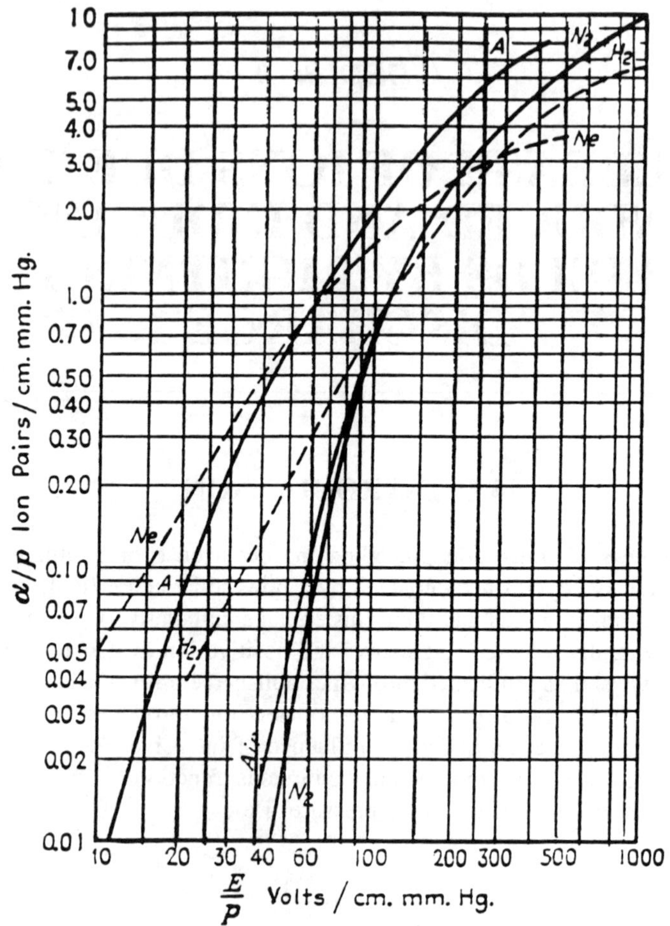

FIGURE 4.1.   Coefficient for field-intensified ionization by electrons.

However, Townsend emphasized clearly that Equation (4.1) is only an approximate estimating criterion that could serve as a useful guide in identifying the role of pressure on the process of field intensified ionization, either by electrons or positive ions.

Table 4.1 lists constants A and B for a number of gases including air for several ratios of E/P. The presence of impurities alters drastically the ratio of γ/P, where such a situation exists in the air continuum in the atmospheric belt during the period preceding voltage breakdown and after breakdown during lightning phenomenon.

Townsend also indicated that errors in computing the ratio of γ/P have been attributed to the presence experimentally of traces of mercury vapor.

The preceding discussion centered on the role of pressure on the process of field ionization by accelerated electrons and positive ions. However, the fact

TABLE 4.1
Constants of the Equation

Gas	A	B	Range of E/P, v/cm/mmHg
Air	14.6	365	150—600
A	13.6	235	100—600
CO_2	20.0	466	500—1,000
H_2	5.0	130	150—400
H_2O	12.9	289	150—1,000
He	2.8	34	20—150

that γ/P is a smooth and continuous function of E/P will be relied upon to establish a new relationship for lightning phenomenon.

4.II EFFECT OF PRESSURE ON LIGHTNING

Due to the huge complexity surrounding the process of cumulative ionization preceding the release of the conductive and convective surges, the role of atmospheric pressure and temperature on lightning phenomenon in air could be shown using the approximate criterion developed by Townsend as indicated by Equation (4.1) where

$$\frac{\gamma}{P} = g\left(\frac{E}{P}\right) \tag{4.3}$$

As explained earlier, γ represents the number of ionizations produced by an accelerated electron or positive ion per meter and P is the pressure in Newtons/m^2 or equivalent atmosphere.

To arrive at a functional form for the number of ionizations or, in effect, the number of electrons released per ionizing agent which could be either electron or positive ion, we will proceed according to the following argument.

Let

$$\overline{P} = \hat{a}_x P_x + \hat{a}_x \cdot P_y + \hat{a}_z P_z \tag{4.4}$$

where $\overline{P}$ is the Poynting vector of propagating lightning surge power in watts/m^2.

Also, it is known that

$$\overline{P} = \overline{E} \times \overline{H} \text{ j/s/m}^2 \tag{4.5}$$

Then, V_i is the inducing voltage in volts or joules/coulomb which has been expressed by Equation (3.38) based on the method of magnetic moment. Also the P vector had been expressed by Equations (3.83), (3.85) and (3.87).

Now let us analyze the following dimensionally:

$$P/V_i \frac{\text{Joules}}{\text{sec} - \text{m}^2} \Big/ \frac{\text{Joules}}{1 \text{ coulomb}}$$

$$\frac{\text{Joules}}{\text{sec} - \text{m}^2} \Big/ \frac{1 \text{ coulomb}}{\text{Joules}}$$

$$\frac{1 \text{ coulomb}}{\text{sec} - \text{m}^2} \tag{4.6}$$

or number of charge carriers/m²/sec.

The number of electrons released per meter is

$$\gamma = \int_o^l \int_o^t \frac{\overline{P}}{V_i} \, dt \cdot dl \tag{4.7}$$

where dl is a three-dimensional space differential vector, dt is the differential time element, and $\overline{P}$ is the Poynting vector in watts/m².

4.III RELEASE OF CHARGE CARRIERS DUE TO INDUCED ELECTRIC FIELD

Consider first the case for the inducing and induced fields based on the concept of the magnetic vector potential and constant current density in return strokes.

Determination of γ_x:

γ_x represents release of charge carriers/meter along the x-axis or in a plane perpendicular to the x-axis.

$$\gamma_{x\text{-induced}} = \int_o^x \int_o^t \frac{\overline{P}_{x1}}{V_{ij}} \, dt \cdot dx \tag{4.8}$$

where $\overline{P}$ is the Poynting vector in the x-axis. From Equations (3.51) and (3.93),

$$= \int_o^x \int_o^t \left[\frac{j4QP(t)}{C^2} \left(\frac{2}{C^2} - \frac{1}{C^3} \sin ct + \frac{1}{2} \delta^2(t)\sin ct \right) \right.$$

$$\left. \sum_{n,m}^{\infty} \sum_{n,m}^{\infty} (Y_{n,m} - Z_{n,m}) \left(\frac{e^{-jX_{n,m}}}{X_{n,m}^2} + \frac{e^{-jX_{n,m}}}{X_{n,m}} \right) dt \cdot dx_{n,m} \right. \tag{4.9}$$

charge carriers/m

Determination of γ_y:

γ_y represents release of charge carriers/m along the y-axis:

$$\gamma_{y\text{-induced}} = \int_o^y \int_o^t \frac{\overline{P}_{y\text{-induced}}}{V_{\text{induced}}} \, dt \cdot dy \tag{4.10}$$

where $\overline{P}_{y\text{-induced}}$ is the Poynting vector in the y-axis. From Equations (3.51) and (3.95),

$$\gamma_{y\text{-induced}} = \int_o^y \int_o^t \left[\frac{j4QP(t)}{C^2} \left(\frac{2}{C^2} - \frac{1}{C^3} \sin ct + \frac{1}{C^2} \delta_o^2(t) \sin ct \right) \right.$$

$$\left. \sum_{m=1}^{\infty} \sum_{n=1}^{\infty} \left(1 + \frac{1}{X_{n,m}} \right) \left(\frac{e^{-jX_{n,m}}}{X_{n,m}^2} + \frac{e^{-jX_{n,m}}}{X_{n,m}} \right) \right] dt \cdot dy_{n,m} \tag{4.11}$$

charge carriers/m

Determination of γ_z:

γ_z represents release of charge carriers/m along the z-axis. Since inducing $H_y = -H_z$

$$\gamma_{z\text{-induced}} = -\gamma_{y\text{-induced}} \tag{4.12}$$

The number of charge carriers released due to the inducing electric and magnetic fields in meter^2 per second or per meter could refer to either the conductive return stroke or convective stroke (time variation of induced bound charges) and of course their combination.

4.IV   RELEASE OF CHARGE CARRIERS DUE TO INDUCING FIELD

We proceed first with the case where induced and inducing fields have been based on the concept of the magnetic vector potential and constant current in return strokes.

$$\gamma_{x\text{-inducing}} = \int_o^x \int_o^t \frac{\overline{P}_{x2}}{V_{i2}} \, dt \cdot dx \tag{4.13}$$

$\gamma_{x\text{-inducing}}$ is measured in number of charge carriers/meter.

$$\gamma_{y\text{-inducing}} = \int_o^y \int_o^t \frac{\overline{P}_{x2}}{V_{i2}} \, dt \cdot dx \tag{4.14}$$

and

$$\gamma_{z\text{-inducing}} = \int_o^z \int_o^t \frac{\overline{P}_{z2}}{V_{i2}} \, dt \cdot dx \qquad (4.15)$$

where

$$\overline{P}_2 = \hat{a}_x P_{x2} + \hat{a}_y P_{y2} + \hat{a}_z P_{z2} \qquad (4.16)$$

$\overline{P}_2$ is the Poynting vector due to the inducing electric and magnetic fields.

$$\overline{P}_2 = \overline{E}_2 \times \overline{H}_2 \qquad (4.17)$$

$\overline{E}_2$ is the inducing electric field which could be obtained from Equation (3.38) and $\overline{H}_2$ is the inducing magnetic field which could be obtained from the relationship

$$\overline{\nabla} \times \overline{E}_2 = - \frac{\partial \overline{B}_2}{\partial t} \qquad (4.18)$$

Solution for $\overline{P}_2$ is left as a problem for the student to solve.

In Sections III and IV, analyses regarding the release of charge carriers due to induced and inducing fields have been based on the magnetic vector potential obtained through the procedure of summation of magnetic moments and where wave shape of current density in the conductive and convective strokes has been treated as a step function. Now, we shall analyze the concept of calculating the release of charge carriers per meter^2 per second with respect to more factual cases for the wave shape of the conductive and convective return strokes.

4.V RELEASE OF CHARGE CARRIERS WHEN CURRENT DENSITY IS A SHARP LINEAR RISE AND A LINEAR SLOW DECAYING TAIL

The set of Equations (2.89) through (2.94) indicates the inducing electromagnetic field at any point in space (ρ, α, τ).

Therefore, inducing the Poynting vector in space is expressed as $P_{\text{inducing}} = \overline{P}_3$.

$$\overline{P}_3 = \overline{E}_3 \times \overline{H}_3 \text{ joules/sec/m}^2 \qquad (4.19)$$

where

$$\overline{E}_3 = \hat{a}_\rho E_\rho + \hat{a}_\alpha E_\alpha + \hat{a}_\tau E_\tau \qquad (4.20)$$

$$\overline{H}_3 = \hat{a}_\rho H_\rho + \hat{a}_\alpha H_\alpha + \hat{a}_\tau H_\tau \qquad (4.21)$$

$$\overline{P}_3 = \begin{bmatrix} \hat{a}_\rho & \hat{a}_\alpha & \hat{a}_\tau \\ E_\rho & E_\alpha & E_\tau \\ H_\rho & H_\alpha & H_\tau \end{bmatrix} \qquad (4.22)$$

$$= \hat{a}_\rho[E_\alpha H_\tau - H_\alpha E_\tau] - \hat{a}_\alpha[E_\rho H_\tau - H_\rho E_\tau] +$$

$$\hat{a}_\tau[H_\alpha E_\rho - H_\rho E_\alpha] \qquad (4.23)$$

Therefore,

$$\overline{P}_{3\rho c} = \hat{a}_\rho(E_\alpha H_\tau - H_\alpha E_\tau)$$

$$\overline{P}_{3\alpha c} = -\hat{a}_\alpha(E_\rho H_\tau - H_\rho E_\tau)$$

and

$$\overline{P}_{3\tau c} = \hat{a}_\tau(H_\alpha E_\rho - H_\rho E_\alpha) \qquad (4.24)$$

Now we can list the following:

$$\overline{P}_{3\rho c} = -\left[\frac{J_c(t)\tau}{2} \sum_{n=o}^{\infty} \frac{1}{\pi\rho^2} J_n(\rho - R_c)H_n^{(2)}(\rho - R_c) - \right.$$

$$\frac{1}{nR_c} H_n^{(2)}(\rho - R_c) \frac{(\rho - R_c)^{n-1}}{2^{n-1}(n - 1)!} \sqrt{\frac{2}{\pi}} (- j)^n + \frac{1}{2}$$

$$J_n(\rho - R_c) \frac{1}{nR_c} \frac{e^{-j(-R_c)}}{(\rho - R_c)^2} + j^{\left(n + \frac{3}{2}\right)} \sqrt{\frac{2}{\pi(\rho - R_c)}}$$

$$\left. e^{-j(\rho - R_c)} \frac{1}{nR_c} \right] \left[\frac{J_c(t)R_c k^2\pi}{2\hat{y}} \sqrt{\frac{j2}{\pi(\rho - R_c)}} \right.$$

$$\left. \sum_{n=0}^{\infty} \frac{j^n(\rho - R_c)^n}{2^n n!} \right] \qquad (4.25)$$

where $\overline{P}_{3\rho c}$ represents the ρ component of Poynting vector due to the conductive stroke. To consider the convective stroke replace A_c by A_v and R_c by R_v. Next,

$$\overline{P}_{3\alpha c} = (H_\rho E_\tau - E_\rho H_\tau)\hat{a}_\alpha \qquad (4.26)$$

since $H_\tau = 0$

$$\overline{P}_{3\alpha c} = (H_\rho E_\tau)\hat{a}_\alpha \qquad (4.27)$$

$$= \left(- \frac{k^2 R_c J_c^2(t)\tau}{2\hat{y}} \right) \left(\frac{1}{(\rho - R_c)} \right) e^{-j(\rho_c - R)} \sum_{n=o}^{\infty} \frac{j^n(\rho_c - R)}{2^n n!}$$

$$\left(\sum_{n=o}^{\infty} \frac{j^n(\rho - R_c)^n}{2^n n!} \right) \tag{4.28}$$

Then,

$$\overline{P}_{3\tau c} = (H_\alpha E_\rho - H_\rho E_\alpha)\hat{a}_\tau \tag{4.29}$$

Expression for $\overline{P}_{3\tau c}$ could be secured by substituting H_α, E_ρ, H_ρ and E_α from Equations (2.89) and (2.90).

Let $\gamma'_{3\alpha c}$ represent the release of the number of charge carriers/m²/sec.

$$\gamma'_{3\alpha c} = \frac{\overline{P}_{3\alpha c}}{V_{i\alpha}} \tag{4.30}$$

where $V_{i\alpha}$ is the inducing voltage along the α vectorial orientation.

$$V_{i\alpha} = -\int_o^{2\pi} \overline{E}_\alpha \cdot \overline{\rho} \, d\alpha \tag{4.31}$$

$$V_{i\alpha} = -\int_o^{2\pi} \frac{jJ_c(t)}{2\hat{y}} \sqrt{\frac{j2}{\pi(\rho - R_c)}} \, e^{-j(\rho - R_c)} \sum_{n=o}^{\infty} \frac{j^n(\rho - R_c)^n}{2^n n!} \rho \, d\alpha$$

$$= -j \frac{\pi J_c(t)}{2\hat{y}} \sqrt{\frac{j2}{\pi(\rho - R_c)}} \, e^{-j(\rho - R_c)} \sum_{n=o}^{\infty} \frac{j^n(\rho - R_c)^n}{2^n n!} \rho \, d\alpha \tag{4.32}$$

From Equations (4.28) and (4.30),

$$\gamma'_{3\alpha c} = -j \frac{tR_c k^2 J_c(t)}{\pi\rho(\rho - R_c)} \sqrt{\frac{(\rho - R)}{j2}} \sum_{n=o}^{\infty} \frac{j^n(\rho - R_c)^n}{2^n n!} \tag{4.33}$$

It is meaningful to examine the condition of field breakdown that will be accompanied by the release of the ionizing avalanche which implies an infinite limit for the inducing current density and also an infinite value for $\alpha_{i\alpha}$, $\alpha_{i\rho}$ and $\alpha_{i\tau}$. For $\gamma'_{3\alpha c} \rightarrow \infty$, from Equation (4.33)

$$\rho(\rho - R_c) \rightarrow \text{zero}$$

$$\rho = 0$$

and

$$\rho = R_c \tag{4.34}$$

However, looking back to Equation (4.33), we see that when $\rho = R_c$, the limiting process has the form of $\frac{0}{0}$. This means that a special limiting process has to be carried out to see whether breakdown will occur.

$$\lim_{\rho \to R_c} \gamma'_{3\alpha c} \to \frac{\partial/\partial\rho}{\partial/\partial\rho} \frac{(\rho - R_c)^{n+1/2}}{\rho(\rho - R_c)}$$

$$\frac{\partial/\partial\rho}{\partial/\partial\rho} \frac{(\rho - R_c)^{n-1/2}}{\rho}$$

Continuing the limiting process,

$$\lim_{\rho \to R_c} \gamma'_{3\alpha c} \to \left(n - \frac{1}{2} \right) \frac{(\rho - R_c)^{n-3/2}}{1} \tag{4.35}$$

$$\to \text{zero}$$

The conclusion is that at $\rho = R_c$, field breakdown will not take place. Therefore, $\gamma'_{3\alpha c} \to \infty$ only when $\rho = 0$ which means at points within the column of the conductive stroke. A similar argument applies regarding the convective stroke.

To find out expression for the release of the number charge carriers per meter, proceed from Equation (4.30), where

$$\gamma_{3\alpha c} = \int_o^{2\pi} \int_o^t \frac{\overline{P}_{3\alpha c}}{V_{i\alpha}} \, dt \cdot d\alpha \tag{4.36}$$

$\gamma_{3\alpha c}$ is with respect to the conductive stroke, which could represent also the convective stroke identified as $\gamma_{3\alpha}$.

It is important to recall that $J_c(t)$ and $J_v(t)$ follow the representation given by Equations (2.95) and (2.96) which refers to a pulse wave shape characterized by a sharp linear rise and a slow linear tail.

4.VI RELEASE OF INDUCED CHARGE CARRIERS WHEN CURRENT DENSITY IS A SHARP LINEAR RISE AND A SLOW DECAYING TAIL

The induced voltage on a transmission line is given by Equations (2.103) to (2.106), where

$$V(x,t) = V_1 + V_2 + V_3 + V_4 \tag{4.37}$$

To find the induced electric magnetic fields, first we take the gradient of $V(x,t)$.

$$\overline{\nabla} V(x,t) = \overline{\nabla} V_1 + \overline{\nabla} V_2 + \nabla V_3 + \nabla V_4 \tag{4.38}$$

From Equations (2.104) to (2.107):

$$\overline{\nabla}V_1(x,t) = \frac{2C^2V_{\max}\delta_o(t)[A_c - A_v]}{V_i(x = L)\delta_o'(t)}\,\overline{\nabla}V_i(x) \tag{4.39}$$

$$\overline{\nabla}V_2(x,t) = \frac{2C^2V_{\max}\delta_o(t - t_1)[A_v - A_c]}{V_i(x = L)\delta_o'(t - t_1)}\,\overline{\nabla}V_i(x) \tag{4.40}$$

$$\overline{\nabla}V_3(x,t) = \frac{2C^2V_{\max}\delta_o(t - t_1)[A_v - A_c]}{V_i(x = L)\delta_o'(t - t_1)}\,\overline{\nabla}V_i(x) \tag{4.41}$$

and

$$\overline{\nabla}V_4(x,t) = \frac{2C^2V_{\max}\delta_o(t - t_1)[A_v - A_c]}{V_i(x = L)\delta_o'(t - t_1)}\,\overline{\nabla}V_i(x) \tag{4.42}$$

where $V_i(x)$ is the inducing voltage given by Equation (2.95). However, $V_i(x)$ will give the inducing electric field vector given by Equation (2.89) and the corresponding inducing magnetic vector is given by Equations (2.90a), (2.90b) and (2.93).

Therefore, to express the induced electric field on a finite transmission line generated by a linear but close to actual lightning pulse, we proceed to express the components in $\overline{\nabla}V_1(x,t)$:

$$\overline{\nabla}V_1(x,t) = \left[\frac{2C^2V_{\max}\delta_o(t)[A_c - A_v]}{V_i(x = L)\delta_o'(t)}\right][\hat{a}_\rho E_\rho + \hat{a}_\alpha E_\alpha + \hat{a}_\tau E_\tau] \tag{4.43}$$

$$\overline{\nabla}V_2(x,t) = \left[\frac{2C^2V_{\max}\delta_o(t - t_1)[A_v - A_c]}{V_i(x = L)\delta_o'(t - t_1)}\right][\hat{a}_\rho E_\rho + \hat{a}_\alpha E_\alpha + \hat{a}_\tau E_\tau] \tag{4.44}$$

$$\overline{\nabla}V_3(x,t) = \left[\frac{2C^2V_{\max}\delta_o(t - t_1)[A_v + A_c]}{V_i(x = L)\delta_o'(t - t_1)}\right][\hat{a}_\rho E_\rho \hat{a}_\alpha E_\alpha + \hat{a}_\tau E_\tau] \tag{4.45}$$

and

$$\overline{\nabla}V_4(s,t) = \left[\frac{2C^2V_{\max}\delta_o(t - t_1)U_{-1}(t)[A_v - A_c]}{V_i(x = L)\delta_o'(t - t_1)}\right][\hat{a}_\rho E_\rho + \hat{a}_\alpha E_\alpha \hat{a}_\tau E_\tau] \tag{4.46}$$

To find the induced magnetic field vector on the finite transmission line, we have to use the following Maxwell field equation:

$$\overline{\nabla} \times \overline{E} = -\frac{\partial B}{\partial t} \tag{4.46a}$$

where $\overline{E}$ is the induced electric field vector given by:

$$\overline{E} = -[\overline{\nabla}V_1(s,t) + \overline{\nabla}V_2(s,t) + \overline{\nabla}V_3(s,t) + \overline{\nabla}V_4(s,t)]$$

$$= (\overline{E}_1 + \overline{E}_2 + \overline{E}_3 + \overline{E}_4) \tag{4.47}$$

$\overline{\nabla}(V_1, V_2, V_3, V_4)$ are given by Equations (4.43) to (4.46). Also note that s replaced x as generalized space vector.

$$\overline{\nabla}x\overline{E} = \overline{\nabla}x(E_1 + E_2 + E_3 + E_4)$$

$$= \overline{\nabla}xE_1 + \overline{\nabla}xE_2 + \overline{\nabla}xE_3 + \overline{\nabla}xE_4 \tag{4.48}$$

From Equation (4.46a)

$$\overline{B} = -\int(\overline{\nabla}x\overline{E})dt \tag{4.49}$$

$$= -\int[\overline{\nabla}x\overline{E}_1 + \overline{\nabla}x\overline{E}_2 + \overline{\nabla}x\overline{E}_3 + \overline{\nabla}x\overline{E}_4]dt$$

$$= [\overline{B}_1 + \overline{B}_2 + \overline{B}_3 + \overline{B}_4] \tag{4.50}$$

where

$$\overline{B}_1 = -\int\overline{\nabla}x\overline{E}_1\, dt$$

$$\overline{B}_2 = -\int\overline{\nabla}x\overline{E}_2\, dt$$

$$\overline{B}_3 = -\int\overline{\nabla}x\overline{E}_3\, dt$$

and

$$\overline{B}_4 = -\int\overline{\nabla}x\overline{E}_4\, dt \tag{4.51}$$

But in reference to Equations (4.43) through (4.46), the fields E_1, E_2, E_3, and E_4 are the same with respect to their space functions, i.e.,

$$\overline{\nabla}x\overline{E}_1(\rho,\alpha,\tau) = \overline{\nabla}x\overline{E}_2(\rho,\alpha,\tau) =$$

$$\overline{\nabla}x\overline{E}_3(\rho,\alpha,\tau) = \overline{\nabla}x\overline{E}_4(\rho,\alpha,\tau)$$

Therefore,

$$\overline{\nabla} \times \overline{E} = \hat{a}_\rho \left[\frac{1}{\rho} \frac{\partial E_\tau}{\partial_\alpha} - \frac{\partial E_\alpha}{\partial \tau} \right] +$$

$$\hat{a}_\alpha \left[\frac{\partial^E \rho}{\partial \tau} - \frac{\partial^E \tau}{\partial \rho} \right] +$$

$$\hat{a}_\tau \left[\frac{1}{\rho} \frac{\partial}{\partial \rho} (\rho E_\alpha) - \frac{1}{\rho} \frac{\partial E_\rho}{\partial \alpha} \right] \tag{4.52}$$

In Equation (4.52), E_ρ, E_α and E_τ are given by Equations (2.89), (2.90a), (2.90b) and (2.93).

$$\overline{\nabla} x \overline{E}_{\text{induced}} = -\hat{a}_\alpha \left[\frac{J_c(t) R_c \pi k^2}{2\hat{y}} \sqrt{\frac{j2}{\pi}} \sum_{n=0}^{\infty} \frac{j^n}{2^n n!} \left(n - \frac{1}{2} \right) (\rho - R_c)^{n-3/2} \right]$$

$$+ \hat{a}_\tau \frac{j J_c(t)}{2\hat{y}} \sqrt{\frac{j2}{\pi}} \frac{1}{\rho} e^{-j(\rho - R_c)} \left[\sum_{n=o}^{\infty} \frac{j^n \left(n - \frac{1}{2} \right)}{2^n n!} \right.$$

$$(\rho - R_c)^{n-3/2} + (1 + j\rho) \sum_{n=o}^{\infty} \left. \frac{j^n (\rho - R_c)^{n-3/2}}{2^n n!} \right] \tag{4.53}$$

We can express the induced magnetic field B_1 according to Equation (4.50):

$$\overline{B}_{1\text{-induced}} = \overline{\nabla} x \overline{E} \left(\frac{-2C^2 V^{\text{max}}}{V_i(x = L)} \right) (A_c - A_v) \int_o^t \frac{\delta_o(t)}{\delta_o(t)} \, dt$$

$$= -\overline{\nabla} x \overline{E} \left(\frac{2C^2 V_{\text{max}}}{V_i(x = L)} \right) (A_c - A_v)[\delta_o^2(t) - \delta_o'(t) U_{-1}(t) - \delta_o(t)] \tag{4.54}$$

where $\overline{\nabla} x \overline{E}$ is given by Equation (4.53)

$$\overline{B}_{2\text{-induced}} = \overline{\nabla} x \overline{E} \left(\frac{-2C^2 V_{\text{max}}}{V_i(x = L)} \right) (A_v - A_c) \int \frac{\delta_o(t - t_1)}{\delta_o'(t - t_1)} \, dt$$

$$= -\overline{\nabla} x \overline{E} \left(\frac{2C^2 V_{\text{max}}}{V_i(x = L)} \right) (A_v - A_c) \left[\delta_o^2(t - t_1) - \right.$$

$$\delta_o'(t - t_1) U_{-1}(t - t_1) - \delta_o(t - t_1)] \tag{4.55}$$

$$\overline{B}_{3\text{-induced}} = \overline{\nabla}x\overline{E}\left(\frac{-2C^2V_{max}}{V_i(x = L)}\right)(A_v - A_c)\int \frac{\delta'_o(t - t_1)}{\delta_o(t - t_1)} \, dt$$

$$= -\overline{\nabla}x\overline{E}\left(\frac{2C^2V_{max}}{V_i(x = L)}\right)(A_v + A_c)[\delta'^2_o(t - t_1) -$$

$$\delta''_o(t - t_1)U_{-1}(t - t_1) - \delta'_o(t - t_1)] \tag{4.56}$$

and

$$\overline{B}_{4\text{-induced}} = \overline{\nabla}x\overline{E}\left(\frac{-2C^2V_{max}}{V_i(x = L)}\right)(A_v - A_c)\int_o^t \frac{\delta_o(t - t_1)}{\delta'_o(t - t_1)} \, dt$$

$$= -\overline{\nabla}x\overline{E}\left[\frac{2CV_{max}}{V_i(x = L)}U_{-1}(t)(A + A_c^2)\right][\delta_o^2(t - t_1)$$

$$- \delta'_o(t - t_1)U_{-1}(t - t_1) - \delta_o(t - t_1)] \tag{4.57}$$

In Equations (4.54) to (4.57) $\overline{\nabla}x\overline{E}$ is the same as expressed in Equation (4.53).

Considering the atmospheric continuum as simple and isotropic, where $\overline{B} = \mu\overline{H}$ and of course μ is a scalar, the induced H field becomes:

$$\overline{H}_{induced} = \mu[\overline{B}_1 + \overline{B}_2 + \overline{B}_3 + \overline{B}_4] \tag{4.58}$$

Now let γ' represent the number of released charge carriers/m²/sec.

$$\gamma'_{induced} = \frac{\overline{P}_{induced}}{V_{induced}} = \frac{\overline{E}_{induced} \times \overline{H}_{induced}}{V_{induced}} \tag{4.59}$$

$$= \frac{(\overline{E}_1 + \overline{E}_2 + \overline{E}_3 + \overline{E}_4) \times (\overline{H}_1 + \overline{H}_2 + \overline{H}_3 + \overline{H}_4)}{(V_1 + V_2 + V_3 + V_4)} \tag{4.60}$$

in terms of ions/m²/sec. $\overline{E}_1$, $\overline{E}_2$, $\overline{E}_3$ and $\overline{E}_4$ are given by Equations (4.43) to (4.46); $\overline{H}_1$, $\overline{H}_2$, $\overline{H}_3$ and $\overline{H}_4$ are given by Equations (4.54) to (4.57); and V_1, V_2, V_3 and V_4 are given by Equations (2.104) to (2.107).

To obtain an expression representing the released number of charge carriers per meter, Equation (4.66) becomes:

$$\gamma_{c,v} = \int_o^x \int_o^t \frac{\overline{P}_{induced}}{V_{induced}} \, dt \cdot dx \tag{4.61}$$

where x is along the transmission line orientation, taking into account coordinate transformation from cylindrical to cartesian systems and c refers to the effect of conductive stroke, while v refers to the convective stroke.

Proceeding to expand Equation (4.60) to some extent,

$$\gamma'_{induced} = \overline{E}_1[\overline{H}_1 + \overline{H}_2 + \overline{H}_3 + \overline{H}_4] +$$
$$\overline{E}_2[\overline{H}_1 + \overline{H}_2 + \overline{H}_3 + \overline{H}_4] +$$
$$\overline{E}_3[\overline{H}_1 + \overline{H}_2 + \overline{H}_3 + \overline{H}_4] +$$
$$\overline{E}_4[\overline{H}_1 + \overline{H}_2 + \overline{H}_3 + \overline{H}_4] \tag{4.62}$$

in terms of charge carriers/m^2/sec or $\gamma'_{induced} = \gamma'_{c,v}$

$$\gamma'_{c,v} = \gamma'_{1c,v} + \gamma'_{2c,v} + \gamma'_{3c,v} + \gamma'_{4c,v} \tag{4.63}$$

where

$$\gamma'_{1c,v} = \overline{E}_1[\overline{H}_1 + \overline{H}_2 + \overline{H}_3 + \overline{H}_4] \tag{4.64}$$

$$\gamma'_{2c,v} = \overline{E}_2[\overline{H}_1 + \overline{H}_2 + \overline{H}_3 + \overline{H}_4] \tag{4.65}$$

$$\gamma'_{3c,v} = \overline{E}_3[\overline{H}_1 + \overline{H}_2 + \overline{H}_3 + \overline{H}_4] \tag{4.66}$$

and

$$\gamma'_{4c,v} = \overline{E}_4[\overline{H}_1 + \overline{H}_2 + \overline{H}_3 + \overline{H}_4] \tag{4.67}$$

Also, $\gamma_{c,v}$ represents the release of induced charge carriers/meter where:

$$\gamma_{c,v} = \gamma_{1c,v} + \gamma_{2c,v} + \gamma_{3c,v} + \gamma_{4c,v} \tag{4.68}$$

Each term is measured in the number of charge carriers/meter where

$$\gamma_{1c,v} = \int_o^x \int_o^t \overline{\gamma}_1 \, dt \cdot dx \tag{4.69}$$

$$\gamma_{2c,v} = \int_o^x \int_o^t \overline{\gamma}_2 \, dt \cdot dx \tag{4.70}$$

$$\gamma_{3c,v} = \int_o^x \int_o^t \overline{\gamma}_3 \, dt \cdot dx \tag{4.71}$$

$$\gamma_{4c,v} = \int_o^x \int_o^t \overline{\gamma}_4 \, dt \cdot dx \tag{4.72}$$

Solutions for Equations (4.64) to (4.72) are left as problems for the student to solve.

4.VII ATMOSPHERIC CONDITIONS

The process of ionization within the clouds continuum is under the influence of the electric field, while that occurring on the transmission line or any other power system installation.

Equation (4.2), attributed to Townsend and repeated below, was based on enclosed process of ionization where the pressure could be controlled and A and B are constants with respect to a particular gaseous continuum.

$$\frac{\gamma}{P} = Ae^{-\frac{B}{E/P}} \tag{4.2}$$

However, in the case of lightning phenomenon, atmospheric pressure cannot be controlled, but this relationship can provide a very approximate or, in effect, guiding mode for the determination of atmospheric pressure as well as temperature conditions.

Based on the principle that the mean free path in the ionization process is inversely proportional to the pressure, and that the average number of mean free paths per unit length is 1/L, where L is the mean free path, Equation (4.2) could be written as:

$$\gamma = APe^{-\frac{AYe}{E/P}} = APe^{-\frac{B}{E/P}} \tag{4.73}$$

where $\overline{E}$ is the prevailing electric field vector in the region, V_e represents an effective ionization potential by electron collision, and γ is the number of ions released per unit length in the analysis that has been carried out in this text.

Taking the natural logarithm of Equation (4.73),

$$\ln\gamma = \ln\left[APe - \frac{AVe}{E/P}\right]$$

or

$$\ln\gamma = \ln AP - \frac{AV_eP}{E}$$

$$= \ln A + \ln P - \frac{Av_eP}{E}$$

Therefore,

$$\ln\frac{\gamma}{A} = \ln P - \frac{AV_eP}{E} \tag{4.74}$$

Expanding the term ln P, where $P > 0$

$$\ln P = 2\left[\frac{P-1}{P+1} + \frac{1}{3}\left(\frac{P-1}{P+1}\right)^3 - + \frac{1}{5}\left(\frac{P-1}{P+1}\right)^5 + - - - - \right] \quad (4.75)$$

To arrive, for the time being, at a simple expression easy for analysis and guidance, only the first term of Equation (4.75) will be substituted in Equation (4.74):

$$\ln \frac{\gamma}{A} \cong 2\frac{P-1}{P-1} - \frac{AVeP}{E} \quad (4.76)$$

Let

$$\ln \frac{\gamma}{A} = k$$

Therefore,

$$k(P + 1) = 2(P - 1) + (P + 1)\frac{AV}{E}P \text{ times}$$

$$EKP + KE = AV_eP^2 + AV_eP + (2P - 2)E$$

Therefore,

$$AV_eP^2 + P(AV_e + 2E - KE) - (2E + KE) = 0 \quad (4.77)$$

Solving for the pressure P from Equation (4.77):

$$P = \frac{1}{AV_e}[(KE - 2E - AV_e) \pm \sqrt{(AV_e + 2E - KE)^2 + 4AV_e(2E + KE)}]$$

$$(4.78)$$

$$\text{newtons/m}^2$$

Therefore,

$$P_1 = \frac{1}{AV_e}[(KE - 2E - AV_e) + \sqrt{(AV_e + 2E - KE)^2 + 4AV_e(2E + KE)}]$$

$$(4.79)$$

$$\text{newtons/m}^2$$

and

$$P_2 = \frac{1}{AV_e} [(KE - 2E - AV_e) - \sqrt{(AV_e + 2E - KE)^2 + 4AV_e(2E + KE)}]$$

(4.80)

$$\text{newtons/m}^2$$

P_1 and P_2 are two alternate solutions for the pressure in terms of effective ionization potential by electron collision and the prevailing electric field in the region. Of course, if more terms will be selected from Equation (4.75), quite different solutions for P will result.

According to Equation (4.73), $AV_e = B$, another constant given for several gases is shown in Table 4.1. For air, $A = 14.6$ and $B = 365$ at a range E/P of 150 to 600 measured in V/cm/mm of Hg.

Therefore, considering air as the dominant gas existing during lightning stroke, Equations (4.79) and (4.80) become:

$$P_1 = \frac{1}{365} [(KE - 2E - 365) + \sqrt{(365 + 2E - KE)^2 + 1460(2E + KE)}] \quad (4.81)$$

and

$$P_2 = \frac{1}{365} [(KE - 2E - 365) - \sqrt{(365 + 2E - KE)^2 + 1460(2E + KE)}] \quad (4.82)$$

where

$$k = \ln \frac{\gamma}{14.6} \tag{4.83}$$

P_1 and P_2 are measured in mm of Hg.

Equations (4.81) and (4.82) indicate that the atmospheric pressure P_1 or P_2 is a direct function of E, where E could be either the magnitude of the inducing electric field or the magnitude of the induced electric field.

A. ATMOSPHERIC PRESSURE THROUGHOUT THE INDUCING FIELD REGION

When the return conductive as well as convective strokes are of the form characterized with a sharp linear rise and relatively slow linearly decreasing tail, the inducing electric field vectors are given by Equation (2.89) where

$$|\overline{E}|_{\text{total}} = \sqrt{E_\rho^2 + E_\alpha^2 + E_\tau^2}$$

$$= \left[[A_c(t - t_1) + jA_c t U_{-1}(t - t_1)J_c(t)]^2 \left(\frac{j(\rho - R_c)}{\pi R_c^2 \hat{y}^2}\right) \right.$$

$$\left[\sum_{n=o}^{\infty} \frac{j^n e^{-j(\rho - R_c)}}{2^n n!} \left(-\frac{(\rho - R_c)}{R_c^n} - (\rho - R_c)^{n-1} - \frac{(\rho - R_c)^{n-3/2}}{\sqrt{2n}} + \frac{j2^n(n-1)!^2}{\sqrt{2}} \right) \right]$$

$$+ \left[\left(\frac{-J_c^2(t)}{2\hat{y}^2} \frac{j}{\pi(\rho - R_c)} \right) e^{-j2(\rho - R_c)} \sum_{n=o}^{\infty} \frac{j^{2n}(\rho - R_c)^{2n}}{2^n n!} \right]$$

$$+ \left. \left[\frac{J_c^2(t)R_c^2 k^4 \pi}{2\hat{y}^2} \frac{j}{(\rho - R_c)} \sum_{n=o}^{\infty} \frac{j^{2n}(\rho - R_c)^{2n}}{2^n n!} \right] \right]^{1/2}$$

$$(4.84)$$

Returning to Equation (4.73):

$$\gamma_{\text{total}} = APe - \frac{BP}{E_{\text{total}}}$$

γ_{total} could be obtained from Equation (4.36) that represents the total release of charge carriers/m, while E_{total} is expressed by Equation (4.84). For A and B usually known, there the pressure P could be computed from Equation (4.73).

Equation (4.73) can be used to find the prevailing atmospheric pressure by considering the release of charge carriers/m along the ρ, α or τ orientation.

E.X. γ_ρ represents released charge carriers/m along the ρ-axis, γ_α represents release of charge carriers along the α-axis, and γ_τ, that along the τ-axis.

Also release of charge carriers/m^2/sec could be in a transversal $\alpha\tau$ plane or along the ρ-axis, $\tau\rho$ transversal plane or along the α axis, and the $\rho\alpha$ transversal plane or along the τ-axis.

The next effect is the prevailing atmospheric pressure in the region of the induced field, where the E in Equation (4.73) is the inducing total electric field and γ_{total} refers to the total released charge carriers in the same region.

The E_{total} for the induced electric field in the region which is, in this case, along the transmission line is given by the set of Equations (4.43) to (4.46).

Another alternative for information about atmospheric pressure in the induced field region is to use the value of the induced electric field individually along the ρ, α or τ directions, and to substitute the corresponding released charges along the same orientation.

4.VIII   SOLVED EXAMPLES

A. In reference to Equation (4.11), carry out the integration process to arrive at a final expression for $\gamma_{y\text{-induced}}$.

Solution

$$\gamma_{y\text{-induced}} = \int_o^y \int_o^t \frac{P_{y\text{-induced dt}}}{V_{\text{induced}}} \cdot \overline{dy} \qquad (A1)$$

Equation (4.11) contains the function $P(t)$ given by Equation (3.50). Therefore,

$$\gamma_{y\text{-induced}} = F(x,y) \int_o^t P(t)^2 \left[\frac{1}{C^2} - \frac{1}{C^3} \sin ct + \frac{1}{C^2} \delta_o^2(t)\sin ct \right] dt \qquad (A2)$$

where

$$F(x,y) = \int_o^y \frac{j4Q}{C^2} \sum_{m=1}^{\infty} \sum_{n=1}^{\infty} \left(1 + \frac{1}{X_{n,m}} \right) \left(\frac{e^{-jX_{n,m}}}{X_{n,m}^2} + \frac{e^{-jX_{n,m}}}{X_{n,m}} \right) dy_{n,m}$$

$$= \frac{j4QY_{n,m}}{C^2} \sum_{m=1}^{\infty} \sum_{n=1}^{\infty} \left(1 + \frac{1}{X_{n,m}} \right) \left(\frac{e^{-jX_{n,m}}}{X_{n,m}^2} + \frac{e^{-jX_{n,m}}}{X_{n,m}} \right) \qquad (A3)$$

Then, from Equation (A2),

$$\int_o^t P^2(t)dt = \int_o^t \left[\frac{2}{C^2} - \frac{1}{C^3} \sin ct + \frac{1}{C^2} \delta_o^2(t)\sin ct \right]^2 dt$$

$$= \int_o^t \left[\frac{4}{C^4} + \frac{\sin^2 ct}{C^6} + \frac{1}{C^4} \delta_o^4(t)\sin^2 ct - \frac{4}{C^5} \sin ct \right.$$

$$\left. + \frac{4}{C^4} \delta_o^2 \sin ct - \frac{2}{C^5} \delta_o^2(t)\sin^2 ct \right] dt$$

$$= \frac{4}{C^4} t + \frac{t}{2C^5} - \frac{1}{4C^6} \sin 2ct + \frac{\delta_o^3}{2C^6} - \frac{3}{4C^7} \delta_o^2(t) \qquad (A4)$$

$$\sin 2ct + \frac{4}{C^6} \cos ct - \frac{2\delta_o^2(t)}{C^5} \sin 2 \, dt +$$

$$\frac{8}{C^5} - \frac{1}{C^4} t\delta_o^2(t) + \frac{1}{2C^5} \delta_o(t)\sin 2ct \qquad (A5)$$

Therefore,

$$\int_o^t P^2(t)dt = t\left(\frac{4}{C^4} - \frac{1}{2C^5} \right) - \frac{1}{4C^6} \sin ct +$$

$$\frac{1}{2C^6} \delta_o^3(t) \cdot - \frac{3}{4C^7} \delta_o^2(t)\sin 2ct - \frac{4}{C^6} \cos ct$$

$$+ \frac{8}{C^5} - \frac{1}{C^4} t\delta_o^2(t) \qquad (A6)$$

From Equations (A1), (A2), (A3) and (A6):

$$\gamma_{y\text{-induced}} = \frac{j4Q}{C^2} \sum_{m=1}^{\infty} \sum_{n=1}^{\infty} Y_{n,m} \left[\left(1 + \frac{1}{X_{n,m}}\right) \right.$$

$$\left(\frac{e^{-jX_{n,m}}}{X_{n,m}^2} + \frac{e^{-jX_{n,m}}}{X_{n,m}}\right)\right]\left[\frac{8}{C^5} + t + \left(\frac{4}{C^4} - \frac{1}{2C^5}\right)\right.$$

$$-\frac{t}{C^4}\delta_o^2(t) - \frac{1}{4C^6}\sin 2ct + \frac{1}{2C^6}\delta_o^3(t) -$$

$$\left. \frac{3}{4C^7}\delta_o^3(t)\sin 2ct + \frac{4}{C^6}\cos ct \right] \tag{A7}$$

$\gamma_{y\text{-induced}}$ represents the number of charge carriers released in the $(x - z)$ plane per meter.

It is interesting to note the time function in $\gamma_{y\text{-induced}}$ at $t = 0^+$:

$$f(t) = \frac{8}{C^5} + \frac{1}{2C^6} + \frac{4}{C^6} \approx 8C^{-5}$$

B. In reference to Equation (4.10), carry out the integration process to express $\gamma_{x\text{-induced}}$ in terms of ions/meter.

Solution

Equation (4.10) contains $F(x,y,z)$ and $f(t)$ where $f(t)$ has been solved in Example A. However, $F(x,y,z)$ must be integrated with respect to x''. Let $F(x,y,z) = G(x,y,z)$:

$$G(x,y,z) = \int (Y_{n,m} - Z_{n,m})\left(\frac{e^{-jX_{n,m}}}{X_{n,m}^2} + \frac{e^{-jX_{n,m}}}{X_{n,m}}\right)dX_{n,m} \tag{B1}$$

$$= (Y_{n,m} - Z_{n,m})\left[-\frac{e^{-jX_{n,m}}}{X_{n,m}^2} - j\int \frac{e^{-jX_{n,m}}}{X_{n,m}}dx_{n,m}\right.$$

$$\left. + j\int \frac{e^{-jX_{n,m}}}{X_{n,m}}dX_{n,m}\right] \tag{B2}$$

Therefore,

$$G(x,y,z) = (Z_{n,m} - Y_{n,m})\frac{e^{-jX_{n,m}}}{X_{n,m}} \tag{B3}$$

and

$$\gamma_{x\text{-induced}} = \frac{j4Q}{C^2}\left[\frac{8}{C^5} + t\left(\frac{4}{C^4} - \frac{1}{2C^5}\right) - \frac{t}{C^4}\,\delta_o^2(t) - \right.$$

$$\frac{1}{4C^6}\sin 2ct + \frac{1}{2C^6}\,\delta_o^3(t) - \frac{3}{4C^7}\,\delta_o^3(t)\sin 2ct$$

$$\left. + \frac{4}{C^6}\cos ct\right]\left[\sum_{m=1}^{\infty}\sum_{n=1}^{\infty}(Z_{n,m} - Y_{n,m})\frac{e^{-jX_{n,m}}}{X_{n,m}}\right]\text{ions/meter}$$

$$(B4)$$

C. Calculate the inducing electric field vector E_z indicated by Equation (4.17) and the associated H_2 field.

Solution

Equation (3.38) shows a solution for the inducing electric potential. The inducing electric field in this case is of course equal to the negative of the gradient of the inducing voltage.

However, the inducing electric field is given by Equation (3.31), where

$$\mu\,\frac{\partial A_y}{\partial t} = -\,\overline{\nabla}V_{\text{inducing}} \qquad (C1)$$

Also from Equation (3.36)

$$-\overline{\nabla}V_{\text{inducing}} = \hat{a}_y 2Q_{c,v}\mu_o\delta_o(t - t_o)\sum_{m=1}^{\infty}\sum_{n=1}^{\infty}e^{-j\sqrt{X_{n,m}^2 + Y_{n,m}^2}}$$

$$= \hat{a}_y E_{2y} \qquad (C2)$$

To find the filed H, we proceed as follows:

$$\overline{\nabla}x\overline{E}_2 = -\,\frac{\partial\overline{B}}{\partial t}$$

$$\overline{\nabla}x\overline{E}_2 = \hat{a}_x\left[\frac{E_{2z}}{y} - \frac{E_{2y}}{z}\right] - \hat{a}_y\left[\frac{E_{2z}}{x} - \frac{E_{2x}}{z}\right]$$

$$+ \hat{a}_z\left[\frac{E_{2y}}{x} - \frac{E_{2x}}{y}\right] \qquad (C3)$$

Therefore,

$$-\frac{\partial \overline{B}_2}{\partial t} = \hat{a}_x\left[-\frac{\partial E_{2y}}{\partial z}\right] + \hat{a}_z\left[\frac{\partial E_{2y}}{\partial x}\right]\ldots$$

$$= \hat{a}_z\frac{\partial}{\partial x}\left[2Q_{c,v}\mu\delta_o(t - t_o)\sum_{n}^{\infty}\sum_{m}^{\infty}e^{-j\sqrt{X_{n,m}^2 + Y_{n,m}^2}}\right] \tag{C4}$$

$$= (\hat{a}_z)\,-j2Q_{c,v}\mu\delta_o(t - t_o)\sum_{m=1}^{\infty}\sum_{n=1}^{\infty}\frac{e^{-j\sqrt{X_{n,m}^2 + Y_{n,m}^2}}}{X_{n,m}\sqrt{X_{n,m}^2 + Y_{n,m}^2}} \tag{C5}$$

Then,

$$\hat{a}_z\frac{\partial B_3}{\partial t} = (\hat{a}_z)[-j2Q_{c,v}\mu\delta_o(t - t_o)]\sum_{n=1}^{\infty}\sum_{m=1}^{\infty}\frac{e^{-j\sqrt{X_{n,m}^2 + Y_{n,m}^2}}}{X_{n,m}\sqrt{X_{n,m}^2 + Y_{n,m}^2}} \tag{C6}$$

Therefore,

$$\hat{a}_z H_{2z} = -j2Q_{c,v}U_{-1}(t - t_o)\sum_{m=1}^{\infty}\sum_{n=1}^{\infty}\frac{e^{-j\sqrt{X_{n,m}^2 + Y_{n,m}^2}}}{X_{n,m}\sqrt{X_{n,m}^2 + Y_{n,m}^2}} \tag{C7}$$

D. Carry out the integration process in Equation (4.36) for the expression representing the number of released charge carriers/meter due to the induced field.

Solution

$\gamma_{3\alpha c}$ represents the number of released charge carriers/meter due to inducing electric field when the return stroke pulse is in the form of a sharp linear rise and the tail is a linear decaying mode.

$$\gamma_{3\alpha c} = \int_o^{2\pi}\int_o^t\frac{\overline{P}_{3\alpha}\,dt}{V_{1\alpha}}\,d\alpha \tag{D1}$$

From Equation (4.33):

$$\gamma_{3\alpha c} = \int_o^{2\pi}\int_o^t(\gamma'_{3\alpha c}\,dt)d\alpha \tag{D2}$$

$$\gamma_{3\alpha c} = -j\int_o^{2\pi}\left[\int_o^t\frac{\tau R_c k^2}{\tau\rho(\rho - R_c)}\sqrt{\frac{\pi(\rho - R_c)}{j2}}\sum_{n=0}^{\infty}\frac{j^n(\rho - R_c)^n}{2^n n!}J_c(t)dt\right]d\alpha \tag{D3}$$

where

$$J_c(t) = A_{ct} \qquad\qquad 0 < t < t_1$$

$$= A_c(t - t_1) \qquad\qquad t > t_1$$

Therefore,

$$\int_o^t J_c(t) = \int_o^{t_1} A_{ct}\, dt\, t + \int_{t_1}^t A_c(t - t_1)dt$$

$$= \frac{1}{2} A_c t_1^2 + A_c\left(\frac{1}{2} t^2 - t_1 t + \frac{1}{2} t_1^2\right)$$

$$= A_c t_1^2 + \frac{1}{2} A_c t^2 - A_c t_1 t \qquad\qquad (D4)$$

From Equations (D3) and (D4):

$$\gamma_{3ac} = -j\left[\frac{2\tau R_c k^2}{\rho(\rho - R_c)} \sqrt{\frac{\pi(\rho - R_c)}{j^2}} \sum_{n=0}^{\infty} \frac{j^2(\rho - R_c)^n}{2^n n!}\right]\left[\frac{1}{2} A_c t^2 + \right.$$

$$\left. A_c t_1^2 - A_c t_1 t\right] \text{ions/meter} \qquad\qquad (D5)$$

γ_{3ac} refers to the conductive return stroke. To represent γ_{3ac} for the convective stroke generated by the time varying bound charges, the parameter A_c is to be replaced by A_v and R_c by R_v, where R_v is the outer radius of the coaxial convective stroke shown in Figure 2.1.

E. Consider the atmospheric case in the air dominated zone where the pressure is 0 mm of Hg. From Equations (4.81) and (4.82), express the field E in volts/cm and probable number for γ.

Solution

From Equation (4.81), let $P_1 = 0$

$$(KE - 2E - 365)^2 = (365 + 2E - KE)^2 + 1460(2E + KE)$$

$$K^2E^2 + 4E + (365)^2 - 4KE^2 - 730KE + 1460E =$$

$$K^2E^2 + 4E^2 + (365)^2 + 1460E - 730KE - 4KE^2$$

$$2920E + 146KE \qquad\qquad (E1)$$

Therefore,

$$2910E + 1460KE = 0 \tag{E2}$$

From Equation (2), $E = 0$ and $K = -2$ but $K = \ln \dfrac{\gamma}{14.6}$; therefore,

$$\ln \frac{\gamma}{14.6} = -2$$

or

$$e^{-2} = \frac{\gamma}{14.6}$$

and

$$\gamma = \frac{14.6}{e^2} = 2 \text{ ions/cm} \tag{E3}$$

at zero electric field implying the presence of one positive and one negative ion, i.e., a neutral state.

4.IX PROBLEMS

1.   Obtain expression for the release of charge carriers/m^{2}sec due to the inducing electric field by lightning discharge in the y-z plane. Fields calculation has been based on the principle of magnetic moment.
2.   Supplement problem 1 for the release of charge carriers/meter.
3.   From the solution obtained in problem 2, solve for the atmospheric pressure for air along the x direction of space continuum.
4.   Repeat problem 1 for the release of ions in the x-z plane.
5.   Repeat problem 2 for the release of ions in along the y-axis.
6.   Repeat problem 3 for the atmospheric pressure along the y direction of space continuum.
7.   Repeat problem 1 for the release of ions in x-y plane.
8.   Repeat problem 2 for the release of ions along the z-axis.
9.   Repeat problem 3 for the atmospheric pressure along the z-axis.
10. Considering space continuum under lightning stroke as constant volume, obtain expressions for the prevailing temperature in problems 3, 6 and 9.
11. Repeat the solution of Example D for the release of ions in the α-τ plane.
12. Repeat problem 11 for the release of ions in the α-ρ plane.
13. Establish a solution for the prevailing atmospheric pressure and temperature, based on information existing in Example D and its solution.

14. Establish a solution for the prevailing atmospheric pressure and temperature, based on information existing in problem 11 and its solution.

15. Establish a solution for the prevailing atmospheric pressure and temperature, based on information existing in problem 12 and its solution.

16. Using expressions for the induced electromagnetic fields on a transmission line, based on a representation of lightning surge in the form of a sharp linear rise and a slow linear tail, establish final expression for the release of ions/m^2/sec in the y-z plane.

17. From the solution obtained in problem 16 proceed to establish a solution for the release of ions/meter along the x-axis.

18. From the solution of problem 17, establish an expression for the prevailing pressure and temperature in the region of the induced fields.

19. Repeat problem 16 for the release of charge carriers in the z-x plane.

20. Repeat problem 17 for the release of charge carriers along the y-axis.

21. Repeat problem 18 for the prevailing conditions of pressure and temperature along the y direction of the induced fields region.

22. Repeat problem 16 for the release of charge carriers in the x-y plane.

23. Repeat problem 17 for the release of charge carriers in the x-axis.

24. Repeat problem 18 for the prevailing conditions of pressure and temperature along the z direction of the induced fields region.

25. In a gaseous dominated zone of water vapor, establish a solution for the atmospheric pressure and temperature when the induced electromagnetic fields due to lightning pull were calculated on the basis of magnetic moment method. Consider effect of total electric and total magnetic field.

26. Repeat problem 25 for a gaseous dominated zone of CO_2.

27. Repeat problem 25 for a gaseous dominated zone of H_2.

28. Repeat problem 25 for a gaseous dominated zone of He.

29. Repeat Example E when the gaseous dominated zone is water vapor under normal atmospheric pressure in mm of Hg.

30. Repeat problem 29 when the gaseous dominated zone is CO_2 under 50% atmospheric pressure.

31. Repeat problem 29 when the gaseous dominated zone is He under 25% atmospheric pressure.

32. Present a qualitative analysis for the effect of ion bombardment due to lightning surge on the phenomenon of corona on the transmission line. Consider linear representation for the lightning surge.

33. Repeat problem 32 when lightning surge representation is based on the method of magnetic moment.

34. Repeat problem 32 when lightning surge representation is a chopped pulse with oscillations.

PRINCIPAL LIST OF SYMBOLS

γ	rate of ionizing collisions/meter
P	pressure in atmospheres
C	velocity of light
v_i	inducing voltage
$a_\rho,\ a_\tau,\ a_\gamma$	unit vectors
t_1	delayed time
$\delta_\rho'(t)$	$\dfrac{\partial \delta_\rho(t)}{\partial t}$
$\overline{A}_c$	conductive magnetic vector potential
$\overline{A}_v$	convective magnetic vector potential

REFERENCES

1. **Cobine, J. D.**, *Gaseous Conductors: Theory and Engineering Applications*, Dover Publ., New York, 1941.
2. **Denno, K.**, Computation of electromagnetic lightning response using moments method, *IEEE Trans., Magnetics,* 20(5) 1953, 1984.
3. **Denno, K.**, Dynamic modelling for the process of inducing and induced voltage surges due to lightning, *J. Electrostat.,* 13, 55, 1982.
4. **Denno, K.**, Mathematical modelling of propagating inducing and induced power due to actual pulse shape of lightning surge, *J. Electrostat.,* 15, 43, 1984.

Chapter 5

INTERACTION OF GASEOUS CONTINUA WITH LIGHTNING

5.I STAGES OF IONIZATION AND VOLTAGE BREAKDOWN AT THE GASEOUS CONTINUUM OF HELIUM

Previous work carried out by this author and presented in Chapter 2 established mathematical model sine time and space domains for the inducting and induced electromagnetic effects at any physical point located on a transmission point in free space due to lightning discharge at another place. Problem objectives took into consideration the conductive current density in the return stroke and the convective current density generated by the state of time variation of bound charges distribution induced through ground surface.

Solutions of electromagnetic field components have been secured with an actual functional form for the conductive and convective electric current densities represented by a sharp linear rise in the pulse front and a lower linear decline in the pulse tail. Also, solutions have been obtained for the propagating inducing and induced power at a space point produced by the corresponding inducing and induced electric magnetic field components.

This chapter presents research centering on investigation of the effects of the propagating lightning electromagnetic phenomena on anisotropic gaseous continua of helium located at some distance from the impact area of lightning surge.

The gaseous continuum is assumed to be under ambient levels of pressure and temperature, an initial neutral state of ionization and of cylindrical configuration.

Calculations in cylindrical coordinates for the effects of electromagnetic interaction between the propagated electromagnetic field generated by lightning and the gaseous continuum resulted in solutions for the ionic as well as breakdown conditions of the gaseous continuum. The analytical picture obtained identifies the state of concentration for the doubly ionized continuum, the induction process as well as microscopic velocities in three dimensions.

Bases in the present analysis are centered on the following given states:

(i)   Doubly ionized cylindrical continuum of He^+ and H^{++} as well as neutrals under standard pressure and temperature, provided with certain mechanism for controlling the ions generation and neutrals

(ii)   The gaseous continuum is anisotropic under the influence of electromagnetic field radiation produced by a lightning surge.

Analytical goals in this section are:

1.   Electrically induced current densities in the anisotropic gas continuum of He^+ and H^{++}

2.   Conditions of electric voltage breakdown within the anisotropic gas continuum of He^+ and He^{++}

3.   Microscopic velocity within the helium ionic continuum

4.   Magnetically induced current densities

5.   Conditions of magnetic voltage breakdown.

A. DESCRIPTION OF He IONIC CONTINUUM

The gas continuum is assumed anisotropic and consists of a mixture of He^+ and He^{++} and neutrals, where

n_1, n_2 = concentration of ion generation for He^+ and He^{++}, respectively
α_1, α_2 = rate of ion generation for He^+ and He^{++}, respectively
β = rate of ion removal due to diffusion
ϵ = ion purity parameter

From previous work carried out by this author, it was established that:

$$n_1 = \frac{\alpha_1(\alpha_1 + \alpha_2)}{(\beta + \alpha_2 - \alpha_1)(\epsilon\beta - 2\alpha_1 - \alpha_2 + \beta)(2\alpha_1 - \beta - \epsilon\beta)\alpha_1} \tag{5.1}$$

and

$$n_2 = \frac{\alpha_1 + \alpha_2}{\epsilon\beta - 2\alpha_1 - \alpha_2 + \beta} - \frac{2\alpha_1 - \beta - \epsilon\beta}{\epsilon\beta - 2\alpha_1 - \alpha_2 + \beta}$$

$$\frac{\alpha_1(\alpha_1 + \alpha_2)}{(\epsilon\beta - 2\alpha_1 - \alpha_2 + \beta) \div (\alpha_1 - \alpha_2 + \beta) + \alpha_1(2\alpha_1 - \beta - \epsilon\beta)} \tag{5.2}$$

The continuum is characterized as anisotropic and tensorial electrical conductivity.

B. ELECTRICALLY INDUCED CURRENT DENSITIES

Interaction between the anisotropic dielectric of the gaseous He continuum and the inducing electromagnetic fields will produce a current density vector,

$$\bar{J} = \sigma \bar{E} \tag{5.3}$$

Since the concentrations of He^+ and He^{++} are constants and macroscopically independent of position as indicated in Equations (5.1) and (5.2), there will be a magnetically induced current contributed only by the inducing electric fields generated by lightning surge. In Equation (5.3) is a rank 2 tensor, in which $\bar{J}$ and $\bar{E}$ are the induced current density and the inducing electric field vectors.

$$\begin{bmatrix} \bar{J}_\rho \\ \bar{J}_\alpha \\ \bar{J}_\tau \end{bmatrix} = \begin{bmatrix} \sigma_T & -\sigma_H & 0 \\ \sigma_H & \sigma_T & 0 \\ 0 & 0 & \sigma \end{bmatrix} \begin{bmatrix} \bar{E}_\rho \\ \bar{E}_\alpha \\ \bar{E}_\tau \end{bmatrix} \tag{5.4}$$

where $\bar{J}_\rho$, $\bar{J}_\alpha$ and $\bar{J}_\tau$ are the three-dimensional components of the induced current density in cylindrical coordinates.

$$\bar{J}_\rho = \sigma_T \bar{E}_\rho - \sigma_H \bar{E}_\alpha \tag{5.5}$$

$$\bar{J}_\alpha = \sigma_H \bar{E}_\rho + \sigma_T \bar{E}_\alpha \tag{5.6}$$

$$\bar{J}_\tau = \sigma \bar{E}_\tau \tag{5.7}$$

Solutions for $\bar{E}_\rho$, $\bar{E}_\alpha$ and $\bar{E}_\tau$ have been given by Equations (2.71) through (2.78).

Also $\bar{J}_\alpha$, $\bar{J}_\rho$ and $\bar{J}_\tau$ could be expressed in another form, in terms of concentration of He^+ and He^{++} charge carriers and local ion motion, as shown in the following equations:

$$\bar{J}_\rho = (n_1 Q^+ + 2n_2 Q^+)\bar{V}_\rho \tag{5.8}$$

$$\bar{J}_\alpha = (n_1 Q^+ + 2n_2 Q^+)\bar{V}_\alpha \tag{5.9}$$

$$\bar{J}_\tau = (n_1 Q^+ + 2n_2 Q^+)\bar{V}_\tau \tag{5.10}$$

where $\bar{V}_\rho$, $\bar{V}_\alpha$ and $\bar{V}_\tau$ are the local velocities along the respective coordinate axes, and Q^+ is the total charge in coulombs carried by clusters in the gaseous cloud.

C. CONDITIONS OF ELECTRICALLY INDUCED VOLTAGE BREAKDOWN

In the dielectric gaseous continuum, potential breakdown corresponds to an infinite current density along any or more than one dimensional flow.

Three criteria for voltage breakdown could be established, one from the set of Equations (5.5), (5.6) and (5.7) and the other from the set of Equations (5.8), (5.9) and (5.10) infinite current density along any orientation of the cylindrical coordinate system.

1.   *The Field Criterion*

This could be obtained from the set of Equations (5.5), (5.6) and (5.7) and expressed below:

$$1/\sigma_H \overline{E}_\rho + \sigma_\tau \overline{E}_\alpha = 0 \tag{5.11}$$

$$1/\sigma_\tau \overline{E}_\rho + \sigma_\tau \overline{E}_\alpha = 0 \tag{5.12}$$

$$1/\sigma \overline{E}_\alpha = 0 \tag{5.13}$$

2.   *The Concentration Criterion*

$$1/\overline{V}_\rho (n_1 Q^+ + 2n_2 Q^+) = 0 \tag{5.14}$$

$$1/\overline{V}_\alpha (n_1 Q^+ + 2n_2 Q^+) = 0 \tag{5.15}$$

$$1/\overline{V}_\tau (n_1 Q^+ + 2n_2 Q^+) = 0 \tag{5.16}$$

3.   *The Field-Concentration Criterion*

This coupled criterion is expressed by equating the corresponding current density vector from the concentration and field equations set as indicated below:

$$\sigma_\tau \overline{E}_\rho - \sigma_H \overline{E}_\alpha = \overline{V}_\rho (n_1 Q^+ + 2n_2 Q) \tag{5.17}$$

$$\sigma_H \overline{E}_\rho + \sigma_\tau \overline{E}_\alpha = \overline{V}_\alpha (n_1 Q^+ + 2n_2 Q) \tag{5.18}$$

$$\sigma \overline{E}_\tau = \overline{V}_\tau (n_1 Q^+ + 2n_2 Q) \tag{5.19}$$

D. LOCAL MICROSCOPIC VELOCITY OF CHARGE CARRIER

It is interesting to note that the coupled field concentration criterion for voltage breakdown within the gaseous He continuum indicated in Equations (5.17), (5.18) and (5.19) could be used to calculate the average velocity of charge, i.e.,

$$\overline{V}_\rho = \frac{\sigma_\tau \overline{E}_\rho - \sigma_H \overline{E}_\alpha}{n_1 Q^+ + 2n_2 Q^+} \tag{5.20}$$

$$\overline{V}_\alpha = \frac{\sigma_\tau \overline{E}_\rho - \sigma_H \overline{E}_\alpha}{n_1 Q^+ + 2n_2 Q^+} \tag{5.21}$$

$$\overline{V}_\tau = \frac{\sigma \overline{E}_\tau}{n_1 Q^+ + 2n_2 Q^+} \tag{5.22}$$

Indeed the average microscopic velocity vectors in Equations (5.20), (5.21) and (5.22) for the doubly ionized mixture could be secured only through the field-concentration criterion for voltage breakdown.

E. MAGNETICALLY INDUCED ELECTRIC FIELDS AND CURRENT DENSITIES

From Equations (2.17) and (2.78), and (5.20), (5.21) and (5.22), the induced electric fields, microscopically, are

$$\begin{bmatrix} \overline{E}_{\rho m} \\ \overline{E}_{\alpha m} \\ \overline{E}_{\tau m} \end{bmatrix} = \frac{1}{\mu_o} \begin{bmatrix} \hat{a}_\rho & \hat{a}_\alpha & \hat{a}_\tau \\ V_\rho & V_\alpha & V_\tau \\ H_\rho & H_\alpha & 0 \end{bmatrix} \tag{5.23}$$

$$\therefore \quad \overline{E}_{\rho m} = \frac{1}{\mu_o} \hat{a}_\rho(-H_\alpha V_\tau) \tag{5.24}$$

$$\overline{E}_{\alpha m} = \frac{1}{\mu_o} \hat{a}_\alpha(-V_\tau H_\rho) \tag{5.25}$$

$$\overline{E}_{\tau m} = \frac{1}{\mu_o} \hat{a}_\tau(V_\rho H_\alpha - V_\alpha H_\rho) \tag{5.26}$$

And the induced current densities are:

$$\text{from} \quad \overline{J} = \sigma \overline{E} \tag{5.27}$$

$$\therefore \quad \overline{J}_{\rho m} = \sigma_\tau \overline{E}_{\rho m} - \sigma_H \overline{E}_{\alpha m} \tag{5.28}$$

$$\overline{J}_{\rho m} = \sigma_H \overline{E}_{\rho m} + \sigma_H \overline{E}_{\alpha m} \tag{5.29}$$

$$\overline{J}_{\tau m} = \sigma \overline{E}_{\tau m} \tag{5.30}$$

The conditions of field microscopic voltage breakdown are:

$$1/(\sigma_\tau \overline{E}_{\rho m} - \sigma_H \overline{E}_{\alpha m}) = 0 \tag{5.31}$$

$$1/(\sigma_\tau \overline{E}_{\rho m} + \sigma_\tau \overline{E}_{\alpha m}) = 0 \tag{5.32}$$

$$1/(\sigma \overline{E}_{\tau m}) = 0 \tag{5.33}$$

Similar modes of microscopically magnetically induced voltage breakdown can be developed for the mode of ion concentration and the coupled field-concentration modes.

In the above equations, $\overline{E}_{\rho m}$, $\overline{E}_{\alpha m}$ and $\overline{E}_{\tau m}$ are the magnetically induced electric fields along ρ, α and τ cylindrical coordinates, respectively. $\overline{J}_{\rho m}$, $\overline{J}_{\alpha m}$ and $\overline{J}_{\tau m}$ are the magnetically induced current densities along the ρ, α, τ cylindrical coordinate system, respectively.

From the preceding sections, we can identify the following conclusions: For an anisotropic gaseous continuum located at a considerable distance from the source of lightning surge where radiation from conduction in the return stroke and time variation in bound charges (convective effect) have been assumed to exist, and with the knowledge that all electromagnetic radiation fields components are known, the following solutions have been established:

1. Induced current density vectors within the gaseous continuum in a three-dimensional coordinate system.
2. Criteria of voltage breakdown (induced electrically and magnetically) along any orientation in the cylindrical coordinate system have been established in three modes: (a) concentration mode; (b) field effect mode; and (c) coupled field-concentration mode.
3. Microscopic average velocities of the anisotropic gaseous double ionized mixture have been secured through the field-concentration criterion of voltage breakdown.

5.II ACCELERATION OF CHARGE CARRIERS

In this section the spectral scope for the accelerating fields generated by the convective current of lightning surge on an anisotropic gaseous continuum is investigated. Charged particles will be time induced electrically and magnetically with distinct velocity spectrum and acceleration.

At the anisotropic gaseous continuum, solutions have been secured for the originated modes of acceleration imposed on the singly and doubly ionized particles, electrons as well as neutrals in terms of their physical and chemical properties and the tensorial elements of their electrical conductivities.

Convective current density by the convective return stroke is expressed by

$$\frac{\partial \overline{D}}{\partial t} = \frac{\partial \overline{D}}{\partial x} \cdot \frac{\partial \overline{x}}{\partial t} \tag{5.34}$$

where $\partial \overline{D}/\partial t$ as the convective current density $\overline{D}$ which is also the electric charge density vector

$$\overline{J}_v = \frac{\partial \overline{D}}{\partial \overline{x}} \cdot \frac{\partial \overline{x}}{\partial t} \tag{5.35}$$

then

$$\frac{\partial \bar{J}_v}{\partial t} = \frac{\partial^2 \overline{D}}{\partial x^2} \cdot \frac{\partial \bar{x}}{\partial t} + \frac{\partial \overline{D}}{\partial x} \cdot \frac{\partial^2 \bar{x}}{\partial x^2} \tag{5.36}$$

Treating the gaseous continuum as anisotropic, the current density $\bar{J}_v$ is written below in terms of the tensorial electrical conductivity and the inducing electric field vector $\overline{E}$:

$$\bar{J}_v = \begin{bmatrix} \sigma_\tau & -\sigma_H & 0 \\ \sigma_H & \sigma_T & 0 \\ 0 & 0 & \sigma \end{bmatrix} \begin{bmatrix} \overline{E}_\rho \\ \overline{E}_\alpha \\ \overline{E}_\tau \end{bmatrix} = \begin{bmatrix} \bar{J}_\rho \\ \bar{J}_\alpha \\ \bar{J}_\tau \end{bmatrix} \tag{5.37}$$

where

$$\bar{J}_{\rho v} = \sigma_\tau \overline{E}_\rho - \sigma_H \overline{E}_\alpha \tag{5.38}$$

$$\bar{J}_{\alpha v} = \sigma_H \overline{E}_\rho + \sigma_H \overline{E}_\alpha \tag{5.39}$$

$$\bar{J}_{\tau v} = \sigma \overline{E}_\tau \tag{5.40}$$

Treating the current density wave in the convective stroke as a sharp impulse,

$$\frac{\partial \bar{J}_v}{\partial t} \rightarrow \delta_o(t) \tag{5.41}$$

Equation (5.36) is rewritten below:

$$\frac{\partial^2 \overline{D}}{\partial x^2} \frac{\partial x}{\partial t} = \delta_o(t) \frac{\partial \overline{D}}{\partial x} \frac{\partial^2 x}{\partial t^2} \tag{5.42}$$

$$\therefore \quad \frac{\partial^2 \overline{D}_\rho}{\partial_\rho^2} \frac{\partial_\rho}{\partial t} = \delta_o(t) \frac{\partial \overline{D}_\rho}{\partial_\rho} - \frac{\partial_\rho^2}{\partial t^2} \tag{5.43}$$

let $\bar{a}_\rho$ represent the ρ oriented acceleration vector,
$\therefore$

$$\bar{a}_\rho = \delta_o'(t) - \frac{\dfrac{\partial^2 \overline{D}_\rho}{\partial_\rho^2} \dfrac{\partial_\rho}{\partial t}}{\partial^2 \overline{D}_\rho / \partial_\rho}$$

or

$$= \delta_o'(t) - \bar{u}_\rho \frac{\partial^2 \overline{D}_\rho / \partial_\rho^2}{\partial^2 \overline{D}_\rho / \partial_\rho} \qquad (5.44)$$

where $\bar{u}_\rho = \partial_\rho / \partial t$.

The convective current density vector $\bar{J}_v$ is written below:

$$\bar{J}_v = \frac{\partial D}{\partial x} \cdot \bar{u}_x$$

or

$$\bar{J}_{\rho v} = \frac{\partial D_\rho}{\partial_\rho} \cdot \bar{u}_\rho \qquad (5.45)$$

From Equations (5.44) and (5.45),

$$\bar{a}_\rho = \delta_o'(t) - \frac{(J_{\rho v})\left(\dfrac{\partial^2 \overline{D}_\rho}{\partial_\rho}\right)^2}{\left(\dfrac{\partial^2 \overline{D}_\rho}{\partial_\rho}\right)^2} \qquad (5.46)$$

Similarly,

$$\bar{a}_\alpha = \delta_o'(t) - \frac{(J_{\alpha v})\left(\dfrac{\partial^2 \overline{D}_\alpha}{\partial_\alpha}\right)^2}{\left(\dfrac{\partial^2 \overline{D}_\alpha}{\partial_\alpha}\right)^2} \qquad (5.47)$$

and

$$\bar{a}_\tau = \delta_o'(t) - \frac{(J_{\tau v})\left(\dfrac{\partial^2 \overline{D}_\tau}{\partial_\tau}\right)^2}{\left(\dfrac{\partial^2 \overline{D}_\tau}{\partial_\tau}\right)^2} \qquad (5.48)$$

Returning to Equations (5.38), (5.39) and (5.40), and if the gaseous medium is characterized as a simple continuum where

$$\overline{D} = \epsilon \overline{E}$$

then

$$\overline{D}_\rho = \epsilon \overline{E}_\rho$$
$$\overline{D}_\alpha = \epsilon \overline{E}_\alpha$$

and

$$\overline{D}_\tau = \epsilon \overline{E}_\tau \tag{5.49}$$

Then if the gaseous continuum is characterized as linear where

$$\overline{D} = \epsilon \overline{E} + \epsilon_1 \frac{\partial E}{\partial_\tau} + \epsilon_2 \frac{\partial^2 E}{\partial_\tau^2} + \cdots \tag{5.50}$$

$$\therefore \quad \overline{D}_\rho = \epsilon \overline{E}_\rho + \epsilon_1 \frac{\partial \overline{E}_\rho}{\partial t} + \epsilon_2 \frac{\partial^2 \overline{E}}{\partial_\tau^2} + \cdots \tag{5.51}$$

$$\overline{D}_\alpha = \epsilon \overline{E}_\alpha + \epsilon_1 \frac{\partial \overline{E}_\alpha}{\partial t} + \epsilon_2 \frac{\partial^2 \overline{E}}{\partial_\tau^2} + \cdots$$

and

$$\overline{D}_\tau = \epsilon E_\tau + \epsilon_1 \frac{\partial \overline{E}_\tau}{\partial t} + \epsilon_2 \frac{\partial^2 \overline{E}}{\partial_\tau^2} + \cdots$$

$$\overline{\overline{\epsilon}} = \begin{bmatrix} \epsilon_{11} & \epsilon_{12} & \epsilon_{13} \\ \epsilon_{21} & \epsilon_{22} & \epsilon_{23} \\ \epsilon_{31} & \epsilon_{32} & \epsilon_{33} \end{bmatrix} \tag{5.52}$$

$$\therefore \quad \overline{D} = \overline{\overline{\epsilon}} \, \overline{E}_\tau$$

$$= \begin{bmatrix} \epsilon_{11} & \epsilon_{12} & \epsilon_{13} \\ \epsilon_{21} & \epsilon_{22} & \epsilon_{23} \\ \epsilon_{31} & \epsilon_{32} & \epsilon_{33} \end{bmatrix} \begin{bmatrix} \overline{E}_\rho \\ \overline{E}_\alpha \\ \overline{E}_\tau \end{bmatrix} \tag{5.53}$$

Also it is useful to remind the reader that $\overline{E}_\rho$, $\overline{E}_\alpha$ and $\overline{E}_\tau$ have been expressed in Equations (2.71), (2.72) and (2.73).

To express $\overline{a}_\rho$, $\overline{a}_\alpha$ and $\overline{a}_\tau$ according to Equations (5.46), (5.47) and (5.48), $\partial \overline{E}_\rho / \partial_\rho$ and $\partial^2 \overline{E}_\rho / \partial_\rho^2$ are given below:

$$\frac{\partial \overline{E}_\rho}{\partial_\rho} = A_{v,c} \sqrt{\frac{j}{\pi}} \sum_{n=o}^{\infty} \frac{j^{n+1}}{2^n n! R_v \hat{y}} e^{-j(\rho - R_v)} \left[\left[\frac{(\rho - R_v)}{R_v} \right]^{n+\frac{1}{2}} \right.$$

$$\left. + (\rho - R_v)^{n-\frac{1}{2}} + \frac{(\rho - R_v)^{n-1}}{2^n} + \frac{j 2^n (n-1)!}{2} (\rho - R_v)^{\frac{1}{2}} \right]$$

$$- \left[\frac{n + \frac{1}{2}}{R^n} (\rho - R_v)^{n - \frac{1}{2}} + \left(n - \frac{1}{2} \right)(\rho - R_v)^{n - \frac{3}{2}} \right.$$

$$\left. + \frac{n - 1}{2^n} (\rho - R_v)^{n - 2} + \frac{j2^n(n - 1)!}{2} (\rho - R_v)^{n - \frac{1}{2}} \right] \tag{5.54}$$

Also $\partial^2 \overline{E}_\rho / \partial_\rho^2$ is given below:

$$\frac{\partial^2 \overline{E}_\rho}{\partial_\rho^2} = A_{v,c} \sqrt{\frac{j}{\pi}} \sum_{n=o}^{\infty} \frac{-j^{n+2}}{2^n n! R_v \hat{y}} e^{-j(\rho - R_v)} \left[\frac{(\rho - R_v)^{n + \frac{1}{2}}}{R^n} \right.$$

$$+ (\rho - R_v)^{n - \frac{1}{2}} + \frac{1}{2^n} + (\rho - R_v)^{n - 1} \frac{j2^n(n - 1)!}{2} (\rho - R_v) \frac{1}{2}$$

$$- \frac{n + \frac{1}{4}}{R^n} (\rho - R_v)^{n - \frac{1}{2}} + \frac{j2^n(n - 1)!}{2} (\rho - R_v)^{n - \frac{3}{2}}$$

$$\left. - \frac{(n - 1)(n - 2)}{2^n} (\rho - R_v)^{n - 3} - \left(n - \frac{1}{2} \right)\left(n - \frac{3}{2} \right)(\rho - R_v)^{n - \frac{5}{2}} \right] \tag{5.55}$$

$\therefore$ All the structural components to account for $\overline{a}$ have been presented as indicated in Equations (5.55), (5.54) and (5.45).

And from Equations (5.47) and (5.48), the reader can notice that

$$\frac{\partial \overline{E}_\alpha}{\partial_\alpha} = \frac{\partial \overline{E}_\tau}{\partial_\tau} = 0$$

Of course mechanical expressing or solutions for $\overline{a}_\rho$, $\overline{a}_\alpha$ and $\overline{a}_\tau$ identified in Equations (5.46), (5.47) and (5.48) are valid if the gaseous continuum is simple where

$$\overline{D} = \epsilon \overline{E}$$

Another aspect of field acceleration is that which could be generated by magnetic induction where, on lumped particle basis,

$$\overline{E}_{mag} = \overline{v} \times \overline{B}$$

$$= \begin{bmatrix} \hat{a}_\rho & \hat{a}_\alpha & \hat{a}_\tau \\ \overline{u}_\rho & \overline{u}_\alpha & \overline{u}_\tau \\ \overline{B}_\rho & \overline{B}_\alpha & \overline{B}_\tau \end{bmatrix} \tag{5.56}$$

$\bar{u}_\rho$, $\bar{u}_\alpha$ and $\bar{u}_\tau$ represent the gaseous charged carrier velocities in cylindrical coordinates. β_ρ, β_α and β_τ represent the induced magnetic induction components within the gaseous continuum. Expressions for $\hat{a}_\rho$, $\hat{a}_\alpha$ and $\hat{a}_\tau$ were given in Equations (2.74) to (2.78).

5.III NITROGEN AND NITROGEN/FREON WITH CCl$_2$F$_2$ MIXTURES

A. BREAKDOWN IN N$_2$

To continue an upward trend in the application of high voltages in power systems, it is necessary to resort to the use of attaching gases having high dielectric strengths such as SF_6 and CCl_2F_2 as insulators. However, because of the high cost of these gases, and to increase breakdown level, the formation of binary/ternary mixture with unitary gases such as N_2, H_2, CO_2 and air is a research goal. This section focuses on the understanding of breakdown characteristics of such mixtures under the impact of lightning and switching impulse voltages. This section will analyze experimentally the response of a gaseous mixture of N_2 with 1-20% of CCl_2F_2 from pressure under a lightning surge (1.2/50 μs) and switching impulse (200/2000 μs) duration, in a continuum gap formed by a rod/plane configuration at pressure up to 5 bar and spacing up to 100 mm.

Results of this work have been obtained by T.V. Babu Rajendran, C. S. Lakshminarasimha, and M.S. Baidu.

The experimental set-up used is shown in Figure 5.1, where electric currents have been recorded in pre-breakdown and breakdown regions as well as time lag for breakdown.

C_g = generator capacitance (0.05, μlF)
R_g = generator resistance
R_s = wavefront resistance
C_1 = load capacitance
R_d = wavetail resistance
R_{se} = divider resistance (1.25 Ω)
R_{sh} = current shunt (0.25 mΩ)
R = terminating resistance (75 Ω)

The cylindrical steel chamber can withstand pressure of up to 14 bar. The high voltage lead was taken in through a bushing that can stand impulse voltage up to 400 kV; for the source of lightning impulse and switching impulse, a Marx-type generator had been used. Measurements as breakdown voltage were taken at gap spacing ranging from 5 to 100 mm, coupled with gradual reduction of pressure.

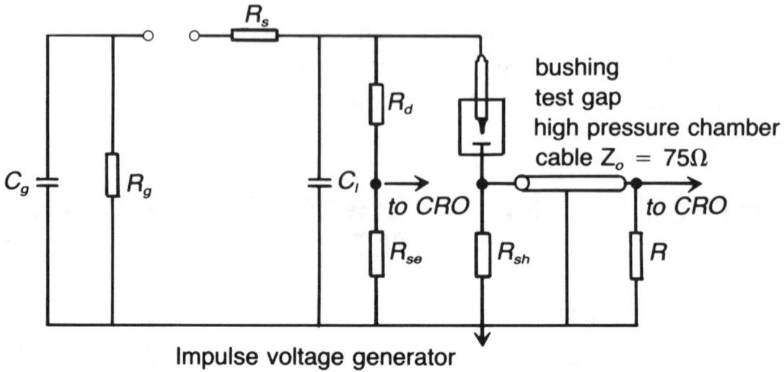

FIGURE 5.1.   Schematic diagram of experimental set-up.

Emphasis in measurement was given to a level of 50% voltage breakdown using conventional probability criteria. Also, throughout all breakdown, rigorous levels of current built have been recorded.

The experimental work conducted by Rajendra, Lakshminarashimha and Naidu continued with N_2 where lightning surges of both polarities over a gap spacing ranging from 5-100 mm, and for switching impulse (200/2000 mm) of both polarities were applied. Breakdown voltage levels for negative polarities were higher for both switching impulses and lightning surges. Results for voltage breakdown characteristics in N_2 for lightning impulses are shown in Figure 5.2 and for switching impulse, in Figure 5.3.

B. BREAKDOWN IN N_2-CCl_1F_2 MIXTURE

Observation of voltage breakdown as conducted in the chamber in which a mixture of N_2-CCl_2F_2 has been injected under pressure conditions varying from 1.3 to 5 bar. Results are shown in Figures 5.4 through 5.9.

It can be seen from the characteristics curves shown in Figures 5.2 to 5.9 that injection of CCl_2F_2 into N_2 will improve markedly the level of voltage breakdown, although a plateau of saturation is developing. Also, behavior of time lag to breakdown is shown in Figure 5.10 for a gaseous continuum of N_2-CCl_2F_2.

The presentation in Section III of this chapter identified extremely useful information for the characteristics of voltage breakdown in pure N_2 gas and in a mixture of N_2-CCl_2F_2 subject to the inception of lightning surge and switching impulse. Results indicate that the addition of impurities will enhance upward the level of voltage breakdown especially for the case of negative polarities. This study also indicates that the 50% of voltage breakdown is substantially higher than the level of corona onset, and that saturation in voltage breakdown will set at a certain percentage addition of CCl_2F_2 impurity.

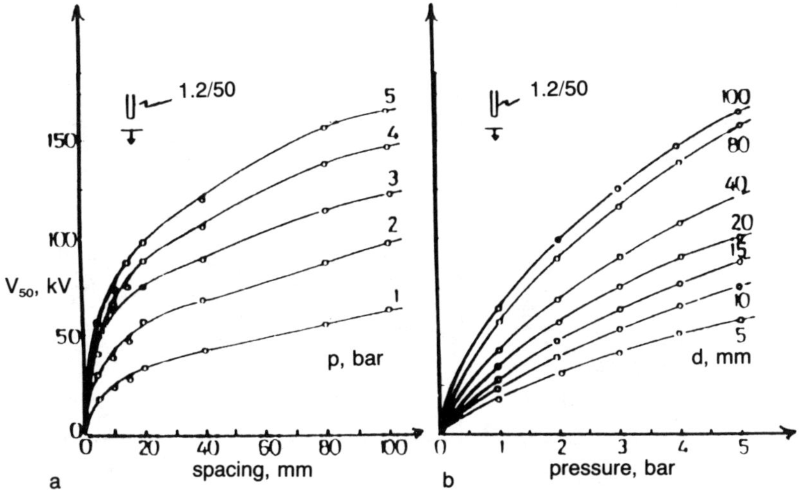

FIGURE 5.2. Impulse breakdown characteristics for positive rod/plane gaps in N_2. (a) Voltage/spacing characteristic. (b) Voltage/pressure characteristics.

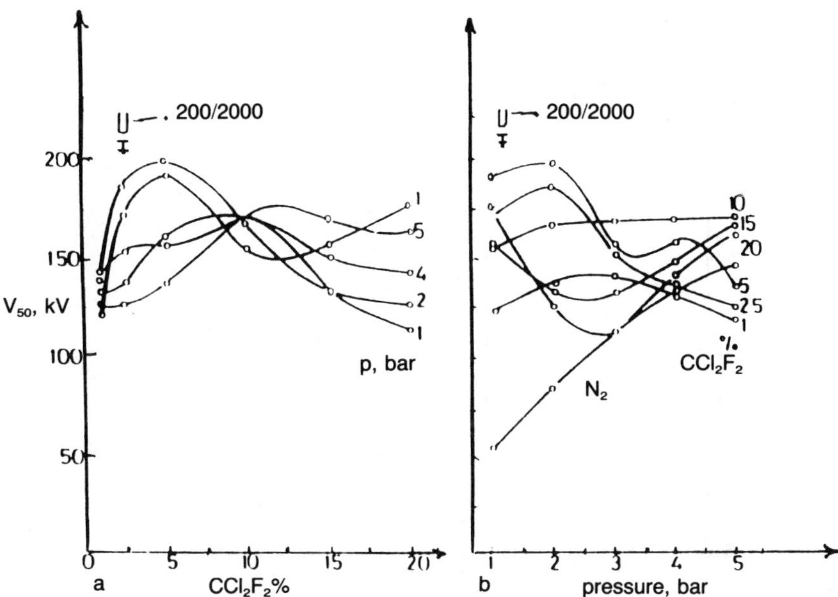

FIGURE 5.3. Switching impulse breakdown voltage characteristics for positive rod/plane gap in N_2/CCl_2F_2 mixtures. $d = 100$ mm.

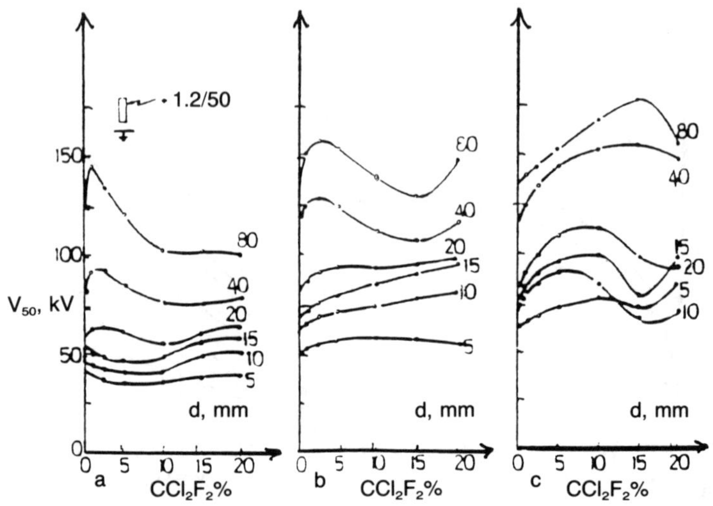

FIGURE 5.4.   Impulse breakdown voltages V_{50} for positive rod/plane gaps in N_2/CCl_2F_2 mixtures. (a) $p = 1$ bar; (b) $p = 3$ bar; (c) $p = 5$ bar.

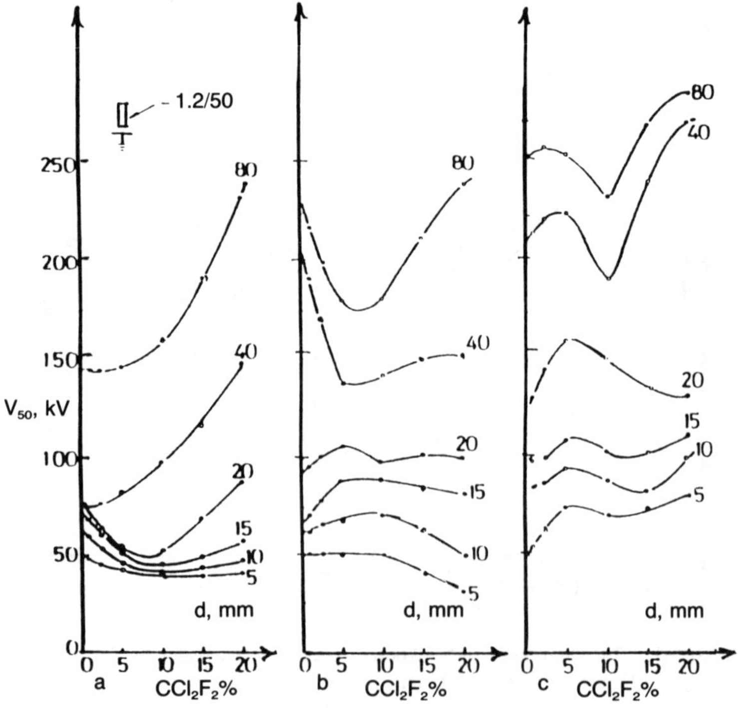

FIGURE 5.5.   Impulse breakdown voltages V_{50} for negative rod/plane gaps in N_2/CCl_2F_2 mixtures. (a) $p = 1$ bar; (b) $p = 3$ bar; (c) $p = 5$ bar.

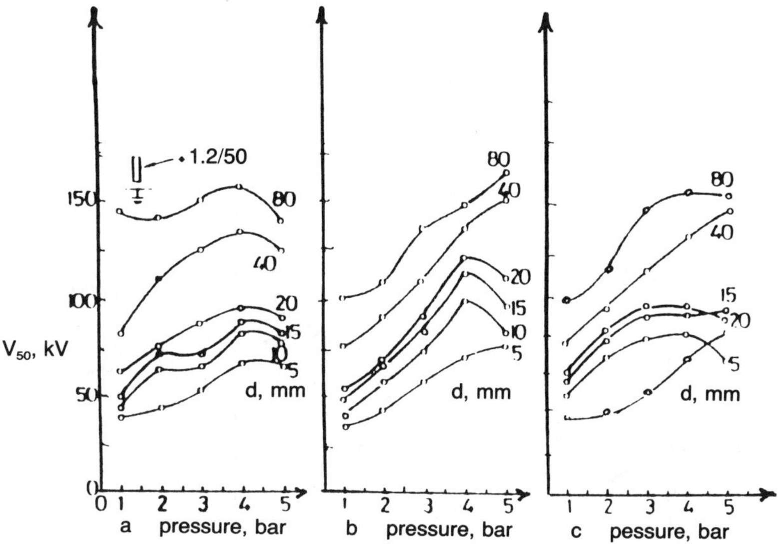

FIGURE 5.6. Impulse breakdown voltage V_{50} against pressure for positive rod/plane gaps in N_2/CCl_2F_2 mixtures. (a) 1% CCl_2F_2; (b) 10% CCl_2F_2; (c) 20% CCl_2F_2.

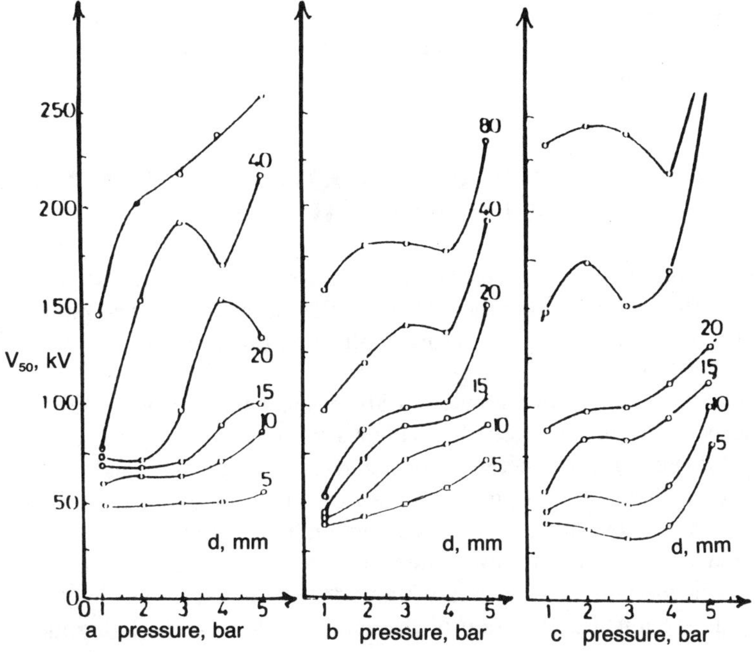

FIGURE 5.7. Impulse breakdown voltage V_{50} against pressure for negative rod/plane gaps in N_2/CCl_2F_2 mixtures. (a) 1% CCl_2F_2; (b) 10% CCl_2F_2; (c) 20% CCl_2F_2.

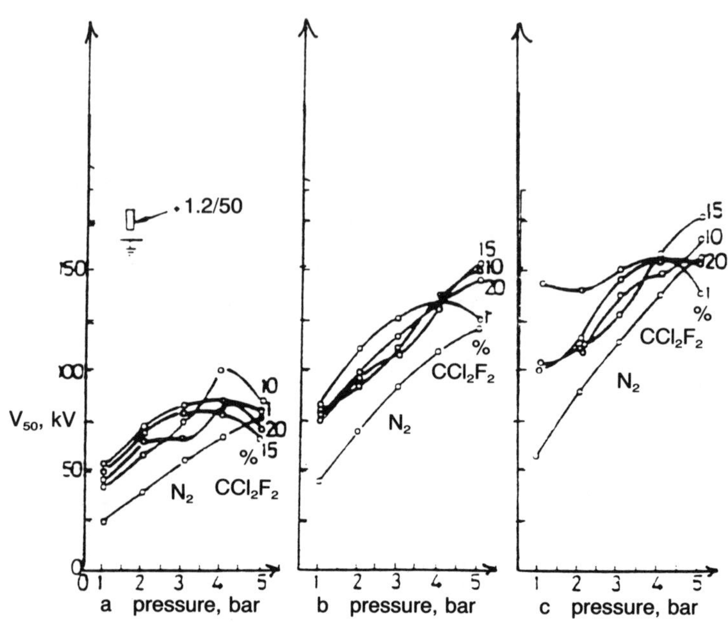

FIGURE 5.8.   Impulse breakdown voltage V_{50} against pressure for positive rod/plane gaps in N_2/CCl_2F_2 mixtures. (a) $d = 10$ mm; (b) $d = 40$ mm; (c) $d = 80$ mm.

5.IV   VOLTAGE BREAKDOWN AND ARCING CHARACTERISTICS OF SF$_6$

Utilization of SF$_6$ gas dielectric in its pure state or through its mixture with oil as an effective insulating medium in cables, transformers and other electric power apparatus is finding increasing application due to various favorable thermal, physical and dielectric properties.

Current research efforts pointed to SF$_6$ gas as having strong electronegative characteristics and clear paternal behavior according to Henry's law with demonstrated linear variations between local conditions of pressure and temperature. Voltage breakdown studies previously carried out showed a strong tendency for the time rate of flow of pure SF$_6$ gas evolution under effective corona, potential stress and specified local temperature coupled with increased acceleration of evolution as the temperature rises without detectable decomposition.

In this section, the intention of the author is that for insulating medium in

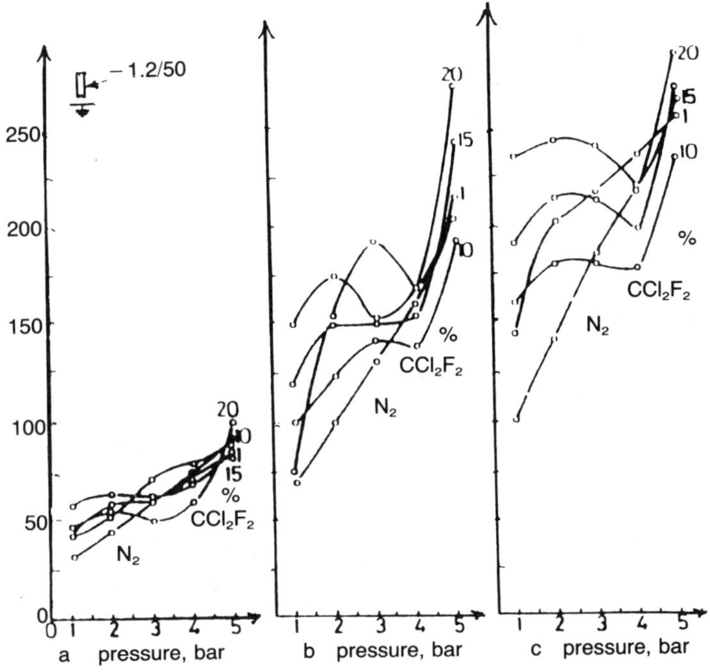

FIGURE 5.9.   Impulse breakdown voltage V_{50} against pressure for negative plane gaps in N_2/CCl_2F_2 mixtures. (a) $d = 10$ mm; (b) $d = 40$ mm; (c) $d = 80$ mm.

power apparatus involving SF_6 gas obeying the paternal behavior according to Henry's law the following analytical information will be established:

1. Solutions for the coefficients of ionizations represented by the first and second Townsend coefficients, the secondary ionization coefficient and the coefficient of photoemission. All those parameters are to be expressed in terms of variable conditions of pressure and temperature, and external electric field.
2. Condition of optimum pressure for arcing and its numerical value.
3. Calculations of sparking potentials under all states of ionizations.
4. Frequency spectrum for the optimum pressure of arcing and the electrons ionization coefficient.

A. IONIZATION COEFFICIENT FOR SF_6

From the work already carried out by Walsh and Kurz, the pressure vs. temperature known as Henry's criterion for SF_6 solution in an enclosed tank

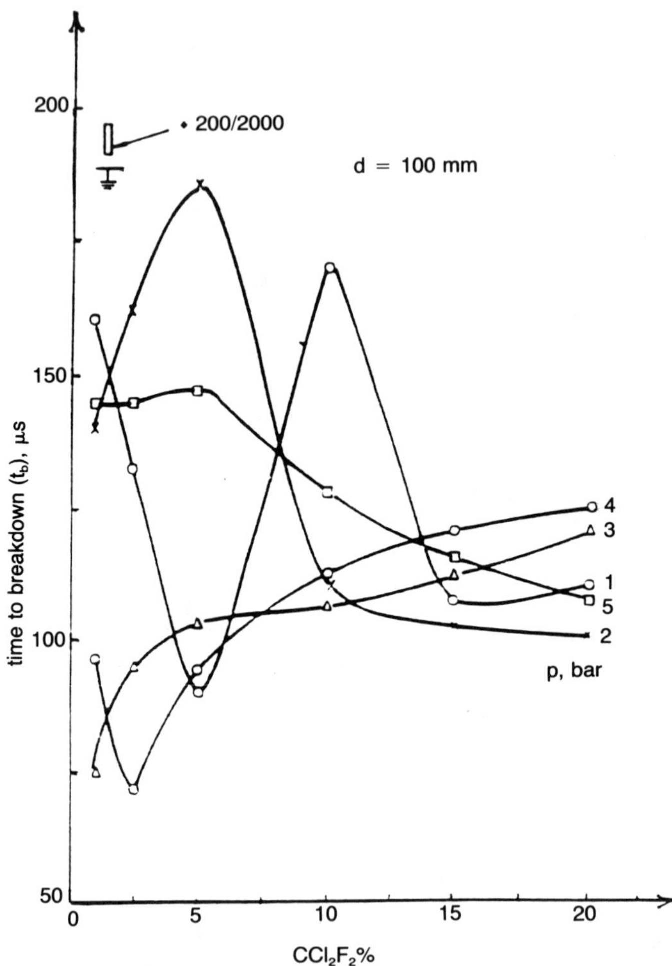

FIGURE 5.10.   Time lag to breakdown for rod/plane gap in N_2/CCl_2F_2 mixtures under switching surges.

with 10% head space and with a pressure of 1.14 atm is shown in Figure 5.1. The linear graphical picture is expressed in the following equation:

$$P = \frac{1}{45}\,(\tau - 10) \tag{5.57}$$

where P is in atm and T in °C.

For controlled ionization, the author would like to discuss the state of ionization by electrons associated with a special list.

The First Townsend Ionization Coefficient. This coefficient designated as τ represents the number of ionizing collisions per cm of path in SF_6 in the direction of an applied external electric field,

$$\tau = \frac{f(EL)}{L} \tag{5.58}$$

where L is the mean free path of electrons in SF_6 and E is the external electric field, and that

$$\alpha_\tau = APe^{-\frac{BP}{E}} \tag{5.59}$$

A and B are functional parameters of pressure, temperature and electric field and the gas ionization potential V_i. From above and Equation (5.57), and also according to the Engel and Steenbeck equation, τ is expressed below:

$$\tau = \frac{600aPV_i}{\sqrt{\pi}2f}\left(1 + \frac{eV_i}{2kTe}\right)e^{-\frac{2\sqrt{2}fV_i}{4\sqrt{\pi}\alpha_{eo}\left(\frac{E}{P}\right)}} \tag{5.60}$$

where V_i = the ionization potential of the gas

$\quad\quad f$ = fractional loss of energy on electron collision

$\quad\alpha_{eo}$ = mfp at 1 mm of Hg

$\quad\quad e$ = electron charge

$\quad\quad a$ = number of ions produced per second per electron in SF_6 with respect to local electron temperature

$\quad\quad T_e$ = electron temperature. $\tag{5.61}$

$\therefore$ From Equations (5.58), (5.59) and (5.60) results

$$A = \left(1 + \frac{eV_i}{2kTe}\right)e - \left(\frac{600\,aV_i}{\sqrt{2}f4\sqrt{\pi}}\right)e\left[45P\ln\frac{\dfrac{45P}{T\text{-}45\text{-}10}}{T\text{-}45P\text{-}10}\right]\cdot$$

$$\left[\frac{2\sqrt{2}fV_i}{4\sqrt{\pi}\alpha_{eo}\dfrac{E}{P}}\right] \tag{5.62}$$

Therefore, with B and A expressed in Equations (5.59) and (5.62), the coefficient τ for SF_6 is solved for explicitly.

Generally the optimum pressure for ionization is dominated by electron collision in SF_6 and hence P_m is expressed as

$$P_m = \overline{E}/\overline{B} \tag{5.63}$$

where B is solved for in Equation (5.61). But with SF_6 following exactly the rule of Henry's law,

$$P_m = \frac{1}{45} \frac{T\text{-}45P\text{-}10}{\ln \dfrac{45P}{T\text{-}10}}$$

$$\text{where } p = \frac{1}{45}(\tau\text{-}10) \tag{5.64}$$

The result of Equation (5.64) is extremely significant because the optimum pressure for field intensified ionization by electrons in SF_6 is indeed independent of the external electric E and temperature.

Ionization by Positive Ions (Second Townsend Coefficient, H).   In general, H is expressed as

$$\eta = Pg\left(\frac{E}{P}\right) \tag{5.65}$$

The current density generated by electrons and positive ions is as below:

$$J = j_o \frac{(\tau - \eta)e^{(\tau - \eta)x}}{\tau - \eta e^{(\tau - \eta)x}} \tag{5.66}$$

Under arcing, the denominator (5.66) is equal to zero.

Solutions for η are obtained as:

$$\eta_1 = \frac{1}{x} \tag{5.67}$$

and

$$\eta_2 = \tau \tag{5.68}$$

where x is any point in the SF_6 container from the cathode, and τ is the first Townsend coefficient.

The Secondary Coefficient of Ionization (γ). This effect is attributed to the phenomenon of ionization by positive ion bombardment of the cathode surface resulting in the release of secondary electrons.

The current density contribution from primary and secondary electron ionization collision is expressed as

$$J = J_o \frac{e^{\tau d}}{1 - \gamma(e^{\tau d} - 1)} \tag{5.69}$$

Under arcing conditions, the denominator of Equation 12 will vanish, and hence,

$$\gamma = 1/e^{\tau d} - 1$$

Solution for τ is already obtained in terms of Equations 3, 4, and 6.

Ionization by photo-emission due to the arrival at the cathode of photons produced in SF_6 gas by electron ionization, represented by the coefficient α, is expressed by the current density J:

$$J = J_o \frac{\tau e^{\tau d}}{\tau - \alpha \xi g e^{(\tau - \mu)d}} \tag{5.70}$$

where μ is the average value of the absorption coefficient for photons in SF_6, g is the geometrical parameter representing the proportion of photons that may reach the cathode, and ξ is the fraction of photons generating electrons at the cathode capable of leaving the surface.

For arcing to occur, the denominator of 14 must vanish, and hence

$$\alpha = \frac{\tau}{\xi g} e^{(\mu - \tau)d} \tag{5.71}$$

and an explicit solution for α is expressed by inserting the value of τ obtained earlier.

B. SPARKING POTENTIAL U_S IN SF_6

Criteria for the enhancement of sparking and the value of sparking potential are expressed below:

a. Due to electron emission by positive ion bombardment, or from Equation (5.59), with

$$\tau d = \ln \frac{1}{\gamma}$$

or

$$\tau = APe^{-\frac{BP}{E}} \text{ and } E = \frac{V_s}{d} \tag{5.72}$$

$$\therefore \quad V_s = \ln \frac{\dfrac{BPd}{APd}}{\ln \dfrac{1}{\gamma}} \tag{5.73}$$

b. Due to emission of positive ions, with

$$\frac{\tau}{\eta} \approx e^{\tau d} \tag{5.74}$$

and

$$V_s = BPd/\ln \frac{AP(1 - \tau_d)}{d} \tag{5.75}$$

A and B solutions for SF_6 were obtained earlier.

c. Due to photoemission from cathode surface, with

$$\alpha \approx \frac{\tau}{\xi g} e^{-\tau d}$$

or

$$\alpha \approx \frac{\tau}{\xi g} (1 - \tau d) \tag{5.76}$$

$$\therefore V_s = ABdP^2/AP - \frac{1 \pm \sqrt{\dfrac{1}{\xi g} - 4\alpha d\xi g}}{2d} \tag{5.77}$$

Again for SF_6, A and B solutions are solved for before as shown in Equations (5.60) and (5.62). Now to look for the minimum value of V_s as a function of the product (Pd), differentiation of V_s with respect to (Pd) could be carried out for each case of those mentioned before and then setting

$$\frac{dV_s}{d(Pd)} = 0$$

An example case is considered for ionization by electrons release from positive ions bombardment, where

$$(Pd) \text{ for } V_{smin} = \frac{2.718}{A} \ln \frac{1}{\gamma} \tag{5.78}$$

Also previous work indicated that the arcing behavior of SF_6 gas mixed with saturated oil as insulating medium in transformers demonstrated clear maximization in arcing energy as well as surge in volume and gas pressure with respect to pure SF_6 gas on one side and SF_6 gas mixed with saturated oil and with arcing impurities on the other. Those analyses proved the occurrence of higher time rate in the volume of SF_6 gas evolution compared to N_2 and C_2F_6 gases under the same arcing conditions.

The section presents another phase of analysis for the control of voltage breakdown and arcing behavior for the dielectrics of SF_6, SF_6 + oil and SF_6 + oil + arcing impurities in cables, transformers and other power apparatus.

Calculations have been carried out to develop closed mathematical relationships for the processes of multi-ionizations in the dielectrics, including solutions for ions generation by electrons and positive ions as well as electron emission by positive ion bombardment in terms of local pressure, temperature and intensity of ionization fields. Results identified conditional criteria for optimum pressure with respect to the level of voltage breakdown for each case of SF_6 purity. Plots have been established for the various coefficients of field-intensified ionization with respect to local pressure and temperature, pointing to clear identification of breakdown conditions. Results also expressed local constraints for the minimum level of voltage breakdown with respect to pressure and temperature as well as ionization field.

For the arcing behavior, optimization study is required for the development of a mathematical model for arcing energy spectra with respect to a wide range of volume, pressure and temperature variations for the three cases mentioned earlier, namely the dielectrics of pure SF_6, SF_6 + oil, and SF_6 + oil + arcing impurities. The developed energy spectrum represented a clear and reliable generalization of arcing energy distribution across the dielectric volume, local pressure and temperature with identification of resonance for the peak arcing energy.

Conditions of arcing instabilities in SF_6 dielectric, dynamic model in the frequency domain are established for arcing energy under different specified local levels of pressure, temperature and the state of solubility in oil, as well as the time rate of gas evolution in the case of SF_6 + oil for any defined period of arcing time.

Voltage breakdown and arcing characteristics presented in this paper point out two important reliable pieces of information for controlling the states of ionization: the rate of absorption and/or gas evolution as well as local conditions of pressure and minimizing the hot spot temperature in SF_6 gas insulated cables, transformers and other electric power equipment.

Regarding Section IV, the author would like to summarize the following conclusions:

1. With the previously proven fact that SF_6 obeys the pattern of Henry's law, where its local pressure follows a linear change with temperature, the following mathematical solutions have been established in this paper:
   a. The ionization coefficient under an intensified external electric field effect by primary electrons, known as the First Townsend Coefficient.
   b. The ionization coefficient under an intensified external electric field effect, by primary positive ions, known as the Second Townsend Coefficient (for arcing condition).
   c. The ionization coefficient for secondary electrons generated by positive ion bombardment at the cathode surface, under arcing conditions.
   d. The ionization coefficient by photoemission generated by electrons and positive ion collisions under arcing conditions.

2. Numerical value for the optimum pressure under which field intensified ionization by electrons will occur is established at 2.22% of 1 atm. The value is very significant since it is independent of the field strength E and temperature.

3. Solutions for the sparking potential are secured for SF_6 under the influence of ionization by secondary electrons and photons, as well as by positive ions.

4. Dynamic solutions in the complex frequency domain are obtained for the ionization process in SF_6, gassed by primary electron collisions, and then frequency spectrum is derived. The absolute value of the frequency spectrum is represented as an even function while the angular spectrum is an odd function with respect to a singular value of 244 rad/s.

5. In reference to previous experimental work done by Walsh and Kurz, pressure for arcing in SF_6 with saturated oil is $^1/_3$ of the optimum pressure in SF_6 gas alone.

5.V BREAKDOWN IN PLASMA SHEATH

Time lag for gas-filled plasma breakdown varies systematically with respect to increasing levels of over-voltage, an indication closely connected with the plasma properties. Variation of ion densities near the cathode electrode will generate adverse regions of space-charge effects that will alter the final level of voltage breakdown. And with further increase of ionization densities due to space charge effects, the gas-filled plasma will transform itself to a sheath. Research carried out by R. M. Clements, Y. Oved and P. R. Smy identified two levels of voltage breakdown. The first breakdown is sheath induced and can vary over a wide range of bulk plasma conditions remaining as a glow discharge over the whole area of the electrode. At higher voltage level, an arc discharge will commence generated by field-intensified ionization. Lowke and Davies showed that sheath ionization will lead to an increase in electrode current, coupled with increase in sheath thickness for gradual step-up of applied voltage. The processes of ionization preceding voltage breakdown encompass secondary emission in addition to the effect of field-intensified ionization. At breakdown, the sheath collapses and the current increases by a large factor.

Applications of voltage breakdown in plasma-filled gap with eventual formation of high pressure sheath include that in a circuit breaker and other mechanisms of high density plasmas.

A. PLASMA SHEATH THEORY

Work carried out by Clements et al.[2] described the formation of a planar sheath by a flow of current from a perfectly conducting plasma to a nonemitting cathode. Prevailing pressure of such continuum practically existing in circuit-breakers is of the order of multi-atmospheres, where ion movement in the form of current density is controlled by their mobility i as expressed below:

$$J_i = \frac{9}{8} \frac{\mu_i \epsilon_o V^2}{X_s^3} \tag{5.79}$$

where J_i is the ion current density through the sheath, V is the voltage across the sheath, ϵ_o is the permittivity of fee space, and X_s is the sheath thickness.

In the absence of an applied electric field, J_i will flow through the plasma column on sheath by effects of diffusion and/or convection. Since the cathode is at a state of depressed potential with respect to the rest of the plasma column, the sheath will expand and the residual ion current J_i will also increase. Recognizing the effect of the secondary emission coefficient (α) which is of the order of 10^{-2}, additional increment in electron current will result. Increase in potential across the sheath will enhance field-intensified ionization, eventually becoming the dominant source for current flow. If the current density at the cathode is J_o, then the emitted electron current density will be j_o and the resultant electron current at the sheath edge will be $\gamma j_o \, e^{\alpha x_s}$, where α is the number of ionizing collisons/unitpath length along the field direction, known as the First Townsend Coefficient.

Therefore, the combined current density involving ions and electrons within the sheath will be equal to $-\alpha\gamma\rho[\exp(\alpha x) - 1]$ which is approximately equal to $-\gamma Je \exp[\alpha xs]$. Consequently,

$$J_i = J_o[1 - \gamma\epsilon^{\alpha x_s}] \tag{5.80}$$

or

$$J_o = \frac{J_i}{1 - \gamma e^{\alpha x_s}} \tag{5.81}$$

If E is the electric field across the sheath thickness,

$$= \frac{V}{x_s} \tag{5.82}$$

However, at low values of (αx_s), x_s varies with the level of the prevailing electric field in the sheath according to the theory of mobility dominated sheath model. As the strength of ionization increases with the field, the point will be reached at which $\gamma e^{\alpha x_s} \rightarrow$ unity. Therefore, x_s will approach the limit of $\frac{1}{\alpha}$ which will be less than an order of magnitude greater than unity. At low electric fields, x_s will increase with E^2 and at some point $(x_s) \rightarrow 1$ will pass through a maximum, and the further increase in the field will decrease according to the relation that αx_s is varying as $\ell n \frac{1}{\gamma}$. Plotting x_s against E for Equation (5.83) could be patterned according to Equation (5.59) as shown in Figure 5.11.

An increase in the electric field will produce an increase in the sheath thickness x_s to a certain limit after which it starts to contract, the sheath voltage falls off and, under constant voltage, the sheath will collapse immediately after the maximum thickness has been reached. Practically, it is very difficult to stabilize a constant voltage across the sheath, hence the state of equilibrium after the sheath collapse will be determined by the external load.

If a low impedance source is available, the sheath theory predicts that the planar sheath will collapse to zero and the electric current will surge to infinity. Kikoshika and Smy had shown that cathode plasma arcs do occur when the electric field is of the order of 3×10^9 V/m.

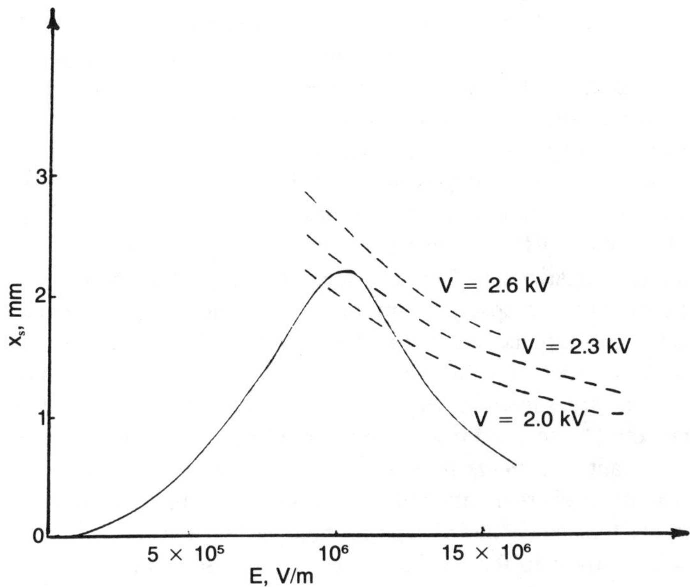

FIGURE 5.11. Calculated sheath thickness x_s against electric field E characteristics for I = 10 A/m^2 and ionization density $\simeq 10^{15}$ m^{-3} (solid line), as well as constant voltage curves for V = 2.0, 2.3 and 2.6 kV (broken lines).

The preceding discussion points to two stages of voltage breakdown: the first is accompanied by expansion of the sheath and then with multi-steps of contractions associated with an increase in current, the equivalent of a glow-to-arc transition with the possibility of another sharp increase in current. The second breakdown is characterized by Mikoshiba and Smy to occur when the sheath electric field will exceed 10^7-10^8 V/m. It should be noted that (g) and (α) are functions of temperature and thus their values must correspond to the appropriate temperature whether in the sheath itself or where the sheath is an integral part of the thermal boundary layers.

B. EXPERIMENTAL PICTURE

The plasma used by Clements, Oved and Smy in their research was produced by burning propane gas with air in a standard meker burner of 3.7 cm diameter. Ionization density was in the range of 10^{15} to 10^{10}/m^3 and was controlled by seeding the flame with a mist of potassium hydroxide.

Spherical platinum electrodes inserted in the flame were used as discharge electrodes. Current-voltage characteristics were measured using the two spherical electrodes inserted in the flame 2 cm above the burner. The separation (D) between the two electrodes was varied, with one electrode charged to $(-V)$ volts, while the second electrode was grounded and the burner was allowed to float. The negative voltage applied at the cathode was generated by a standard (RX) circuit and triggered by a spark gap.

As the applied voltage was increased, the first breakdown occurred accompanied by an increase in current from few μamp to several mamp, while the voltage level of the first breakdown remained stable. Further increase in voltage drew more current up to the inception of the second breakdown, where an arc was initiated between the two electrodes with clear indication that plasma properties became immaterial. It was also noted that between the first and second breakdowns, the plasma impedance was lower by 50% than at first breakdown. Figure 5.12 shows (VI) characteristics of this prepared plasma. Also, Figure 5.13 illustrates variations of first and second voltage breakdowns with respect to ionization density. A special cylindrical probe was used by Clement et al. to sweep the horizontal plane containing the two electrodes to map the equipotential lines for voltages above the first breakdown. The equipotential lines obtained in this experiment are shown in Figure 5.14. Plasma conductivity after the first breakdown could be secured from the measured equipotential curves on the basis that it is constant in a plane normal to the axis joining the centers of the two spheres. The researchers mentioned also identified the presence of electron heating in the plasma, since their findings indicate that the ratio of electron temperature with current to that without current increased from one to a factor of two as the current was increased from 5 to 20 m-amp.

This section summarizes research efforts conducted by Clements, Oved, and Smy. Results permit observation of plasma sheath at high voltages and adverse ionic densities. The outer layer of the sheath is seen as a virtual anode, and precise values for sheath thickness and voltage at which sheath collapse takes place cannot be obtained without much error. This study clearly presented a new model for the transition of glow discharge throughout the plasma after the first breakdown and then to arc discharge after the second breakdown. The model also indicates a process of plasma expansion followed by contraction or collapse to a thickness determined by the external load. Measurement of the impedance of a plasma filled gap indicates that reduction by a factor of as much as 100 occurred due to the effect of external loading. The ionization process before and after the first breakdown involved a field intensification effect by electron secondary emission due to electron bombardment as well as ionization by positive ions and their secondary emission.

5.VI SOLVED EXAMPLES

A. Given a helium ionic cloud of total concentration with He$^+$ of cluster density 10^{20} ions/cluster, express the e energy in joules carried out per cluster in terms of radial vector acceleration, $\partial \overline{D}_\rho / \partial_\rho$ and $\partial \overline{D}_\rho / \partial_\rho^2$.

Solution

Let the mass per cluster $= $ Mkg, with W_{He}^+ as the energy/cluster

$$= \frac{1}{2} M |\overline{u}_\rho|^2 \text{ joules} \tag{A1}$$

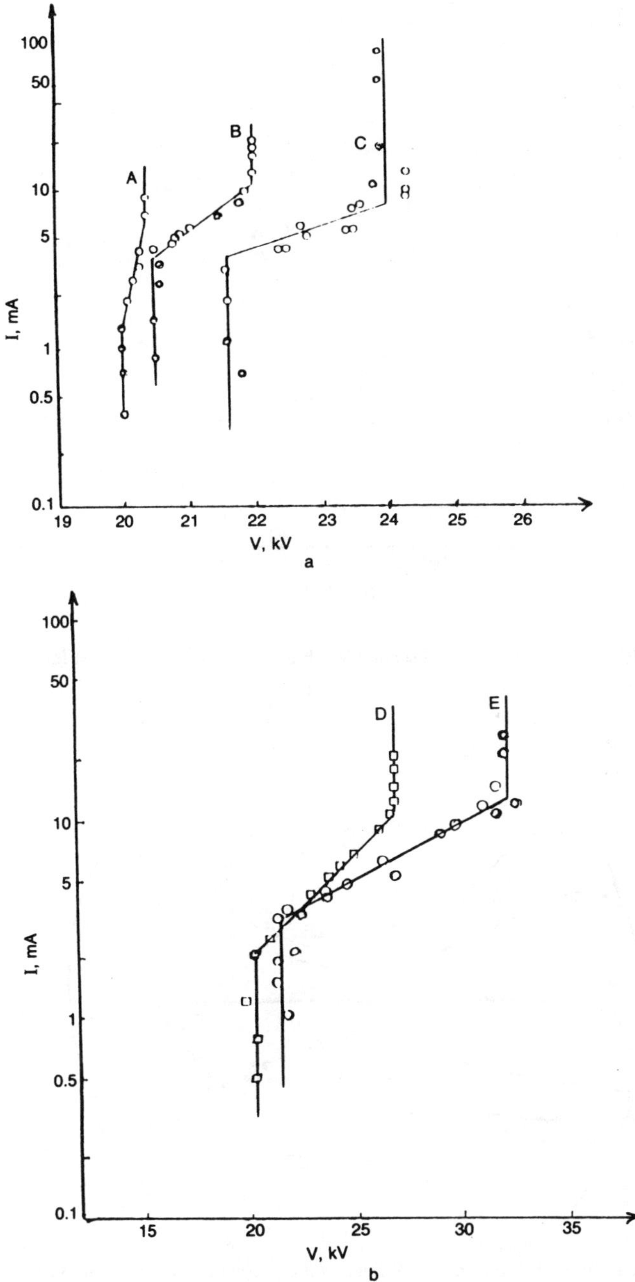

FIGURE 5.12. Current voltage characteristics for different separations. The distance (D) between the centers of the spheres is: (a) A = 1.0 cm, B = 1.3 cm, C = 1.5 cm; (b) D = 1.7 cm, E = 2.0 cm. I_o, the current just before the first breakdown, $\approx 10^{-6}$ A.

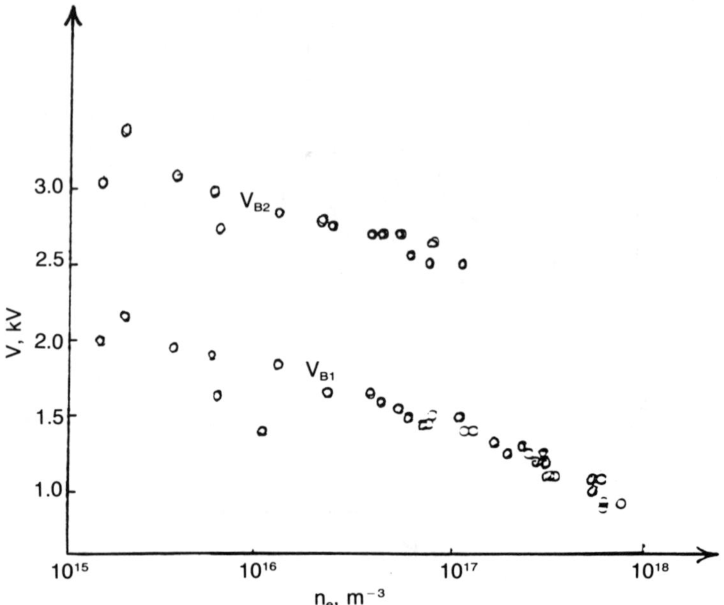

FIGURE 5.13.   First V_{B_1} and second V_{B_2} breakdown voltages against ionization density n_e.

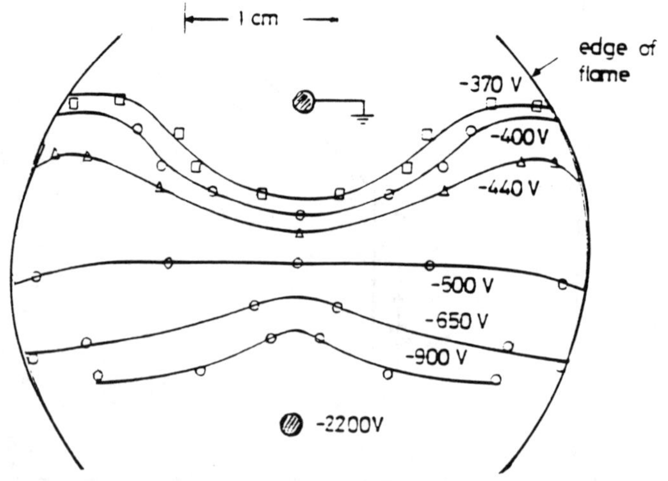

FIGURE 5.14.   Equipotential lines in the plasma for a voltage above the first breakdown.

From Equation (5.44),

$$\bar{u}_\rho = (\delta_o'(t) - \bar{a}_\rho) \frac{\partial \overline{D}_\rho / f_\rho}{\partial^2 \overline{D}_\rho / {}^2_\rho} \tag{A2}$$

or since

$$\delta_o'(t) = \frac{\partial \bar{J}\Delta}{\partial t}$$

$$\therefore \quad \bar{u}_\rho = \frac{\dfrac{\partial \bar{J}_v}{\partial t} - \bar{a}_\rho}{\dfrac{\partial \overline{D}_\rho}{\partial_\rho} \bigg/ \dfrac{\partial^2 \overline{D}_\rho}{\partial^2_\rho}} \quad \text{m/s} \tag{A3}$$

$$\therefore \quad W_{He}^+ = \frac{1}{2} M \left| \frac{\dfrac{fJv}{\partial t} - \bar{\alpha}_\rho}{\dfrac{\partial D_\rho}{\partial_\rho} \bigg/ \dfrac{\partial^2 D_\rho}{\partial^2_\rho}} \right|^2 \quad \text{joules per cluster} \tag{A4}$$

It is of interest to find out some space points at which W_{He}^+ is maximum. From Equation (A4) above, W_{We}^+ is maximum for

$$\frac{\partial \overline{D}_\rho}{\partial_\rho} \to 0 \tag{A5}$$

If the gaseous continuum is simple, implying

$$\overline{D}_\rho = \epsilon \overline{E}_\rho$$

$\dfrac{\partial \overline{E}_\rho}{\partial_\rho}$ is given by Equation (5.54)

$$\therefore \quad \text{for } \frac{\partial E_\rho}{\partial_\rho} \to 0 \tag{A6}$$

case points are:

$$R_v \to 0 \tag{A7}$$

$$e^{-j(\rho - R_v)} \to 0 \tag{A8}$$

Condition (A7) indicates location just at the core of the released convective stroke. Condition (A8) implies a location at the outer cylindrical boundary of the convective stroke.

B. A helium cloud exhausted from the diverter of a Tokamak fusion reactor has the following parametric values:

$$\alpha_1 = 10^5/\text{s}$$

$$\alpha_2 = 0.1_{\alpha 1}$$

$$\beta = -1\alpha_2$$

$$\epsilon = 0$$

Also given the induced electric field vector $\overline{E}$,

$$\overline{E} = (\hat{a}_\rho + \hat{a}_\alpha + \hat{a}_\tau)10 \text{ kV/cm}$$

$$\sigma = \sigma_H = \sigma_t = 1000 \text{ mho/meter}$$

Calculate the local microscopic velocities of the charge carriers. Assume the helium cloud contains ion clusters each carrying a charge of 10^{20} coulombs.

Solution

Using Equations (5.1) and (5.2) to calculate n_1 and n_2:

$$n_1 = \frac{1.1 \times 10^{10}}{(-10^5)(-2 \times 10^5 - 0.2 \times 10^5)(2 \times 10^5)(2 \times 10 - 0.1 \times 10^5)10^5}$$

$$= 0.277$$

$$n_2 = \frac{1.1 \times 10^5}{-1.8 \times 10^5} - \frac{2.1 \times 10^5}{-1.8 \times 10^5}(2.1 \times 10^{10})/$$

$$(-2.2 \times 10^5)10^5 + 10^5(2.1 \times 10^5)$$

$$n_2 = -25.012/\text{S} \tag{B1}$$

From Equations (5.20), (5.21) and (5.22),

$$\overline{v}_\rho = 0$$

$$v_\alpha = \frac{1000(2.10^6)}{-1.9 \times 10^{-19}(0.277 - 50.024) \times 10^{20}}$$

$$= 2.1 \times 10^5 \text{ m/s}$$

$$u_\tau = \frac{1000 \times 10^6}{-1.9 \times 49.947 \times 10^{-19} \times 10^{20}}$$

$$= 1.05 \times 10^5 \text{ m/s} \tag{B2}$$

C. The set of Equations (5.20), (5.21) and (5.22) describes the velocities in cylindrical coordinates of He^+ and He^{++} ions. Indicate the state of compressibility of a He ionic cloud.

Solution

In hydrodynamics, compressibility or otherwise is based on the criterion

$$\overline{\nabla} \cdot \overline{v} = 0 \quad \text{incompressible} \tag{C1}$$

where $\overline{v}$ is the local velocity vector of helium ionic cloud.

Also in Equations (5.20) to (5.22), n_1 and n_2 are independent of position. However $\overline{E}_\rho$, $\overline{E}_\alpha$ and $\overline{E}_\tau$ are space dependent according to Equations (2.71) to (2.73).

$$\overline{\nabla}\overline{v} = \frac{1}{2}\frac{\partial}{\partial_\rho}(\rho v_\rho) + \frac{1}{\rho}\frac{\partial v_\alpha}{\partial_\alpha} + \frac{\partial v_\tau}{\partial_\tau} \tag{C2}$$

$$\rho v_\rho = \rho\left[\frac{\sigma_\tau \overline{E}_\rho - \sigma_H \overline{E}_\alpha}{n_1 e^+ + 2n_2 e^+}\right] \tag{C3}$$

$$\rho v_\rho = \rho\,\frac{\rho(t)}{n_1 e^+ + 2n_2 e^+}\left[\sigma_\tau\sqrt{\frac{j}{\pi}}\,\frac{\rho\sqrt{\rho - R}}{R\hat{y}}\sum_{n=o}^{\infty}\frac{j^n e^{-j(\rho - R)}}{2^n n!} - \left[\frac{\rho - R}{R^n}\right.\right.$$

$$\left. - (\rho - R)^{n-1} - \frac{(\rho - R)^{n-\frac{3}{2}}}{\sqrt{2n}} + \frac{j2(n-1)!}{\sqrt{2}}\right] -$$

$$\sigma_H\rho\left[\frac{j}{2\hat{y}}\sqrt{\frac{2j}{\pi(\rho - R)}}\,e^{-j(\rho - R)}\frac{j^n(\rho - R)^n}{2^n n!}\right]\right] \tag{C4}$$

where

$$\pounds(t) = [A_c,ut - t_1) + jAc,utV_1(t - t_1] \tag{C5}$$

R is first R_c for conduction stroke column radius and then R_v for convective coaxial cylindrical strip.

$$u_\alpha = \frac{\sigma_H E_\rho + \sigma_\tau E_\alpha}{n_1 e^+ + 2n_2 e^+} \tag{C6}$$

E_o, however is independent of α, and E_α is also dependent of α according to Equation (2.72).

$$\therefore \quad \frac{1}{\rho}\frac{\partial u_\alpha}{\partial_\alpha} = 0 \tag{C7}$$

$$u_\alpha = \frac{\rho \overline{E}_\tau}{n_1 e^+ + 2n_2 e^+} \tag{C8}$$

$$\frac{\partial v_\tau}{\partial_\tau} = 0 \tag{C9}$$

It appears now that compressibility will be determined by the process of Equation (C2), where $\overline{\nabla} \cdot \overline{v} = 0$ or not,

$$\overline{\nabla} \cdot \overline{v} = \frac{1}{\rho}\frac{\partial}{\partial}(\rho v_\rho) \tag{C10}$$

The author had carried out the process of $\dfrac{\partial_\rho v_\rho}{\partial_\rho}$ and determined that $\overline{\nabla} \cdot \overline{v} = 0$. Thus, the helium ionic gas as an anisotropic cloud is compressible.

D. Prove that optimum pressure of arcing for SF_6 under controlled ionization is as indicated in Equation (5.64) supplemented by Henry's law that states $P = \dfrac{1}{45}(t - 10)$.

Solution

Since

$$t = APe^{-\frac{B_\rho}{E}} \tag{D1}$$

and

$$B = 45E\,\frac{\ln \dfrac{45P}{T - 10}}{T - 45P - 10} \tag{D2}$$

then for ionization dominated by electrons, optimum pressure is expressed by

$$P_m = E/B \tag{D3}$$

From Equations (D1), (D2) and (D3)

$$P_m = \frac{1}{45} \left[\frac{T - 45P - 10}{\ln \dfrac{45}{T - 10}} \right] \tag{D4}$$

but

$$P = \frac{1}{45} (T - 10)$$

$$P_m = \left[\frac{T - T + 10 - 10}{\ln \dfrac{T - 10}{T - 10}} \right] \frac{1}{45}$$

$$= \left(\frac{0}{0} \right) \frac{1}{45} \rightarrow \text{limit} \rightarrow \frac{1}{45} \tag{D5}$$

According to limits theorem

In this case a limiting process has to be applied on the equation by taking the differentiation of the numerator and denominator separately with respect to T. However, it had been determined that a successive limiting process resulted in obtaining the fractional form of 0/0 which implies an undetermined situation.

In this case, where ionization is controlled mainly by electrons, the author concludes that optimum pressure P_m is as expressed in Equation (D4) subject to the condition that

$$P = \frac{1}{45} (T - 10)$$

E. Refer to Equation (5.79) expressing the ion current density through plasma sheath of thickness x_s. Obtain a relationship for the average velocity of ions through the plasma sheath in terms of J_i and other relevant plasma elements.

Solution

From Equation (5.79)

$$J_i = \frac{9}{8} \frac{\mu i \epsilon_o V^2}{x_s^3} \tag{E1}$$

the mobility

$$\mu i = \frac{\bar{v}_d}{E} \tag{E2}$$

where $\bar{v}_d$ is the ion drift velocity, and

$$\bar{v}_d = \frac{Q^+ L}{M_i \overline{C}} \tag{E3}$$

where Q^+ is the charge carried by the cluster of positive ions, L is the mean free path, $\overline{C}$ is the average ions velocity, and M_i is the mass of an ion cluster provided that the plasma is a homogeneous continuum.

Therefore, from Equations (E1) and (E2)

$$J_i = \frac{9}{8} \left[\frac{\bar{v}_d \epsilon_o V^2}{E x_s^3} \right]$$

$$= \left(\frac{9}{8}\right) \frac{\bar{v}_d \epsilon_o V^2}{x_s^4} \tag{E4}$$

where

$$E = \frac{V}{x_s}$$

Then, from Equation (E3)

$$\overline{C} = \frac{8 J_i M_i x_s}{9 \epsilon_o V^3 Q^+ L} \text{ m/s} \tag{E5}$$

5.VII PROBLEMS

1. A helium cloud exhausted from the diverter of a Tokamak fusion reactor has the following parametric values: $\alpha_1 = 20_{\alpha 2}$; $\alpha_2 = 10^4$ ions/s; $\beta = -4_{\alpha 2}$; $\epsilon = 0.5$. Calculate the concentration of He^+ and He^{++}.

2. The helium cloud specified in problem 1 is subjected to an electric field generated by a lightning surge given by:

$$\overline{E} = \left(\hat{a}_\rho \frac{1}{\rho} + \hat{a}_\alpha \, \alpha + \frac{a_\tau}{\tau} \right) 20 \text{ kV/cm}$$

ρ, α and τ are in cylindrical coordinates. Calculate the local microscopic velocities of He^+ and He^{++}. Assume helium ion cluster for He^+ carries a charge of 10^{20} e^+ and for He^{++} 2×10^{22} e^+.

3. A helium gas having He^+ and He^{++} concentrations has values for the following parameters: $\alpha_1 = 10^8$/s; $\alpha_2 = 10^6$/s; $\beta = 0$. Calculate a set for n_1 and n_2 under these conditions, and plot each against $\epsilon = 0, 0.1, 0.2, 0.3, 0.4, 0.5, 0.6, 0.7, 0.8, 0.9$ and 1.0.

4. A helium gas having He^+ and He^{++} concentrations possesses values for the following parameters: $N_2 = 22.5$; $N_2/N_1 = 0.29$; $\beta = 0$. Calculate values for and α_1, α_2 knowing that $0 \leqslant \epsilon \leqslant$ where $\epsilon = 0$ implies perfect diverter function, and $\epsilon = 1$ no diverter role.

5. From data and solutions of problems 1 and 2, and given the gas electrical conductivity as tensor of rank 2 given by

$$\overline{\overline{\sigma}} = \begin{bmatrix} 400 & -300 & 0 \\ -300 & 400 & 0 \\ 0 & 0 & 500 \end{bmatrix} \text{mho/M}$$

solve for the induced current density vector in the helium gas.

6. Using the solution for the induced current density vector obtained in problem 5, solve for the induced magnetic field intensity vector $\overline{H}$.

7. From the solution for the field $\overline{H}$ in problem 6 and considering that helium gaseous media are linear, obtain a solution for the induced magnetic induction.

8. A helium gas dominated by He^+ cluster ions carries a charge of $10^{20}\ e^+$ and mass of $10^{10}\ m_e$. The convective current density is a step function of magnitude 10 Mamp and if the induced electric field vector $\overline{E}_\rho$ is given by ρ

$$\overline{E}_\rho = \hat{k}_\rho \frac{2.5 \times 10^6}{\rho^2 + 1} \text{ V/m}$$

where $\hat{k}_\rho$ is a unit vector. Calculate the acceleration vector $\hat{\alpha}_\rho$ in m/s² if the energy gained for each cluster = 10 joules.

9. Regarding problem 8, if the electric field vector is given by

$$\overline{E} = \hat{k}_\rho \frac{3 \times 10^6}{\rho + 1 \text{ V/m}} + \hat{k}_\alpha (3 \times 10^6) \frac{\sin \alpha n}{\pi} + \hat{k}_\gamma \frac{3 \times 10^6}{\tau^2}$$

$n = 0,1,2,3\ldots$

calculate the acceleration vector if the energy carried by each cluster = 10 joules.

10. Referring to Equations (5.54) and (5.55), calculate the acceleration vector $\overline{a}_\rho$ at $\rho = R_v$ and if $\partial \overline{J}/\partial t = 2 \times 10^6\ \delta'_o(t)$. Consider the medium of lightning impact as a simple continuum.

11. Repeat problem 10 if the medium of lightning impact area is linear, and $J = A_v t$ for $0 < t < t_1$ or $J = A_v(t - t)_1$ for $t > t_1$, independent of r, θ and z.

12. Repeat problem 10 if the medium of lightning impact area is an anisotropic reflecting tensorial mathematical form of ϵ and μ, the permittivity and permeability parameters respectively, and $\bar{J}_v = A_v U_1(t)$.

13. Refer to Example C where the value of $\overline{\nabla} \cdot \overline{v}$ determines compressibility of gaseous clouds under the impact of lightning stroke. Carry out systematic calculation to determine the result of $\partial_\rho v_\rho / \partial_\rho$.

14. Refer to Example D and in particular Equation (D5). Examine the possibility that optimum pressure in SF_6 may approach in the limit $\dfrac{1}{45}$ atm.

15. Referring to Figure 5.2a, make a plot for V_{50} in kV with respect to pressure in bar for a separation of electrodes at 40, 60, 80 and 100 mm. Examine the possibility of deriving an empirical equation for V_{50}, v_s, ρ at a certain separation.

16. Referring to Figure 5.2b, obtain a plot for V_{50} in kV with respect to separation between the two electrodes in mm, when operation pressures are at 2, 3, 4 and 5 bar. Examine equation for V_{50}, v_s, d at a certain pressure.

17. Referring to Figure 5.3a, obtain a plot for V_{50} i kV with respect to pressure (in bar) at CCl_2F_2 concentrations of 5%, 10%, 15% and 20%. Examine the possibility of deriving an empirical equation for V_{50}, v_s pressure at a certain concentration for CCl_2F_2.

18. Refer to Figure 5.3b, obtain a plot for V_{50} in kV with respect to a concentration of CCl_2F_2 at operating pressures of 1, 2, 3, 4 and 5 bar. Examine the possibility of deriving an empirical equation for V_{50}, v_s concentration at a certain pressure at parameter.

19. Referring to Figure 5.4a, obtain a plot for V_{50} in kV with respect to electrode separation (d) at CCl_2F_2 concentrations of 5%, 10%, 15%, and 20%. Examine the possibility of deriving an empirical equation for V_{50}, v_s, d at a certain parametric concentration of CCl_2F_2.

20. Repeat problem 19 with respect to Figure 5.4b.

21. Repeat problem 19 with respect to Figure 5.4c.

22. Repeat problem 19 with respect to Figure 5.5a.

23. Repeat problem 19 with respect to Figure 5.5b.

24. Repeat problem 19 with respect to Figure 5.5c.

25. Referring to Figure 5.6a, obtain a plot for V_{50} in kV with respect to electrode separation (d) at gas pressures of 1, 2, 3, 4 and 5 bar. Examine the possibility of deriving an empirical equation for V_{50}, d with gas pressure as constant parameter.

26. Repeat problem 25 with respect to Figure 5.6b.

27. Repeat problem 25 with respect to Figure 5.6c.

28. Referring to Figure 5.7a, obtain a plot for V_{50} with respect to separation (d) between electrodes with gaseous pressure as constant parameter at 1, 2, 3, 4 and 5 bar. Examine the possibility of deriving an empirical equation for V_{50}, v_s, d with percentage pressure as constant pressure.

29. Repeat problem 28 with respect to Figure 5.7b.
30. Repeat problem 29 with respect to Figure 5.7c.
31. Referring to Figure 5.9b, obtain a plot for V_{50} with respect to percentage concentration of CCl_2F_2 with the gas pressure as constant parameter at 1, 2, 3, 4 and 5 bar. Examine the possibility of securing an empirical equation for V_{50}, v_s percentage concentration with time pressure as a constant parameter.
32. Referring to Figure 5.10, obtain a plot for the time lag to breakdown with respect to gas pressure with the percentage concentration of CCl_2F_2 as constant parameter.
33. Equation (5.83) shows expression of the ion current J_i as a function of position with the cross-section of plasma sheath. Using the well known Poisson's equation, obtain a solution for the potential and the electrode field as a function of X.
34. Referring to Figure 5.12a,b, obtain a plot for sheath current with respect to separation between the two spheres, with sheath voltage as constant parameter at $V_{sh} = 20, 21, 24, 25, 30$ and 35 kV. Examine the possibility of establishing an empirical equation for (d) as sheath voltage for constant parameter.

PRINCIPAL LIST OF SYMBOLS

$\bar{J}$	current density vector
$\bar{\bar{\sigma}}$	electrical conductivity tensor
$\bar{E}$	electrical field intensity vector
μ_o	magnetic permeability for free space
$\bar{J}_v$	convective current density vector
$\bar{J}_c$	conductive current density vector
$\bar{D}$	electric displacement vector
$\bar{a}$	acceleration vector
$\bar{\bar{\epsilon}}$	electric permittivity tensor

REFERENCES

1. **Babu, D. V. Lakshminarasimha, C. S., and Naidu, M. S.,** Lightning and switching impulse breakdown of rod/plane gaps in nitrogen and nitrogen/freon (CCL_2F_2) mixtures, *IEE Proc.,* 130(3), 134, 1983.
2. **Clements, R. M., Oved, Y., and Smy, P.R.,** High-pressure sheath to glow transition in a plasma-filled gap, *IEE Proc.,* 127(7), 447, 1980.
3. **Chugh, M. K. and Denno, K.,** Application of SF_6 Gas in Power Distribution and Transmission Systems, M.Sc. degree project, New Jersey Institute of Technology, 1980.
4. **Cobine, J. D.,** *Gaseous Conductors,* Dover Publ., New York, 1941, 1958.

5. **Denno, K.,** Voltage Breakdown and Arcing Characteristics of SF_6 Dielectric in Power Apparatus, IEEE, Canadian Communications and Power Conference Record, Montreal, Canada, October 1990, pp. 22-25.

6. **Denno, K.,** Modes of Acceleration at the Anisotropic Gaseous Continuum Due to Lightning Convective Surge, Record of 3rd Conference on Electrostatics, Cracow, Poland, September 1985, pp. 23-25.

7. **Denno, K.,** Stages of Ionization and Voltage Breakdown at Gaseous Continuum Produced by Lightning Surge, Conference Record of 1984 IEEE International Symposium on Electrical Insulation, Montreal, Canada, June 1984, pp. 226-228.

8. **Garzon, R. D.,** Rate of change of voltage and current as functions of pressure and nozzle area in breakers using SF_6, *IEEE Trans. Power Apparatus and Systems,* 95(5), Oct. 1976.

9. **Strasser, H., Schmidt, K. D., and Hogg, P.,** Effects of arcs in enclosures filled with SF_6, *IEEE Trans. Power Apparatus and Systems,* 94(3), June 1975.

10. **Walsh, E. J. and Kurz, R. A.,** Investigation of SF_6/Oil and C_2F_6/Oil as Transformer Dielectric Media, proceedings of the 14th Electrical/Electronics Insulation Conference, Oct. 1979.

11. **Harrington, R. F.,** *Time-Harmonic Electromagnetic Fields,* McGraw-Hill, New York, 1961.

Chapter 6

TRANSFORMER BEHAVIOR UNDER LIGHTNING SURGE

6.I ELECTROMAGNETIC FIELD MODEL

Electromagnetic characterizations for power transformers have been in continuous formulation for a long period of intensive research because of the extremely important role such static inductive devices occupy in a wide scope of applications. Calculation of electromagnetic inductive phenomenon within the multi-winding layers of the power transformer for the purpose of formulating proper but approximate mechanisms of this machine's operational parameters was still centered on the assumption of two-dimensional field distribution on projected front view plane, and differential separation among the multi-layers.

Consider also the fact that the two-dimensional calculations carried out previously ignored the state of field variations along the dimensional change perpendicular to the front view plane. As a result of forced two-dimensional calculations of inductive electromagnetic field distribution, in many situations, vague results for the sequential operational parameters led to inaccurate solutions for the transformer response to the incidence and propagation of lightning as well as switching surges inside the transformer windings.

The author undertakes the task of development of a mathematical model of the power transformer in a cylindrical three-dimensional coordinate system. Physical structure of the power transformer is visualized to consist of (M) multi-winding layers spread between the inner ferromagnetic material core and the outer grounded tank. Separating gaps among the (M) layers are sizable enough to establish real and distinct changes of electromagnetic modes in each of them.

The objectives are:

1. Mathematical model in three-dimensional cylindrical coordinates for all electromagnetic field components within regions of multi-gaps.
2. Surface impedances and Poynting vector distribution
3. Spectrum of velocity of propagation and cut-off frequencies.

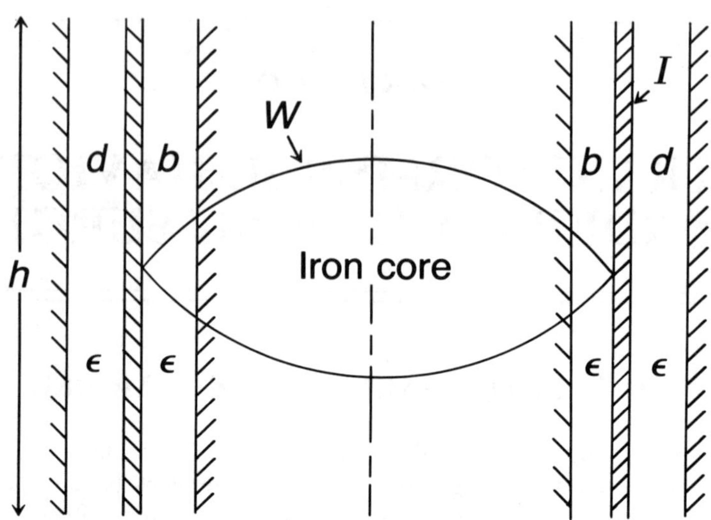

FIGURE 6.1.   Cross-section of multi-winding layer transformer.

A. MATHEMATICAL MODEL USING HELMHOLTZ RADIATION FUNCTION

With the existence of (M) multi-winding layers, arranged in ascending order from the central core, as shown in Figure 6.1 each layer is visualized as a symmetrical current sheet in time variation as expressed below:

$$J_1 = J_1(r_1,\theta,Z,t)$$

$$J_2 = J_2(r_2,\theta,Z,t)$$

$$\vdots$$

$$\vdots$$

$$\vdots$$

$$\vdots$$

$$J_m = J_m(r_m,\theta,Z,t) \tag{6.1}$$

where r_m, θ and Z are the source coordinates of the mth winding layer, $m = 1, 2, 3, \ldots M$, $d_1, d_2 \ldots d_m$ and $J_1, J_2 \ldots J_m$ represent the diameter and current density for the first, second and mth winding layer, respectively, starting from the tank toward the core.

Let $\hat{\epsilon}_m$, $\hat{\mu}_m$ represent the complex permittivity and permeability, respectively; ω is the frequency, and $\hat{y}_m$ and $\hat{Z}_m$ are the admitivity and impedivity of the spatial gap, respectively

$$\hat{y}_m = \sigma_m + j\omega_m\hat{\epsilon}_m \text{ and } \hat{Z}_m = j\omega\hat{\mu}_m$$

where

$$k_m = \sqrt{-\hat{y}_m \hat{Z}_m} \tag{6.2}$$

Then the solution for the Helmholtz scalar wave equation that was obtained in previous work detailed in Chapter 3, in cylindrical coordinates at a point ρ_m, valid for cylindrical current shell is presented below.

$$\phi_m = \frac{A_m r_m}{2} \sum_{n=0}^{\infty} \sum_{m=1}^{M} \frac{1}{n} J_{n-m}(k_m|\rho_m - r_m|)(H_{n-m}^{(2)}(K_m|\rho_m - r_m|) \tag{6.3}$$

The solution of the wave function ϕ_m expressed in Equation (6.3) is valid for the outward direction from the transformer core toward the tank, where r_m represents the mth radius of any layer $\rho_m > r_m$, A_m is the arbitrary current density at the mth layer, J_{n-m} is the Bassel function of the first kind with respect to the mth gap, and $H_{n-m}^{(2)}$ is the Hankel function of the second kind with respect to the mth gap.

Solution of all electromagnetic field components at any point ρ_m, α_m, τ_m in a cylindrical system could be secured from the following set:

$$E_{\rho_m} = \frac{1}{\hat{y}_m}\left(\frac{\partial^2 \phi_m}{\partial \rho_m \partial \tau}\right)$$

$$E_{\alpha_m} = \frac{1}{\hat{y}_m \rho_m}\left(\frac{\partial^2 \phi_m}{\partial_\alpha \partial \tau}\right)$$

$$E_{\tau_m} = \frac{1}{\hat{y}_m}\left(\frac{\partial^2}{\partial \tau_m^2} + k_m^2\right)\phi_m \tag{6.4}$$

$$H_{\rho_m} = \frac{1}{\rho_m}\left(\frac{\partial \phi_m}{\partial \rho_m}\right)$$

$$H_{\alpha_m} = \frac{\partial \phi_m}{\partial \rho_m}$$

$$H_{\tau_m} = 0$$

From Equations (6.3) and (6.4):

$$E_{\rho_m} = \sum_{n=0}^{\infty} \sum_{m=1}^{M} A_m \sqrt{\frac{j}{\pi} \frac{\rho_m - r_m}{r_m \hat{y}_m}} \frac{j^n e^{-j(\rho_m - r_m)}}{2^n n!}$$

$$\frac{(\rho_m - r_m)^n}{r_m^n} - (\rho_m - r_m)^{n-1} - \frac{(\rho_m - r_m)^{n-\frac{3}{2}}}{\sqrt{2n}}$$

$$+ \frac{j2^n(n-1)!}{2} \tag{6.5}$$

$$E_{\rho_m} = \sum_{n=0}^{\infty} \sum_{m=1}^{M} \frac{jA_m}{2\hat{y}_m} \sqrt{\frac{j2}{\pi(\rho_m - r_m)}} \, e^{-(j(\rho_m - r_r))} \frac{j^n(\rho_m - r_m)^n}{2^n n!} \qquad (6.6)$$

$$E_{\tau_m} = \sum_{n=0}^{\infty} \sum_{m=1}^{M} \frac{A_m r_m k_m^2 \pi}{2\hat{y}_m} \sqrt{\frac{j2}{\pi(\rho_m - r_m)}} \frac{j^n(\rho_m - r_m)^n}{2^n n!} \qquad (6.7)$$

$$H_{\rho_m} = \sum_{n=0}^{\infty} \sum_{m=1}^{M} \frac{jA_m \tau_m}{2} \sqrt{\frac{j2}{\pi(\rho_m - r_m)}} \, e^{-j(\rho_m - r_m)} \frac{j^n(\rho_m - r_m)^n}{2^n n!} \qquad (6.8)$$

$$\begin{aligned}
H_{\alpha_m} = \sum_{n=0}^{\infty} \sum_{m=1}^{M} \Bigg[& \frac{A_m \tau_m}{2} \frac{1}{n^2 \rho_m^2} J_n(\rho_m - r_m) H_n^{(2)}(\rho_m - r_m) \\
& - \frac{1}{n r_m} H_n^{(2)}(\rho_m - r_m) \frac{(\rho_m - r_m)^{n-1}}{2^{n-1}(n-1)!} + \\
& \sqrt{\frac{2}{\pi}} (-j)^{n+\frac{1}{2}} J_n(\rho_m - r_m) \frac{1}{n r_m} \frac{e^{-j(\rho_m - r_m)}}{(\rho_m - r_m)^2} \\
& + j^{n+\frac{2}{3}} \frac{2}{\pi(\rho_m - r_m)} e^{-j(\rho_m - r_m)} \frac{1}{n r_m} \Bigg]
\end{aligned} \qquad (6.9)$$

$$H_{\tau_m} = 0 \qquad (6.9\text{a})$$

where at large ρ_m:

$$J_n(\rho_m - r_m) \approx \frac{1}{n!} \frac{(\rho_m - r_m)^n}{2^n}$$

$$H_n^{(2)}(\rho_m - r_m) \approx \frac{j2}{\pi(\rho_m - r_m)} j^n e^{-j(\rho_m - r_m)} \qquad (6.10)$$

B. SURFACE IMPEDANCE SPECTRUM Z_{sm}

Basically Z_{sm} could be secured from the ratio of E to H. At any space point within the multi-gaps among the m layers, the spectrum of Z_{sm} is expressed below.

(1)   Impedance along the positive directions of the system coordinates:

$$Z_{\rho_\alpha}^+ = \overline{E}_{\rho_m} / \overline{H}_{\alpha_m}$$

$$Z_{\alpha_\rho}^+ = \overline{E}_{\alpha_m} / \overline{H}_{\rho_m}$$

$$Z_{\rho_\tau}^+ = \overline{E}_{\rho_m} / \overline{H}_{\tau_m}$$

$$Z_{\tau_\rho}^+ = \overline{E}_{\tau_m}/\overline{H}_{\rho_m}$$

$$Z_{\alpha_\tau}^+ = \overline{E}_{\alpha_m}/\overline{H}_{\tau_m}$$

$$Z_{\tau_\alpha}^+ = \overline{E}_{\tau_m}/\overline{H}_{\alpha_m} \qquad (6.11)$$

(2) Impedances along the negative direction of the system coordinates:

$$Z_{\rho_\alpha}^- = \overline{E}_{\rho_m}/\overline{H}_{\alpha_m}$$

$$Z_{\alpha_\rho}^- = \overline{E}_{\alpha_m}/\overline{H}_{\rho_m}$$

$$Z_{\rho_\tau}^- = \overline{E}_{\rho_m}/\overline{H}_{\tau_m}$$

$$Z_{\tau_\rho}^- = -\overline{E}_{\tau_m}/\overline{H}_{\rho_m}$$

$$Z_{\alpha_\tau}^- = -\overline{E}_{\alpha_m}/\overline{H}_{\tau_m}$$

$$Z_{\tau_\alpha}^- = \overline{E}_{\tau_m}/\overline{H}_{\alpha_m} \qquad (6.12)$$

The two subspectra of Z_{sm} could be expressed completely by substituting the respective electromagnetic field components as indicated above with the corresponding values from Equations (6.6) to (6.10).

C. POYNTING VECTORS

Power transfer per unit area across the m multi-winding layers is expressed below in terms of magnetic field components and the corresponding surface impedances,

$$\overline{P}_\rho = \frac{1}{2} [Z_{\alpha_\tau}^+ H_\tau H_\tau^* + Z_{\tau_\alpha}^+ H_\alpha H_\alpha^+]$$

$$= -\frac{1}{2} [Z_{\alpha_\tau}^- H_\tau H_\tau^* + Z_{\tau_\alpha}^- H_\alpha H_\alpha^*] \qquad (6.13)$$

$$\overline{P}_\alpha = \frac{1}{2} [Z_{\tau_\rho}^+ H_\rho H_\rho^* + Z_{\rho_\tau}^+ H_\tau H_\tau^+]$$

$$= -\frac{1}{2} [Z_{\tau_\rho}^- H_\rho H_\rho^* + Z_{\rho_\tau}^- H_\tau H_\tau^*] \qquad (6.14)$$

$$\overline{P}_\tau = \frac{1}{2} [Z_{\rho_\alpha}^+ H_\alpha H_\alpha^* + Z_{\alpha_\rho}^+ H_\rho H_\rho^+]$$

$$= -\frac{1}{2} [Z_{\rho_\alpha}^- H_\alpha H_\alpha^* + Z_{\alpha_\rho}^- H_\rho H_\rho^*] \qquad (6.15)$$

where the upper index * indicates conjugate.

D. VELOCITY OF PROPAGATION SPECTRUM

Let s_m represent the longitudinal mth winding pitch where

$$\frac{S_m}{\rho_m} = -\overline{E}_{\tau_m}/\overline{E}_{\rho_m} \tag{6.16}$$

v_t represents the axial velocity, where

$$v_{\tau_m} = \frac{v_o}{\sqrt{\epsilon_m \mu_m}} \frac{S_m}{\sqrt{\rho_m^2 + S_m^2}} \tag{6.17}$$

v_o is the velocity of light.

$$\tag{6.18}$$

(1)   For standing wave solution inside the multi-winding set, the frequency spectrum ω_m is expressed:

$$\omega_m = \frac{\lambda_m}{[C_m l_m + \gamma_m \rho_m W_m^2 \lambda_m^2]^{1/2}} \tag{6.19}$$

where l_m and C_m represent the self-inductance and capacitance to ground of the mth winding, and λ_m is the wave density $= \omega_m/v_{\tau_m}$. Subscript m is the inter-turn capacitance of the mth winding.

$$W_m = \sqrt{\rho_m^2 + S_m^2} \tag{6.20}$$

and the critical cut-off frequency v_m,

$$v_m = 1/W_m \sqrt{\gamma_m l_m} \tag{6.21}$$

(2)   For traveling wave solution spectrum

$$v_{\tau_m} = \sqrt{\frac{1}{l_m C_m} - \frac{r_m}{C_m} \omega_m^2 \omega_m^2} \tag{6.22}$$

$$\omega_m = 2\pi Q v_{\tau_m}/L \tag{6.23}$$

where L is the winding length, and $Q = 1, 3, \ldots$

From the preceding analytical presentation for the multi-layers transformer using Helmholtz field radiation function, we can list the following conclusions.

For power transformer of m multi-layers of winding, the following solutions have been expressed using the concept of Helmholtz field radiation function:

1. All components of three-dimensional electric and magnetic fields throughout multi-gap regions in the transformer structure
2. Spectrum of surface impedances in three dimensions along the outward as well as the inward directions
3. Spectrum of Poynting vectors in three dimensions along the outward and inward directions
4. Spectrum of propagating velocities and cut-off frequencies.

6.II TRANSFORMER RESPONSE TO LIGHTNING SURGE

Exact modeling of a multi-layer winding transformer has been developed by this author in a three-dimensional system which took into consideration space variations of electromagnetic components along a transverse axis with respect to the two-dimensional plane used before as the approximate model for the transformer in calculating the operational performance parameters.

The three-dimensional model established previously by this author secured solutions for the transformer surface impedance, the Poynting vector, equivalent ground distance and the operational frequency spectrum as well as the cut-off range.

This author presented in Chapter 2 solutions for all electromagnetic components generated by lightning surge at any field impact area in space. The lightning surge was represented by an actual pulse waveform and included the convective as well as the conductive effects.

In this section, an exact equivalent circuit model for the multi-layer winding transformer is presented which took into consideration the three space dimensional variations of all electromagnetic field components coupled with the fact of sizable separation among the in-winding layers. The circuit model developed is in terms of the m layers, their separation from the inner core structure and the outer grounded tank, field frequency, the surface impedance, the equivalent ground distance, and relevant geometrical parameters.

The second phase of results that will be secured in this section is the steady-state and time varying models for the transformer response due to the incidence and propagation of an actual lightning pulse wave shape. From the basic response equations, a unified response model due to the incidence of lightning surge is established, which identifies threshold boundaries of protection for sustained operational performance for the transformer. The response model also identifies parametric conditions of transformer operation in the domain under short-circuit and open-circuit conditions, when subjected to the program of intense lightning surge. Another aspect pointed out by the response model is the effect of the discrete frequency spectrum of the lightning pulse on the peak expected impact on the multi-winding layer structure.

In Chapters 2 and 3, solutions for the inducing voltage developed by approximate and lightning strokes have been established.

In this section, a solution will be sought for the steady-state and time varying form of the induced transmitting voltage in the transformer as well as representation for the surge impedance offered by the transformer to the incident voltage surge due to lightning.

A. SOLUTION OF THE INDUCED VOLTAGE SURGE[1-7]

The following four order differential represents the proper vehicle to solve for the induced voltage surge, inflicted on a relatively tightly wound long multi-layer transformer.

$$\frac{\partial^2 e}{\partial X^2} - lc\frac{\partial^2 e}{\partial t^2} + \omega l\frac{\partial^4 e}{\partial X^2 \partial t^2} = lc\frac{\partial^2 v}{\partial t^2} \tag{6.24}$$

where

τ = transformer inter-turn capacitance/unit length
c = transformer external capacitance to ground/unit length of coil conductor
l = transformer self-inductance/unit length
ω = one complete turn length of conductor
e = the induced voltage surge
v = the induced incident voltage surge

Taking the Laplace transform of Equation (6.24) with respect to t:

$$\frac{\partial^2 E}{\partial X^2} - lcS^2 E + l\gamma\omega^2 S^2 \frac{\partial^2 E}{\partial X^2} = cS^2 V \tag{6.25}$$

where E is a function of S, X; V is a function of S, X; and S is Laplace transform variable. The complementary solution of Equation (6.25) is

$$E_c(X,S) = A_2 e^{-\alpha X} \tag{6.26}$$

where

$$\alpha = \frac{S\sqrt{lc}}{\sqrt{1 + S^2 l\gamma\omega^2}} \tag{6.27}$$

However since $lc \ll \tau\mu$, μ = inter-turn mutual inductance implying approximately that $\alpha \to 0$

$$E_c(X,S) = A_2 \tag{6.28}$$

where $E_c(X,S)$ is the complementary solution part.

However, from Equation (6.26):

$$E_c(X,S) = A_2\, e^{\frac{-S\sqrt{lc}}{\sqrt{1 + S^2/\gamma\omega^2}}x}$$

$$E_{\text{total}}(X,S) = E_c + E_p \tag{6.29}$$

where E_p is the particular integral part of the total solution.

$$E_{\text{total}} = A_2 e^{-\alpha X} + A_3 G(X,S) \tag{6.30}$$

$$G(X,S) = Lg(X,t) \tag{6.31}$$

$G(X,S)$, taking into consideration the actual pulse shape of lightning stroke, presented in Chapter 2 is given by:

$$G(X,S) = \left[\frac{A_c - A_v}{S^2} - \frac{(A_c - A_v)}{S} e^{-St_1} \right] f(P) \tag{6.32}$$

where $f(P)$ the associative space-dependent function expressed by:

$$f(P) = \left(\frac{j}{\pi} \right)^{1/2} \frac{1}{R} \sum_{n=o}^{\infty} \frac{j^n}{2^n n!} \left[\frac{\rho^{n+\frac{1}{2}} e^{-jn\rho}}{jn^2 R} - \right.$$

$$\frac{n + \frac{1}{2}}{jn^2 R} \int \rho^{n-\frac{1}{2}} e^{-jn\rho}\, d\rho + \frac{\rho^{n-1} e^{-nn\rho}}{jn}$$

$$\left. - \frac{n-1}{jn} \int \rho^{n-2} e^{-jn\rho}\, d\rho + \frac{j2\rho^{n-1}(n-1)!}{3\,2} \rho^{3/2} \right] \cdots \tag{6.33}$$

Now referring to Equation (6.30), the solution for the induced electric potential is given below:

$$E_{\text{total}}^{(X,S)} = A_2 e^{-\frac{S\sqrt{lc}X}{\sqrt{1 + S^2/\gamma\omega^2}}} +$$

$$A_3 \left[\frac{A_c - A_v}{S^2} - \frac{A_c - A_v}{S} e^{-St_1} \right] F(X) \tag{6.34}$$

The function $f(X)$ is replaced by $F(P)$, since $x = \rho \cos\alpha$, and x is along the extended opened full winding of the transformer, where $\alpha = 0$; hence $X = P$, where

γ = capacitance/unit length between adjacent pancake layers
c = external capacitance to ground unit length of conductor

$$\gamma \gg c \tag{6.35}$$

l = self inductance/unit length of conductor
ω = one complete turn length of conductor.

Also, since $\dfrac{\gamma}{c} \gg \dfrac{l}{\mu}$, which implies $\mu\gamma \gg lc$, then

$$E(X,S) = A_2 + A_3\left[\frac{A_c - A_v}{S^2} - \frac{A_c - A_v}{S}e^{-St_1}\right]F(X)$$

$$E(X,t) = A_2\delta(t) + A_3[(A_c - A_v)t - (A_c - A_v)U(t - t)]f(X)_1 \tag{6.36}$$

Of course t in Equation (6.36) cannot exceed a few microseconds, where $f(X)$ is the associative space function of $V(X)$ which is $f(P)$. A_c and A_v represent the conductive and the convective amplitude of the lightning current stroke. t_1 is the time delay for the singularity function included for $v(X,t)$.

To find A_2 and A_3, the following initial and boundary conditions are used:

$$e(X,t) \quad 0 \text{ as } X \to \infty$$

$$e(X,t) \quad E_o \text{ at } t = 0, X = 0$$

E_o is the nominal transformer terminal voltage. Values of A_2 and A_3 are expressed below:

$$A_2 = E_o \frac{\delta(t) - U_1(t - t_1) - U_1(t_1)/c + t}{\delta(t) - U_1(t - t_1) + t}$$

and

$$A_3 = \frac{E_o}{A_c - A_v} \tag{6.37}$$

where $\delta(t)$ is the impulse Delta-Dirac function and $U_1(t - t_1)$ is the delayed step function by t_1.

Therefore, the final solution for the induced voltage in time and space variation $E(X,t)$ is given by:

$$E(x,t) = \left[A_2\delta(t) + A_3(A_c - A_v)[t - U(t - t_1)]\right]f(x) \tag{6.38}$$

B. IMPACTS OF $E(X,t)$?

$$\lim_{t \to 0^+} E(x,t) = \lim_{t \to 0^+} A_2\delta(t) + A_3(A_c - A_v)\text{times}$$

$$[t - U(t - t_1)]f(x) \tag{6.39}$$

$$\lim A_2\delta(t) = E_o$$

$$\lim A_3(A_c - A_v)[t - U(t - t_1)]$$

$$= (A_c - A_v)\frac{E_o}{A_v - A_c}[O - U(-t_1)]$$

$$= U(t_1)E_o \tag{6.40}$$

$$\therefore \quad \lim_{t \to O^+} E(X,O^+) = 2E_o \tag{6.41}$$

At $X = O^+$ $\qquad E(O^+,O^+) = E_o$

Any X and $t = O^+$, $E(x,O) = 2E_o \tag{6.42}$

At any t:

$$A_2 \to E_o$$

$$A_3 \to \frac{E_o}{A_c - A_v}$$

$\therefore$

$$E(x,t) \to \left[E_o\delta(t) + \frac{E_o}{A_c - A_v}(t - U(t - t_1)) \right]f(x) \tag{6.43}$$

$$E(x,t) \to E_o\delta(t)f(t) \tag{6.44}$$

C. SOLUTION OF INDUCED SURGE CURRENT

From the solution of the induced voltage surge expressed in Equation (6.38) and considering the transformer could be represented by a single lumped self-inductance (L), the induced current surge could be obtained from the following:

$$i(X,t) = \frac{1}{L}\int e\,dt \tag{6.45}$$

$$= \frac{1}{L}\left[A_2 U(t) + A_3(A_c - A_v)\left[\frac{1}{2} t^2 - tU(t - t_1)\right]\right] \text{ times}$$

$$[J_c(t) - J_v(t)]\left(\frac{j}{\pi}\right)^{1/2} \frac{1}{R\hat{y}} \sum_{n=o}^{\infty} \frac{j^n}{2^n n!} \left[\frac{-X^{n+\frac{1}{2}} e^{-jnX}}{jn^2 R}\right.$$

$$-\frac{n + \frac{1}{2}}{jn^2 R} \int X^{n-\frac{1}{2}} e^{-jnX} dX + \frac{X^{n+\frac{1}{2}} e^{-jnX}}{j^n}$$

$$\left. \frac{n - 1}{j^n} \int X^{n-2} e^{-jnX} dX + \frac{j2^{n-1}(n - 1)!}{3\,2} X^{3/2}\right] \tag{6.46}$$

where

$$R = R_c \text{ for conductive state}$$

$$= R_v \text{ for convective state}$$

$$J_c = A_c t, O < t < t_1$$

$$= A_c(t - t_1)U_1(t - t_1), \text{ for } t > t_1$$

$$J_v = A_v t, O < t < t_1$$

$$= A_v(t - t_1)U_1(t - t_1), \text{ for } t > t_1 \tag{6.47}$$

D. DISTRIBUTION OF THE SURGE IMPEDANCE

$$Z(X,t) = e(X,t)/i(X,t) \tag{6.48}$$

From Equations (6.38) and (6.46)

$$Z(X,t) = L \frac{A_2 \delta(t) + A_3(A_c - A_v)[t - U(t - t_1)]}{A_2 U(T) + A_3(A_c - A_v)[{}^1\!/_2 t^2 - tU(t - t_1)]} \tag{6.49}$$

Equation (6.49) demonstrates an important fact that the surge transformer impedance due to lightning surge is time dependent as illustrated in Figure 6.2.

$$Z \rightarrow L\delta(t)$$

$$\text{and at } t \rightarrow \infty$$

$$Z \rightarrow \text{zero} \tag{6.50}$$

Distribution of imposed surge impedance as a result of lightning voltage incidence on a transformer could be represented as shown in Figure 6.2.

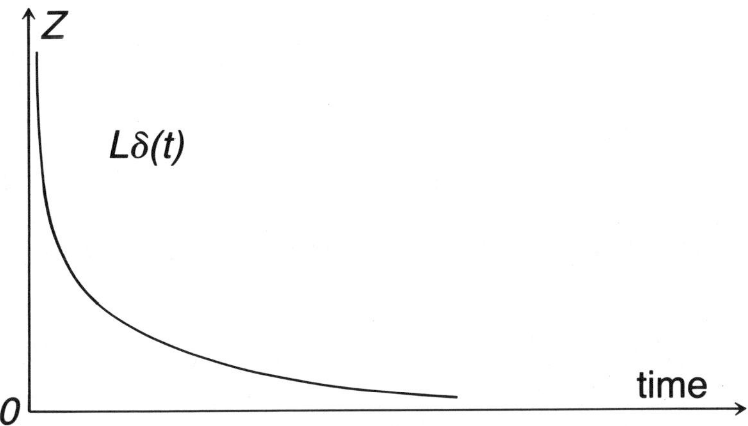

FIGURE 6.2. Surge impedance distribution.

From the preceding Section II, mathematical models in the form of closed form solutions have been secured successfully resulting from the incidence of propagating inducing voltage produced by lightning conductive return stroke and the convective time varying bound charges in the bound charges in the ground terrain. The following have been secured:

1. Distribution of the induced voltage surge, taking into account effects of the transformer self-inductance and the external capacitance to ground but neglecting any resistive effects.
2. Distribution of the induced current transmitted surge.
3. Time pattern of the surge impedance imposed on the transformer structure.

It is established that at the instant of voltage surge incidence, the surge impedance is an impulse, while the steady-state value approaches a vanishing inductance.

6.III CONDITIONS OF INDUCED VOLTAGE BREAKDOWN[1-10]

At breakdown, the induced current flow in the transformer structure will approach an infinite level.

Breakdown could be described by:

$$i(x,t) = \frac{1}{L} \int e(x,t)dt \rightarrow \infty \qquad (6.51)$$

or from Equation (6.38),

$$\frac{\partial e(x,t)}{\partial x} \rightarrow 0 \qquad (6.52)$$

Condition (6.52) implies that

$$\frac{\partial f(x)}{\partial x} = 0 \tag{6.53}$$

$$i(x,t) \to \infty \text{ if } R \to 0 \tag{6.54}$$

Turning to Equation (6.53), and performing the differentiation with respect to X,

$$\frac{\partial f(x)}{\partial X} = k\left[\frac{l}{jRn^2}(n+1)X^{n-\frac{1}{2}}e^{-jnX} + (-jn)e^{-jnX}X^{n+\frac{1}{2}}\right]$$

$$+ \frac{n+\frac{1}{2}}{jRn^2}X^{n-\frac{1}{2}}e^{-jnX} + \frac{1}{jn}[(n-1)X^{n-2}e^{-jnX} +$$

$$(-jn)e^{-jnX}X^{n-1}] - \frac{n-1}{jn}X^{n-2}e^{-jnX} +$$

$$\frac{j2^{n-1}(n-1)!}{3\sqrt{2}}\frac{3}{2}X^{1/2}\right] \tag{6.55}$$

where

$$k = \frac{\sqrt{j}}{\sqrt{\pi R\hat{y}}}\frac{jn}{2^n n!} \tag{6.56}$$

Therefore,

$$\frac{\partial f(x)}{\partial X} = k\left[\frac{1}{jRn^2}\left[(n+1)e^{-jnX}X^{n-\frac{1}{2}} - jne^{-jnX}X^{n+\frac{1}{2}}\right] +\right.$$

$$\cdot \frac{n+\frac{1}{2}}{jRn^2}X^{n-\frac{1}{2}}e^{-jnX} + \frac{n-1}{jn}e^{-jnX}X^{n-2} - e^{-jnX}X^{n-1}$$

$$\left. - \frac{n-1}{jn}X^{n-2}e^{-jnX} + \frac{j2^{n-1}(n-1)!}{\sqrt{2}}X^{1/2}\right] = 0 \tag{6.57}$$

From Equation (6.57), adding similar terms, and letting $X^{1/2} = y$, hence $X = y^2$, it follows that:

$$\frac{2n + \dfrac{3}{2}}{jRn^2}\, y - \frac{1}{Rn}\, y^3 + \frac{j2^{n-1}(n-1)!}{2}\, e^{jnX}2ny^3 = 1$$

$$\frac{2n + \dfrac{3}{2}}{jRn^2}\, y^{-2} - \frac{1}{Rn} + \frac{j2^{n-1}(n-1)!}{2}\, y^{2n}e^{jny^2} = 1 \tag{6.58}$$

$$\frac{2n + \dfrac{3}{2}}{jRn^2} - \frac{y^2}{Rn} + \frac{j2^{n-1}(n-1)!}{2}\, y^{2n+2}e^{jny^2} = y2 \tag{6.59}$$

Using:

$$\lim_{\substack{n\to\infty \\ y\to\infty}} e^{jny}\Big|^{2} = \cos ny^2 + j \sin ny^2 = 1 \tag{6.60}$$

Therefore,

$$f(x) = \frac{2n + \dfrac{3}{2}}{jRn^2} - \frac{y^2}{Rn} + \frac{j2^{n-1}(n-1)!}{2}\, y^2 y^{2n} = y2$$

$$\frac{2n + \dfrac{3}{2}}{jRn^2}\, y^{-2} - \frac{1}{Rn} + \frac{j2^{n-1}(n-1)}{2}\, y^{2n} = 1$$

$$\frac{j2^{n+1}(n-1)!}{2}\, (y^2)^n + \frac{2n + \dfrac{3}{2}}{jRn^2}\, y^{-2} = 1 + \frac{1}{nR} \tag{6.61}$$

let $y^2 = Z$

Equation (6.61) becomes:

$$\frac{j2^{n+1}(n-1)!}{2}\, Z^{n+1} - \frac{1 + nR}{nR}\, Z = \frac{2n + \dfrac{3}{2}}{jRn^2} \tag{6.62}$$

$$Z^{n+1} Z + Z^2 + Z^3 + Z^4 + \ldots$$

and with

$$\sum_{n=o}^{\infty} Z^{n+1} \gg Z$$

$$n = 0, 1, 2, 3, 4 \ldots \infty$$

(6.63)

$$\frac{j2^{n+1}(n-1)!}{2} Z^{n+1} \frac{2n + \dfrac{3}{2}}{j2^{n+1}(n-1)!}$$

(6.64)

or

$$Z^{n+1} = \frac{2n + \dfrac{3}{2}}{jRn^2} \frac{\sqrt{2}}{j2^{n+1}(n-1)!}$$

(6.65)

Therefore,

$$\sum_{n=o}^{\infty} Z^{n+1} = \frac{-1}{R2} \sum_{n=0}^{\infty} \frac{2n + \dfrac{3}{2}}{n^2(n-1)!}$$

(6.66)

or

$$Z^{n+1} = \frac{-1}{2R} \left[\frac{2n + \dfrac{3}{2}}{n^2(n-1)!} \right]$$

(6.67)

and

$$y^{2n+2} = \frac{-1}{2R} \left[\frac{2n + \dfrac{3}{2}}{n^2(n-1)!} \right]$$

(6.68)

Returning to $y^2 = X$

$$\therefore \quad X^{n+1} = \frac{-1}{2R} \left[\frac{2n + \dfrac{3}{2}}{n^2(n-1)!} \right]$$

(6.69)

or

$$X = \frac{1}{\sqrt{2}\, R_{c,v}} \left[\frac{2n + \dfrac{3}{2}}{n^2(n-1)!} \right]^{\frac{1}{nH}} \qquad (6.70)$$

where $R_{c,v}$ in Equation (6.70) represents the radius of the conductive lightning return stroke, or that for the convective stroke, i.e., the subscript (c) is for conduction and (v) for convection.

Table 6.1 shows correlations for a range of discrete locations of voltage breakdown on a cylindrical transformer for a set of limits on the index n.

Values of X described in Equation (6.70) for voltage breakdown thresholds on the transformer structure represent discrete locations with respect to the range of the series index (n). However, after proceedings with the index (n), it is recognized that X will approach almost zero value for $n > 10$.

Figure 6.3 illustrates the discrete spectrum of X with respect to the source location of lightning surge. In the figure, R is the radius of the conductive or convective surge and X represents the geometrical distance from source of lightning surge.

As for the breakdown condition identified by Equation (6.17), for $R_v \rightarrow 0$, it is implied that the radius of the conductive stroke is infinitesimal and hence the conductive current is an impulse, while the convective stroke current will become a thin surface current sheet whose radius approaches infinite values as shown in Figure 6.4.

Conditions of voltage breakdown threshold subjected on a cylindrical multi-winding power transformer due to conductive and convective effects of lightning stroke are established under the criterion that total current flow induced in the transformer will approach infinite value.

Two conditions have been established:

1. Geometrical location of the transformer with respect to the place of initiation of lightning surge. It is established, as shown in Figure 6.3, that certain specific locations will subject the transformer with more certainty of destructive breakdown than others. These locations are physically identifiable.
2. Where the radius of the conductive lightning surge is extremely small, and the corresponding convective part is very large, it is implied that the conductive stroke becomes an impulse and the convective part a current sheet of infinite radius. This situation will create an extremely destructive breakdown effect.

6.IV   REFLECTION OF INDUCED VOLTAGE

It is of special interest to see reflection of the transformer induced voltage due to lightning surge on the other winding layers, in the form of the equivalent circuit representation.

TABLE 6.1

Series index (*n*)	Point along winding (*X*)
0	∞
1	$\pm\, 2\sqrt{\dfrac{7}{2\,2\,R}}$
2	$\dfrac{3}{2}\,3\sqrt{\dfrac{1}{2\,R}}$
3	$+\, 4\sqrt{\dfrac{15}{3\,2\,R}}$
4	$\pm\, \dfrac{1}{2}\,5\sqrt{\dfrac{19}{3\sqrt{2}\,R}}$
5	$+\, \dfrac{1}{2}\,6\sqrt{\dfrac{23}{2\,\sqrt{2}\,R}}$
6	$7\sqrt{\dfrac{7}{2160\,2\,R}}$
7	$+\, 8\sqrt{\dfrac{32}{1764\,2\,R}}$
8	$\pm\, 9\sqrt{\dfrac{5}{23040\,2\,R}}$
9	$10\sqrt{\dfrac{13}{120960\,2\,R}}$
10	$\pm\, 11\sqrt{\dfrac{0.0043}{23144\,2\,R}}$

Let the induced voltage effect first the load winding side of a multi-phase transformer, expressed by Equation (6.38), with a turns ratio of (a) with respect to the supply side. Therefore, the induced voltage on the supply side $E(x,t)$ becomes $aE(x,t)$, while the current flow is equal to $i(x,t)/a$. Also, all ohmic elements including R_L and jx_L representing the load winding side series resistance and leakage reactance will be multiplied by (a^2) in their reflection to the supply winding side.

The equivalent circuit for the transformer under the effect of lightning surge is shown in Figure 6.5 in terms of one phase, where the basic structural elements are the induced voltage surge, the induced current surge and the appropriate surge impedance.

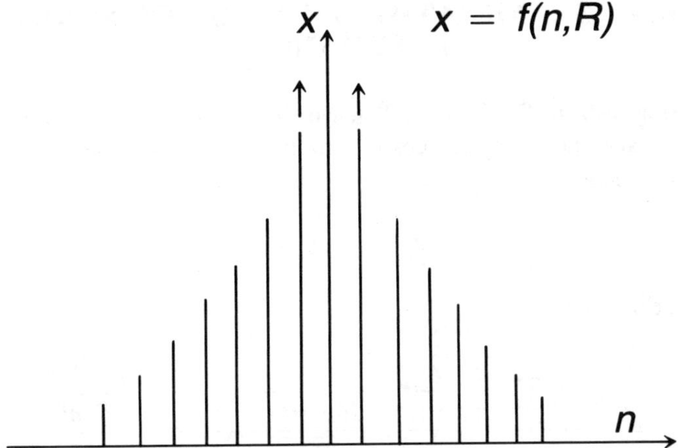

FIGURE 6.3. Discrete distribution of voltage breakdown on transformer due to lightning.

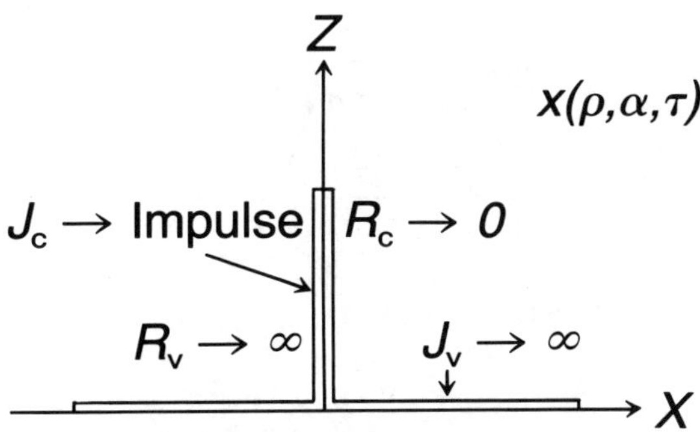

FIGURE 6.4. Extreme condition of voltage breakdown.

FIGURE 6.5. Transformer circuit under lightning surge.

6.V EXAMINATION OF SURFACE SURGE IMPEDANCES

Referring back to Equation (6.11), we shall proceed to inspect some of the vectorial surface surge impedances in a multi-layer transformer.

Let us consider:

$$Z_{\alpha\rho}^{+} = -E_\alpha/H_\rho$$

From Equation (6.5):

$$Z_{\alpha\rho-m} = -\frac{\displaystyle\sum_{n=0}^{\infty}\sum_{m=1}^{\infty}\frac{jA_m}{\hat{y}_m}\sqrt{\frac{j2}{\pi(\rho_m - r_m)}}\,e^{-j(\rho_m - r_m)}\frac{j^{n(\rho_m - r_m)^n}}{2^n n!}}{\displaystyle\sum_{n=0}^{\infty}\sum_{m=1}^{\infty}\frac{jA_m\tau_m}{2}\sqrt{\frac{j2}{\pi(\rho_m - r_m)}}\,e^{-j(\rho_m - r_m)}\frac{j^n(\rho_m - r_m)^n}{2^n n!}} \qquad (6.71)$$

Then from Equation (6.71)

$$Z_{\alpha\rho-m}^{+} - \frac{1}{\hat{y}_m\tau_m} \qquad (6.72)$$

$$= -\frac{1}{\tau_m(\sigma_m + j\omega_m\hat{\epsilon}_m)} \qquad (6.73)$$

where as indicated in Equation (6.2)

$$\hat{y}_m = \sigma_m + j\omega_m\hat{\epsilon}_m$$

Therefore, $Z_{\alpha\rho-m}$ could be written also as:

$$= -\frac{\hat{\sigma}_m - j\omega\hat{\epsilon}_m}{\tau_m(\sigma_m^2 + \omega_m^2\hat{\epsilon}_m^2)} \qquad (6.74)$$

It can be easily seen that $Z_{\alpha\tau}^{+}$ and $Z_{\rho\tau}^{+}$ are infinity since $H_{\tau m} = 0$. Now consider $Z_{\tau\rho}^{+}$:

$$Z_{\tau\rho-m}^{+}$$

$$= \frac{\displaystyle\sum_{n=0}^{\infty}\sum_{m=1}^{\infty}\frac{A_m r_{m\tau}}{2\hat{y}_m}\sqrt{\frac{j2}{\pi(\rho_m - r_m)}}\,e^{-j(\rho_m - r_m)}\frac{j^{n(\rho_m - r_m)^n}}{2^n n!}}{\displaystyle\sum_{n=0}^{\infty}\sum_{m=1}^{\infty}\frac{jA_m\tau_m}{2}\sqrt{\frac{j2}{\pi(\rho_m - r_m)}}\,e^{-j(\rho_m - r_m)}\frac{j^n(\rho_m - r_m)^n}{2^n n!}}$$

$$= \sum_{m=1}^{M} -j \frac{\pi r_m k_m^2}{\tau_m \hat{y}_m} e^{j(\rho_m - r_m)} \tag{6.75}$$

While we notice that $Z_{\rho\tau-m}^{+} = \overline{E}_{\rho m}/\overline{H}_{\tau m} \rightarrow \infty$ (an open-circuit); hence $H_{\tau\mu} = 0$.

Inspection of $Z_{\rho-m}$ which is in terms of complex quantities $\hat{\sigma}_m$ and $\hat{\epsilon}_m$ can be written as follows:

$$\hat{\sigma}_m = a_m \pm jb_m \tag{6.76}$$

$$\hat{\epsilon}_m = K_{1m} + jK_{2m} \tag{6.77}$$

$\therefore$

$$Z_{\alpha\rho-m} = \frac{jW^{(k_{1m} \pm jk_{2m}) - (a_m \pm jb_m)}}{\tau_m(a_m \pm jb_m)^2 + \omega^2(k_{1m} \pm jk_{2m})^2} \tag{6.78}$$

Clearly, $Z_{\alpha\rho-m}$ is a complex impedance which involves a resistive component and either an inductive or capacitive reactance. Similarly, looking at $Z_{\tau\rho-m}$ with $k_m = \sqrt{\hat{y}_m \hat{Z}_m}$ also indicates a complex form.

Regarding $Z_{\alpha\rho-m}^{+} = E_{\rho m}/H_{\alpha m}$, we'll consider this ratio for a relatively large argument for $J_n(x)$ and $H_n^{(2)}(x)$, where, as indicated in Equation (6.10),

$$J_n(x) \approx \frac{1}{n!} \frac{X^n}{2^n} \quad \text{and}$$

$$H_n^{(2)}(x) \approx \sqrt{\frac{j2}{\pi X}} \, j^n e^{-jX}$$

First proceeding to write H for a large argument,

$$\overline{H}_{\alpha m} = \sum_{n=0}^{\infty} \sum_{m=1}^{M} \frac{A_m T_m}{\rho^2} \frac{1}{n^2 \rho_m^2} \frac{1}{n!} \frac{(\rho_m - r_m)^n}{2^n} \sqrt{\frac{j2}{\pi(\rho_m - r_m)}} \, e^{-j(\rho_m - r_m)}$$

$$\frac{1}{nr_m} \sqrt{\frac{j2}{\pi(\rho_m - r_m)}} \, j^n e^{-j(\rho_m - r_m)} \frac{(\rho_m - r_m)^{n-1}}{2^{n-1}(n-1)!} +$$

$$\sqrt{\frac{2}{\pi}} (-j)^{n+\frac{1}{2}} \frac{1}{n!} \frac{(\rho_m - r_m)^n}{2^n} \frac{1}{nr_m} \frac{e^{-j(\rho_m - r_m)}}{(\rho_m - r_m)^2} +$$

$$+ j^{\frac{n}{3}} \sqrt{\frac{2}{\pi(\rho_m - r_m)}} \, e^{-j(\rho_m - r_m)} \frac{1}{nr_m} \tag{6.79}$$

$E_{\rho m}$ from Equation (6.5)

$$= \sqrt{\frac{j}{\pi}} \sum_{n=0}^{\infty} \sum_{m=1}^{M} A_m \frac{\sqrt{\rho_m - r_m}}{r_m \hat{y}_m} \frac{j^n e^{-j(\rho_m - r_m)}}{2^n n!} \left[\frac{(\rho_m - r_m^n)}{r_m^n} \right.$$

$$\left. - (\rho_m - r_m)^{n-1} - \frac{(\rho_m - r_m)^{n-\frac{3}{2}}}{2n} + \frac{j 2^n (n-1)!}{2} \right] \tag{6.80}$$

$Z_{\rho\alpha - m}$ which is the ratio of $E_{\rho m}$ to $H_{\alpha m}$ forms a somewhat involved expression of unclear picture. The author would prefer to examine the form of $Z_{\rho\alpha - m}$ for $\rho_m = r_m$.

$$\lim E_{\rho m} \to 0$$

as

$$\rho_m = r_m$$

$$\lim H_{\alpha m} \to \infty$$

as

$$\rho_m = r_m \tag{6.81}$$

Consequently Equation (6.81) indicates tht $Z_{\rho\alpha - m} \to 0$ is a short circuit condition. The last surface surge impedance to examine is $Z_{\tau\alpha - m}$:

$$Z_{\tau\alpha - m} = \frac{E_{\tau - m}}{H_{\alpha m}} \tag{6.82}$$

From Equation (6.7)

$$\overline{E}_{\tau - m} = \sum_{n=0}^{\infty} \sum_{m=1}^{M} \left[\frac{A_m r_m k_m^2 \pi}{2 \hat{y}_m} \sqrt{\frac{j2}{\pi(\rho_m - r_m)}} \frac{j^{n(\rho_m - r_m)}}{2^n n!} \right]$$

Also, it is clear that for $\rho_m = r_m$,

$$E_{\tau - m} \rightarrow 0$$

$$\text{since } H_{\alpha - m} \rightarrow \infty \qquad (6.83)$$

$Z_{\tau\alpha - m} \rightarrow 0$ is a short-circuit condition.

Other conditions of interest to examine include a point at a grounded tank surrounding the mth layer where $P_m \rightarrow \infty$ and a point at the central axis of the transformer where $P_m = 0$.

6.VI THE SURGE IMPEDANCE IN COMPLEX FORM

Examination of transformer surge impedance derived in Section II of this chapter shows that it could be a pure inductance reactance, dependent on time, having its maximum value initially then decaying to zero afterwards. Analysis carried out in Section V indicates that vectorial surface surge impedance, when in a finite form, is a complex number.

In other published works, from the text by Reinhold Rudenberg entitled: *Electrical Shock Waves in Power Systems,* it was indicated that the steady state form and its nature are totally resistive and is of the order of 5000.

In this section, we shall discover whether a complex form of a lumped surge impedance could be identified which includes resistance and inductive effects. Refer to Equation (6.43), where the induced voltage as a function of space along the transformer winding length has been expressed, and rewritten below.

$$E(x,t) = E_o\delta(t) + \frac{E_o}{A_cA_v}[t - U_1(t - t_1)]F(x)$$

R represents the total resistive part of the transformer surge impedance, and L is the total self-inductance for the transformer. Hence,

$$E(x,t) = R(x,t)I(x,t) + L(x,t)\frac{dI(x,t)}{dt} \qquad (6.84)$$

The problem is to solve for $I(x,t)$ using a combination of circuit theory and field theory concepts, after which the lumped surge impedance could be secured from the ratio of $E(x,t)/I(c,t)$.

In Equation (6.84), $E(x,t)$ is the induced surge voltage, $I(x,t)$ is the induced surge current, $L(x,t)$ is the transformer self-inductance at very high frequency, and $R(x,t)$ is the transformer resistance at very high frequency.

The intent now is to solve for $I(x,t)$ from Equation (6.84) as shown below. Let $I_c(x,t) = $ the complementary solution, where

$$I_c(x,t) = A_1(x)e^{-\frac{R}{L}t} \tag{6.85}$$

while the particular integral $I_p(x,t)$ is:

$$I_p(x,t) = A_2(x)\psi(x,t) \tag{6.86}$$

where, from Equation (6.84),

$$I_p(x,t) = \frac{E(x,t)}{L(x,t)} \tag{6.87}$$

Equation (6.84) could be rewritten as:

$$\frac{dI(x,t)}{dt} + I(x,t)\frac{R(x,t)}{L(x,t)} = \frac{E(x,t)}{L(x,t)}$$
$$= \psi(x,t) \tag{6.88}$$

Total solution for $I(x,t)$ becomes

$$I(x,t) = A_1(x)e^{-\frac{R}{L}t} + A_2(x)(x,t) \tag{6.89}$$

Upon substitution of $I(x,t)$ and $dI(x,t)/dt$ into Equation (6.88), the following are obtained:

$$A_2(x) = \psi(x,t)\Big/\left[\frac{d\psi(x,t)}{dt} + \psi(x,t)\frac{R(x,t)}{L(x,t)}\right]$$
$$A_1(x) = 0 \tag{6.90}$$

Therefore,

$$I(x,t) = \psi^2(x,t)\Big/\left[\frac{\partial(x,t)}{dt} + \frac{R(x,t)}{L(x,t)}(x,t)\right] \tag{6.91}$$

or

$$I(x,t) = E^2(x,t)/[E(x,t)R(x,t) + L(x,t)E(x,t) - E(x,t)L'(x,t)] \tag{6.92}$$

The surge impedance $Z(x,t)$ becomes:

$$Z(x,t) = E(x,t)/I(x,t)$$

$$= [R(x,t) - L'(x,t)] + L'(x,t)\frac{E'(x,t)}{E(x,t)} \qquad (6.93)$$

$E(x,t)$ has been given by Equation (6.43) and

$$E'(x,t) = E_o\delta'(t) + \frac{E_o}{A_c - A_v}[U_1(t) - \delta(t - t_1)]f(x) \qquad (6.94)$$

Therefore, from Equation (6.93),

$$Z(x,t) = R(x,t) + \frac{dL(x,t)}{dt} + L(x,t)$$

$$\left[\frac{E_o\delta'(t) + \dfrac{E_o}{A_c - A_v}[U_1(t) - \delta(t - t_1)]}{E_o\delta(t) + \dfrac{E_o}{A_c - A_v}[t - U_1(t - t_1)]}\right] \qquad (6.95)$$

Looking at $Z(x,t)$ from Equation (6.95), we can see that at $t \to \infty$

$$Z(x,t) \to R(x,t) + \frac{dL(x,t)}{dt} \qquad (6.96)$$

6.VII SWITCHING AND LIGHTNING IMPULSE STRENGTH FOR POWER TRANSFORMERS

This section refers to a study carried out by H. Studinger and K. H. Weck. External clearances for components of the three-phase power transformer such as bushing terminals and oil compensation tank are determined by the switching impulse withstand voltage and lightning impulse withstand voltage levels. For transformers with rated voltages in kilovolts $\geqslant 245$, air clearances are determined by the rated SI and LI, phase-to-neutral and phase-phase voltages, with the positive polarity as the determining base. The SI and LI withstand voltages depend largely on the gap geometry and the application of the minimum air clearances to the clearances of power transformers using controlled-potential bushings. Figure 6.6 shows the main dimensions of air clearance for a power transformer including the transformer tank, the oil compensating tank and bushings, intended for a phase-to-earth clearance and phase-phase clearance, while Figure 6.7 shows a plot for U_{10} with respect to gap(s). To obtain the measured strength of transformer air clearances between phases involving controlled-po-

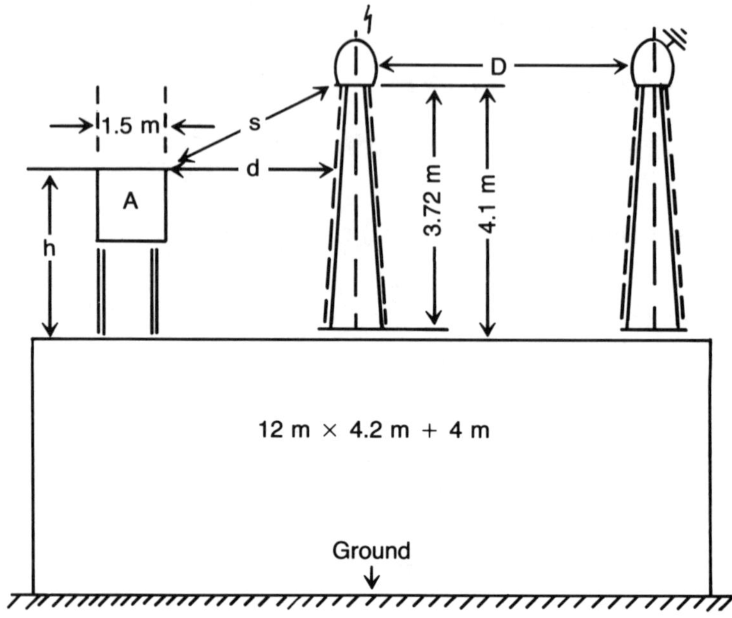

FIGURE 6.6.   Test set up for the test of the phase-to-earth clearance.

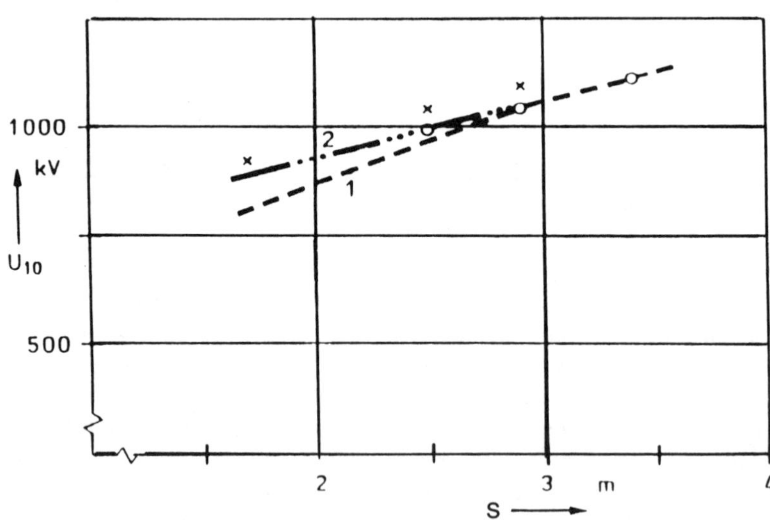

FIGURE 6.7.   Switching impulse withstand voltage V_{10} dependent on the gap clearance between bushing and oil compensation tank, positive polarity. x, lowest test results, dry; ○ lowest test results, wet; 1, calculated with gap factor model; 2, calculated with leader inception model.

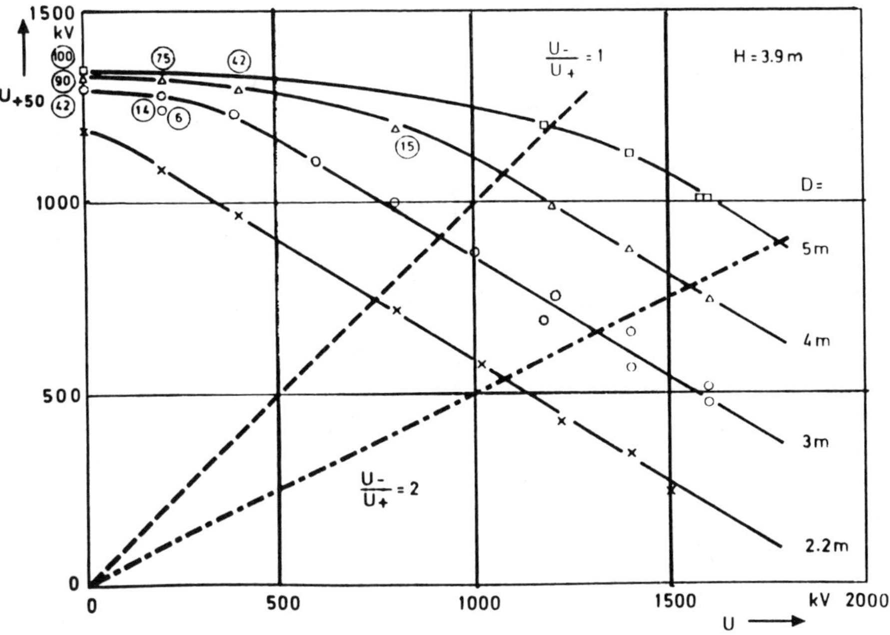

FIGURE 6.8. 50% SI flashover voltage phase-to-phase, positive component against negative component. 2- capacitor stacks, dry.

tential bushings with normal air clearances, tests had been conducted by Studinger and Weck replacing the bushings by post-insulator stacks. Controlled potential bushings used in the experimental tests included two-capacitor stacks (500 PF) and three-capacitor stacks (330 PF). Transformer clearances have to be divided into two main groups: (1) clearances within the transformer, and (2) clearances between live parts of the transformer and surrounding objects, such as structures, buildings, etc. Three categories of clearances must be dealt with: (1) clearances between high voltage terminals of the bushing and the transformer tank, (2) clearances between the high voltage terminal of the bushing and earthed structures installed on the tank such as oil compensation tank, and (3) clearances between the high voltage terminals of the bushing.

A. CLEARANCE BETWEEN LIVE PARTS AND EARTH-DRY

The study conducted by Studinger and Weck pointed to the conclusion that the application of positive voltage stress at one bushing and earthing of the other as actual representation for the actual voltage stress in practice created a situation of dielectric strength that has direct dependence on the clearance between phases with a limiting condition that H/D ≤ unity. This practically means that distances between phases must be at least equal to the separation between bushing terminals and the transformer tank, as indicated in Figures 6.8 and 6.9. Also Figure 6.10

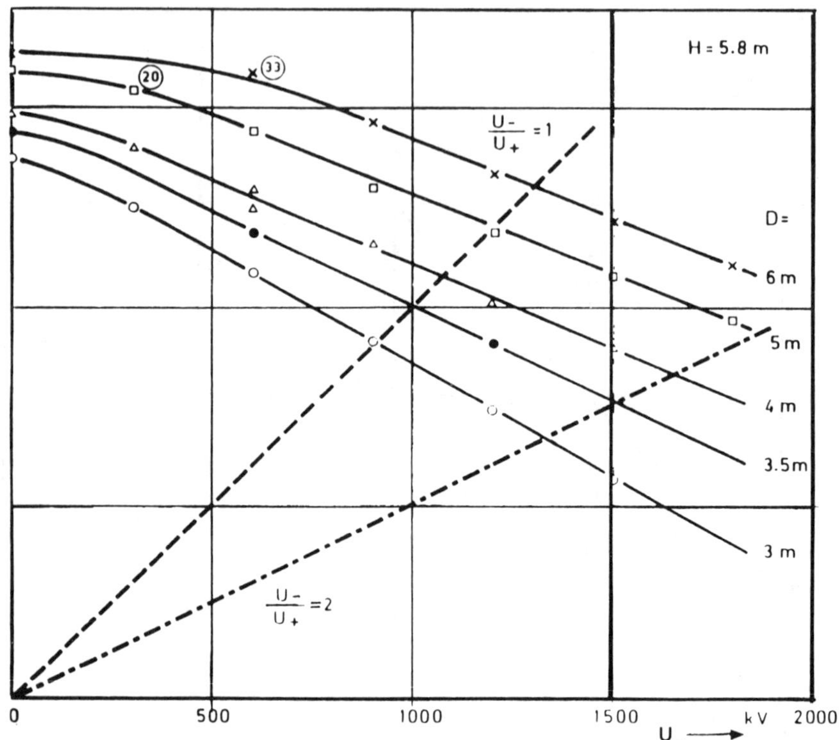

FIGURE 6.9.   50% SI flashover voltage phase-to-phase, positive component against negative component. 3-capacitor stacks, dry.

shows phase-to-earth clearance (*H*) with respect to SI withstand voltage for dry condition. An empirical equation derived by Studinger and Weck is expressed below:

$$U_{10} = 530\, H^{0-6} \text{ kV} \tag{6.97}$$

where the factor (6.30) corresponds to the gap factor of 1.15, and U_{10} is 10% of SI or LI limits.

B. CLEARANCE BETWEEN LIVE PARTS AND EARTH-WET

The wet condition refers to the case of rain which results in a reduction of the dielectric strength. The study of Studinger and Weck indicated that the minimum clearance $H_{\min}$ follows the expression below (also shown in Figure 6.10):

$$H_{\min} = \left[\frac{U_{r\omega}}{490}\right]^{1.6} \text{ meters} \tag{6.98}$$

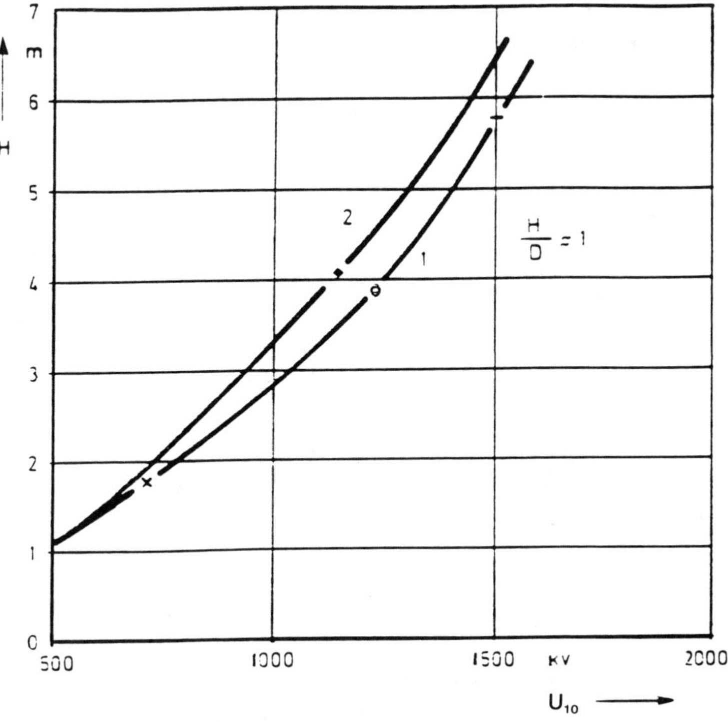

FIGURE 6.10.   Phase-to-earth clearance H between bushing terminal and transformer tank dependent on the SI withstand voltage, positive. 1, Capacitor stacks, dry; x, single capacitor; ○, 2-capacitor stack; +, 3-capacitor stack; 2, bushing type 2, wet.

Numbers in circles in Figure 6.10 give the percentage of flashovers occurring from the positive center capacitor to the earthed bank. This applies also to Figure 6.9 where $U_{r\omega}$ = rated SI withstand phase-earth voltage.

C. CLEARANCE TO EARTHED OBJECTS ON TRANSFORMER TANK

This refers to clearance between the bushing terminal and earthed object on the transformer tank, such as the oil compensation tank. The study by Stundinger and Weck emphasized that for such objects to be treated differently, they must be of small dimensions compared with the transformer tank. Figure 6.10 illustrates clearances (S) with respect to the required withstand voltage U_{10} for two conditions, namely $h < H \leq 0.5$ and $h/H < 1$. Expression for S is given below:

$$S = AH - \frac{U_B - U_{10}}{E} \tag{6.99}$$

where U_B is the withstand voltage bushing to transformer tank without structure. A and E are constants given by

$$A = 0.84, \ E = 112 \text{ kV/m, for } \frac{h}{H} \leq 0.5$$

$$A = 0.90, \ E = 130 \text{ kV/m, for } \frac{h}{H} = 1$$

Also a minimum S is given by

$$S_{\min} = \left[\frac{U_{r\omega}}{540}\right]^{1.67} \text{meter} \tag{6.100}$$

D. MINIMUM PHASE-PHASE CLEARANCES

The minimum clearance between phases according to the work of Studinger and Weck is given by

$$D_{\min} = \left[\frac{U_{r\omega\rho}}{810}\right]^{1.67} \text{meters} \tag{6.101}$$

where $U_{r\omega\rho}$ = rated SI withstand phase-phase voltage.

D_{min} expressed above has a gap factor of 1.75 to account for the application of positive and negative components of equal levels. It has been mentioned that $D_{\min}$ is practically independent of the height of the bushing terminal above the transformer tank.

As a matter of interest, for sinusoidal time variation, $\partial/t = jW$; then

$$Z(x,t) = R(x,t) + jWL(x,t) \tag{6.102}$$

The result indicated in Equation (6.102) affirms the soundness of the present approach of analysis.

6.VIII   SOLVED EXAMPLES

A. Equation (6.36) shows a solution for the induced voltage in a cylindrical multi-winding transformer due to incident lightning surge. Identify $E(x,O)$ and $E(x,T)$, where T is the end of lightning stroke duration.

Solution

From Equation (6.36) at $t = 0$

$$E(x,O) = A_2\delta(o) \tag{A1}$$

an impulse A_2 is given by Equation (6.37).

$$A_2 \text{ at } t = O$$

$$= E_o \frac{\delta(o)}{\delta(o)} \tag{A2}$$

$$\therefore \quad E(x,O) = E_o\delta(o) \tag{A3}$$

Then

$$E(x,t) = \frac{E_o}{A_c - A_v} [(A_c - A_v)T - (A_c - A_v)U_1(T - t_1)]f(x)$$

$$= E_o t_1 f(x) \tag{A4}$$

B. Calculate the total transformer surge impedance that an induced Poynting vector can face in the $\tau - \rho$ plane.

Solution

From Equation (6.75),

$$Z_{\tau\rho - m} = \sum_{m=1}^{M} \frac{-j\pi r_m k_m^2}{T_m \hat{y}_m} e^{j(\rho_m - r_m)} \; \Omega/m^2$$

$$dA = d\rho_m \, dT_m \tag{B1}$$

$$\therefore \quad Z'_{T\rho - m} = \int_o^{\rho_o} \int_o^{T_o} Z_{T\rho - m} dd_{\rho m} \tag{B2}$$

$$= \int_o^{\rho_m} \int_o^{\tau_m} \frac{-j\pi r_m k_m^2}{T_m \hat{y}_m} e^{j(\rho_m - r_m)} d\rho_m d_{Tm}$$

where $Z'_{\tau\rho - m}$ is the total surge impedance in $\tau\gamma$ plane.

$$= \int_o^{\rho_m} \frac{-j\pi r_m k_m^2}{\hat{y}_m} e^{j(\rho_m - r_m)} ln T_m d_{\rho m}$$

$$= - \frac{\pi r_m k_m^2 ln T_m}{\hat{y}_m} e^{j_m(\rho_m - r)} \; \Omega \tag{B3}$$

$$= - \frac{\pi r_m k_m^2 ln T_m}{\hat{y}_m} [\cos(\rho_m - r_m) + j \sin(\rho_m - r_m)] \tag{B4}$$

C. Obtain solution for the transformer surge impedance under lightning pulse. Consider effects of power loss and capacitive domination.

Solution

An approach for the calculation of a transformer surge impedance containing resistive and capacitive effects is based on the visualization of a lumped resistance across a lumped capacitance, subjected as a shunt combination to the induced lightning surge voltage $E(x,t)$.

$$I(x,t) = \frac{E(x,t)}{R} + C\frac{dE(x,t)}{dt} \tag{C1}$$

From Equation (6.94),

$$\frac{dE(x,t)}{dt} = E_o\delta'(t) + \frac{E_o}{(A_c - A_v)}[U_1(t) - \delta(t - t_1)]f(x) \tag{C2}$$

Therefore,

$$I(x,t) = \frac{E_o}{R}\,\delta(t) + \frac{E_o}{R(A_c - A_v)}[t - U_1(t - t_1)] +$$

$$CE_o\delta(t) + \frac{E_oC}{A_c - A_v}[U_1(t) - \delta(t - t_1)] \tag{C3}$$

$$Z(x,t) = E(x,t)/I(x,t)$$

$$= \frac{1}{R} + C\frac{E'(x,t)}{E(x,t)}$$

$$= \frac{1}{R} + C\,\ln E(x,t)$$

$$= \frac{1}{R(x,t)} + C\,\ln\left[E_o\delta(t) + \frac{E_o}{A_c - A_v}[t - U_1(t - t_1)]\right]$$

D.   Given $U_{r\omega} = 540$ kV, find H required in the design of minimum clearance between bushing terminal and an earthed object on a transformer tank. Consider a case where $h/H = 1$.

Solution

From Equation (6.100),

$$S_{\min} = \frac{U_{r\omega}^{1.6}}{540} = (1)^{1.6} = 1 \tag{D1}$$

Equation (6.99)

$$S = AH - \frac{U_B - U_{10}}{E} \tag{D2}$$

From Figure 6.10 for $S = 1m - U_{10} = 800$ kV, and $U_B = 1140$ kV

$$1 = 0.9H - \frac{1140 - 800}{130} \tag{D3}$$

Therefore, $H = 4m$.

6.IX PROBLEMS

1. Using electromagnetic field components of the induced field in a cylindrical transformer of M multi-winding layers expressed by Equations (6.5) through (6.10), obtain an expression for the Poynting vector P_ρ according to Equation (6.13).
2. Repeat problem 1 for the Poynting vector P_α.
3. Repeat problem 1 for the Poynting vector P_τ.
4. Using Equation (6.17), secure a solution for $V_{\tau m}$ (the axial velocity of propagation) in a cylindrical transformer.
5. Using winding pitch information and appropriate induced electromagnetic components, secure an expression for the radial velocity of propagation in a cylindrical transformer.
6. Repeat problem 5 to secure an expression for the component of the velocity of propagation.
7. Refer to Equation (6.34) expressing $E(x,s)$. Obtain a solution for $E(x,t)$ when the condition $\mu \gg lc$ is not valid.
8. Voltage breakdown conditions due to lightning surge could be guided by the criterion:

$$\frac{\partial E(x,t)}{\partial t} \to 0$$

From the solution of $E(x,t)$ secured in problem 7, establish the new condition of breakdown.
9. A multi-layer cylindrical transformer subjected to the incidence of lightning surge with an induced voltage secured in problem 7 has a lumped representation of a self-inductance L. Obtain a solution for the induced surge current and the resulting surge impedance.
10. Repeat problem 9 if the transformer has a lumped representation of $L(x,t)$ and the $R(x,t)$ in series.

11. Repeat problem 9 if the cylindrical transformer has a lumped representation of $R(x,t)$ in shunt with $C[c,t]$.
12. Repeat problem 9 if the cylindrical transformer has a lumped representation of $R(x,t)$, $L(x,t)$ and $C[x,t]$ in a shunt combination.
13. From solution data secured in problem 9, establish conditions for induced voltage breakdown due to lightning surge using the criteria $i(x,t) \to \infty$ and $\partial E(x,t)/\partial X \to 0$.
14. With respect to solution data secured in problem 10, establish conditions for induced voltage breakdown due to lightning surge using the criteria $i(x,t) \to \infty$ and $\partial E(x,t)/\partial X \to 0$.
15. With respect to solution data secured in problem 11, establish conditions for induced voltage breakdown due to lightning surge using the criteria $i(x,t) \to \infty$ and $E(x,t)/\partial X \to 0$.
16. With respect to solution data secured in problem 12, establish conditions for induced voltage breakdown due to lightning surge using the criteria $i(x,t) \to \infty$ and $\partial E(x,t)/\partial X \to 0$.
17. Using the solution of voltage breakdown obtained in problem 13, established a table similar to Table (6.1) showing a range for discrete locations along the transformer windings in terms of the power series index (n). Also make a plot for $X = F(n,R)U_s(n)$.
18. Repeat problem 17 regarding solution information obtained in problem 15.
19. Repeat problem 17 regarding solution information obtained in problem 16.
20. Establish expression for the vector surface surge impedance $Z_{\rho\alpha - m}$ for all ranges of argument. Reduce the expression toward a possible simplest form.
21. Calculate the total transformer surge impedance that an induced Poynting vector faces in the α-ρ plane.
22. Repeat problem 21 for the total transformer surge impedance in the $\rho - \alpha$ plane.
23. Modify the total surface surge impedances calculated in problems 20, 21 and 22 to take into account a grounded tank.
24. Given the nature of the induced lightning surge voltage as a chopped pulse, obtain an expression for the induced current through a cylindrical transformer. Consider that the transformer has a lumped representation of only a unified self-inductance. Also secure a solution for the time and space varying surge impedance.
25. Repeat problem 24 if the transformer has lumped representation of self-inductance in series with a resistance.
26. Repeat problem 24 if the inducing lightning voltage surge is an oscillatory waveform having a sharp rising wave front and decaying tail.
27. Using information established in problem 26 obtain a solution for the induced current and the time-space varying surge impedance, if the transformer has a lumped representation of self-inductance and resistance.

28. Repeat problem 24 if the inducing lightning voltage surge is an over-damped wave shape.
29. Repeat problem 24 if the inducing lightning voltage surge is a critically damped wave shape.
30. Repeat example D if $h/H = 0.3$.
31. Referring to Figure 6.9, derive an empirical relationship for U_{50} with respect to D at $U- = 1000$ kV; consider $H = 2.5$ m instead of 3.9 m indicated on Figure 6.9.
32. Referring to Figure 6.10, derive an empirical relationship for D with respect to U_{10}, for the case of bushing type 1, wet, and then for bushing type 1, dry.

PRINCIPAL LIST OF SYMBOLS

Z_m	impedivity of mth layer
Y_m	admitivity of mth layer
$\overline{Z}_{\rho\alpha}$	surface surge impedance in $\rho - \alpha$ plane
ν_m	cut-off frequency at the mth layer
$i(x,t)$	induced surge current
A_c	max. value of conductive current surge
A_v	max. value of convective current surge
$U_{r\omega}$	rated (SI) withstand phase voltage
$U_{r\omega p}$	rated (SI) withstand line-line voltage

REFERENCES

1. **Denno, K.,** Dynamic modeling for the process of inducing and induced voltage surges due to lightning, *J. Electrostat.*, 13, 55, 1982.
2. **Denno, K.,** Mathematical modeling of propagating inducing and induced power due to actual shape of lightning surge, *J. Electrostat.*, 15, 43, 1984.
3. **Denno, K.,** Three dimensional electromagnetic field model of power transformer with multi-layers winding, *Proc. Int. J. Energy Systems*, p. 10, 1985.
4. **Garabedian, P. R.,** *Partial Differential Equations*, John Wiley, New York, 1984.
5. **Harrington, R. F.,** *Time-Harmonic Electromagnetic Fields*, McGraw-Hill, New York, 1961.
6. **Rudenberg, R.,** *Electric Shock Waves in Power Systems*, Harvard Press, 1968.
7. **Denno, K.,** Steady-state and transient response of model of multi-winding layers transformer due to lightning surge, *J. Mathematical Modeling*, 1985.
8. **Denno, K.,** Computation of electromagnetic lightning response using moments method, *IEEE Trans. Magnetics,* p. 1953, Sept. 1984.
9. **Studinger, H. and Weck, K. H.,** Switching-impulse strength of the external insulation of three phase power transformers, *IEE Proc.*, 133, (8), 534, 1986.
10. **Denno, K.,** Induced voltage distribution and breakdown level in power transformer due to lightning surge, *Proc. 3rd Conf. on Electrostatics*, Technical University Press, Technical University of Wroclaw, Poland, Sept. 1985, pp. 265–289.

Chapter 7

EFFECTS OF LIGHTNING SURGES ON TOWERS

This chapter discusses the effects of incident lightning scores on transmission line towers or ground wire, relying on experimental and theoretical research carried out by Gary, Garabedian, Zhang, Cristescu, and Enache. Results of their research indicated that when a lightning surge impacts a tower or ground wire, the resulting overvoltage is identified in its having a considerably shorter fail than the induction lightning surge. Their study also indicated that the back flash-over voltage regarding the withstand level of insulator strings is much higher than that expected from conventional limit resulting from the standard impulse of lightning test surge characterized by durations of 1.2/50 μs.

The problem of improvement in insulation coordination along a transmission line system has direct dependence on the withstand voltage level of insulation string, lightning-induced voltages on line conductors, and the combined effect on line-line as well as line-ground. The Gary et al. study also presented ways to minimize the shielding failure flash-over rate and the proper location of ground wire to allow passage of nondangerous current surges while safeguarding against back flash-over that may occur due to the unification of the descending negative lightning stroke and the positive streamer from towers and ground wires. Table 7.1 shows data for the number of ground wires, and the rate of lightning strokes on towers and at a point on a span between towers taken during several years of measurements.

7.1 SHIELDING FAILURE

Shielding failure due to insulation breakdown stress has been associated with the calculation of induced voltage at the strike point of lightning surge using a simple circuit model involving the lightning current channel surge impedance and the line surge impedance, which may change due to corona effect. This circuit model uses either a time varying current generator $l'_g(t)$ with a characteristic impedance for the lightning channel Z_{ch} or a voltage generator $e_g(t)$ in connection with the characteristic impedances for the line and light

$$e_g(t) = l'_g(t)\frac{Z_0}{2} \tag{7.1}$$

TABLE 7.1
Stroke Number Distribution

Impact point	Ground wire number			
	0	**1**	**2**	**3**
Tower, 0/0	55	35	20	10
Span, 0/0	45	65	80	90

or

$$e_g(t) = l'_g(t) \frac{Z_{ch}Z_0}{2Z_{ch} + Z_0} \tag{7.2}$$

where Z_0 is the characteristic impedance of the phase conductor.

The improvement made in the effectiveness of the impedance circuit model mentioned above was in the introduction of variable surge impedance of the transmission line due to the propagation of an incident wave whose amplitude exceeds the corona threshold limit U_0. Consequently, in the study conducted by Gary et al.,

$$Z_d = \sqrt{\frac{L}{C_d}} \tag{7.3}$$

where

$$C_d = \frac{dQ(t)}{dV(t)}$$

which depends on line geometry and stroke polarity; Q is the conductor charge at time (t), and the instantaneous impedance Z_i is given by

$$Z_i = V \Big/ \int_0^v \frac{dV}{Z_d} \tag{7.4}$$

The over-voltage level calculated in this way takes into account the clearance between line conductor and ground for a wave shape that is identical to the lightning impulse characterized as of 1.2/50 μs (front rise and tail). Such a line-ground over-voltage will be subjected across the insulator string of the transmission line tower. Calculation of over-voltage stress can be secured from

$$V_{ins}(t) = (1 - K_{cd}) - V_a(t) \tag{7.5}$$

where K_{cd} represents the dynamic coupling coefficient and $V_a(t)$ is the over-voltage stress on the line conductor. Also,

$$K_{cd} = K_g \frac{Z_g}{Z_i} \tag{7.6}$$

where K_g is the geometric coupling coefficient and Z_g is the geometric characteristic impedance; and

$$K_{cd}(t) = K_g \left[1 - \left(1 - \sqrt{\frac{C_g}{C_d}} \right) \frac{V_0}{V(t)} \right] \sqrt{\frac{C_d}{C_g}} \tag{7.7}$$

where V_0 is the voltage from the bottom of the insulation string to the ground and $V(t)$ is the over-voltage induced at the top of the tower. It is reported in the study by Gary et al.[1] that an increase in the coupling coefficient due to corona within the dynamic impedance model will lead to a decrease in the probability of flash-over by about 7% for positive polarity and 5.5% for negative polarity when shielding failure does occur. Their observation indicated that impulse corona will lengthen the duration of over-voltage front wave, but will not affect the tail duration.

7.II STRESS DUE TO LIGHTNING IMPULSE AT TOWERS AND GROUND WIRES

In order to reduce the rate of shielding failure and the rate of transmission line outages, an effective design of ground wire has to be implemented coupled with detailed analysis for amplitude and wave shape for the expected over-voltage wave in order to avoid a back flash-over.

Calculations for the inducing electromagnetic field effects, including induced current densities in towers and ground wires, involve complicated mathematical problems that take into consideration the coupling among neighboring towers, their spans and the assembly of ground wire, besides transmission lines placed at the joints of insulator strings. Such complications had been shown in preceding chapters for cases of transmission lines, power transformer, and gaseous continua. However, the study conducted by Gary et al. presented the following approach for the present situation.

For a cylindrical tower, its characteristic impedance, $Z_i(t)$, is given by:

$$Z_{t-ch}(t) = 60\ln \frac{C_t}{r_t} \tag{7.8}$$

where C is the velocity of light and r_t is the effective tower radius, and for conical tower,

$$Z_{t-con} = \ln \frac{\sqrt{2}}{\sin\alpha} \tag{7.9}$$

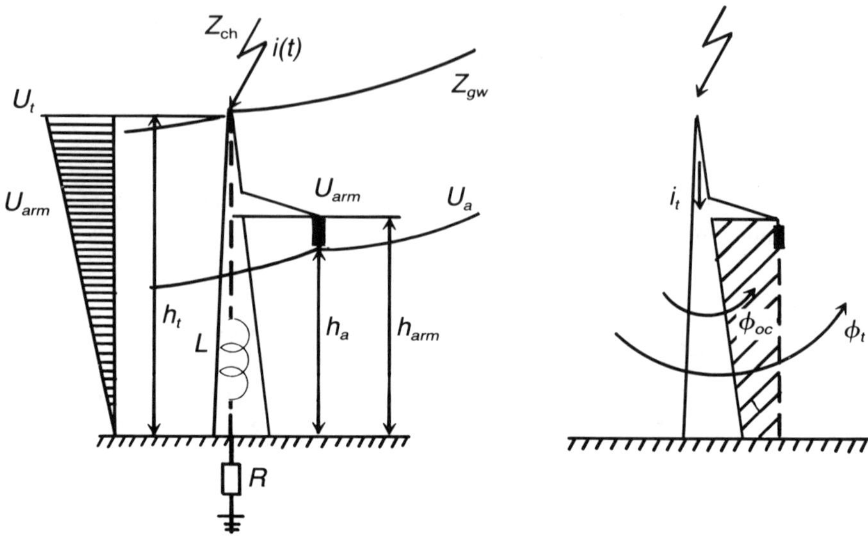

FIGURE 7.1 Equivalent diagram of a tower, used for the calculation of the overvoltages across the insulator string.

where α is half the cone peak angle; the tower self-inductance, L, is given by:

$$L = Z_t T = Z_t \frac{h_t}{C} \tag{7.10}$$

where T is the travel time through the tower while Z_t is the tower characteristic impedance and h_t is the tower height.

It is relevant to mention according to this study that the net voltage that will be subjected on the insulator string is the difference between the induced voltage the tower due to lightning voltage impact at the top of the tower or at any point on the line span and the line voltage due to normal electromagnetic coupling in the transmission line system.

A lumped circuit clement representation with the basic concept of traveling wave theory had been used to account for acceptable results as shown in Figure 7.1, in which R and L represent the lumped foot resistance and self-inductance of the tower, where the principle of linearity could be used. Therefore,

$$U_{drm} = U_t \frac{h_{arm}}{h_t} \tag{7.11}$$

Induced voltage on the ground wire will lead to a corresponding voltage on the line of K_i, V_t due to electromagnetic coupling between any two line conductors which takes into account corona effect.

In the Gary et al. study the principle of field superposition involving the scalar electric potential due to electric charge carried out by the ascending lightning stroke and the magnetic vector potential produced by the released stroke electric current were used to calculate the total voltage stress subjected on the insulator string whereby

$$V(P) = \frac{1}{4\pi f_0} \int_{\Delta Q_0}^{Q} \left[Q\!\left(X, t - \frac{v}{c}\right) \Big/ r \right] dx \tag{7.12}$$

and

$$\overline{A}(P) = \frac{\mu}{4\pi f_0} \int_{\Delta l_0}^{l} \left[I\!\left(X, t - \frac{v}{c}\right) \Big/ r \right] dx \tag{7.13}$$

where P is a point distant (r) from lightning channel element (dx)

$$Q = \frac{I}{v} \text{ coulombs} \tag{7.14}$$

v is the return stroke velocity in m/s.

According to work that was carried out at Saint Privat d'Allier, the overvoltage on insulator string is given by:

$$U_{ins} = \frac{h_{arm}}{h_t} U_t t - U_{induced} \neq U_{ac} \tag{7.15}$$

$$= \frac{h_{arm}}{h_t} U_t t - [K_{cd} U_t + U_{lm} - U_{ls}] + 0.5\, U_{ac} \tag{7.16}$$

where V_{ac} is the conventional power system voltage.

The expression in Equation (7.16) includes effects of electromagnetic coupling such as between the line conductor and ground wire (K_{cd}) and between the stroke channel and line conductor identified by K_{is} and K_{lm}. It also includes

$$U_{lm} = M \frac{dl'\, t}{dt} \tag{7.17}$$

where M is the mutual inductance between the tower as a vertical conductor and the transmission line as the horizontal conductor, and $i(t)$ is the current through the tower.

The tower self-inductance L is expressed by:

$$L = \frac{Q_t}{I} \qquad (7.18)$$

where Q_t is the total magnetic flux through the tower structural surface and ground.

The specific inductance per meter L_0 of tower height h_t is given by

$$L = L_0 h_t \qquad (7.19)$$

The study by Gary and associates also observed that the magnetic flux Q_{dc} passing through a surface containing the tower axis, tower arm, line conductor, and the tower vertical axis is half that of Q_t, and the mutual coupling between the tower vertical axis and transmission line conductor is expressed by

$$M = 0.5L \, \frac{h_a}{h_t} \qquad (7.20)$$

where h_a is the arm height of the insulation suspension point. Therefore, from Equations (7.17) and (7.20):

$$U_{lm} = 0.5[L_0 h_a] \, \frac{dl'\,t}{dt} \qquad (7.21)$$

Consequently, expression for the insulation string voltage becomes:

$$U_{ins}(t) = \frac{h_{arm}}{h_t} \, U_t(t) - \left[K_{cd}(t)U_t(t) + 0.5 \, L_0 h_e \, \frac{di(t)}{dt} \, U_{ls}(t) - 0.5U_{ac} \right] \qquad (7.22)$$

7.III OVER-VOLTAGE AT TOWER OR GROUND WIRE STROKE

The study conducted by Gary et al. relied on parametric representation through digital and analog simulation for the problem of predetermination of induced over-voltage due to a lightning stroke on a tower span between towers and ground wire. Simplifying assumptions had been used including discarding of the parameter (M) and accounting for the effects of neighboring towers, their spans, and ground wire by an empirical correcting factor of 10 Ω in the parametric circuit representation as shown in Figure 7.2. Table 7.2 shows values of all parameters used in the study.

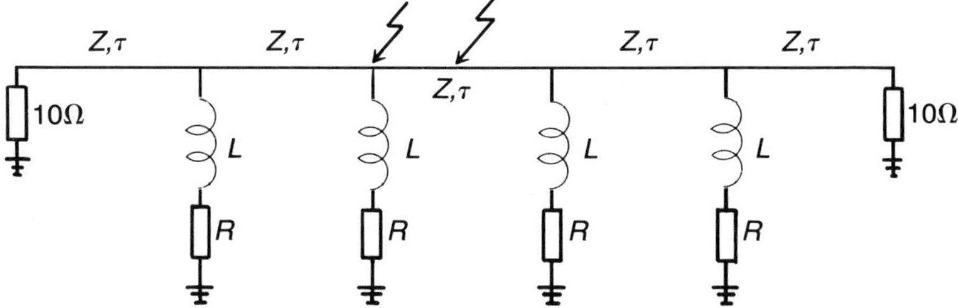

FIGURE 7.2   Equivalent diagram of a line used for the computation of overvoltages U_{arm}. L, $R =$ inductance and footing resistance of the struck tower and of adjacent towers; Z, $\tau =$ surge impedance and propagation time related to a span of the ground wire.

The study carried out by Gary et al. reported the solution for the induced voltage at the tower as:

$$U(t) = \frac{i(t)Z_{gw}}{2} - Ak_1\left\{\sum_{i=2}^{n} \tau(\theta_1)k_1^{i-2}\right.$$

$$\times \left\{e^{-k_2\theta_1}k_1^{i-2}\sum_{k=1}^{i-1}\left[\frac{1}{(k_2 - l)^k} - \frac{1}{(k_2 - m)^k}\right]\right.$$

$$\times \frac{\theta_1^{i-k-1}}{(i - k - 1)!} + \frac{e^{-m\theta_1}}{(k_2 - m)^{i-1}} - \frac{e^{-l\theta_1}}{(k_2 - l)^{i-1}}\right\}$$

$$+ \sum_{j=2}^{n'} \tau(\theta_2)k_1^{j-2}\left\{e^{-k_2\theta_2}\right.$$

$$\times \sum_{r=1}^{j-1}\left[\frac{1}{(k_2 - l)^r} - \frac{1}{(k_2 - m)^r}\right]\frac{\theta_2^{j-r-1}}{(j - r - 1)!}$$

$$\left.\left.+ \frac{e^{-m\theta_2}}{(k_2 - m)^{j-1}} - \frac{e^{-l\theta_2}}{(k_2 - l)^{j-1}}\right\}\right\} \tag{7.23}$$

$$k_1 = \frac{Z_{ch}Z_{gw}^2}{2L(2Z_{ch} + Z_{gw})}$$

$$k_2 = \frac{Z_{gw} + 2R}{2L} \tag{7.24}$$

$$\theta_1 = t - 2(i - 1)\frac{\alpha d}{c}$$

$$\theta_2 = t - 2(j - 1)\frac{(1 - \alpha)d}{c} \tag{7.25}$$

where A, l, and m are parameters of the incident impulse $i(t)$:

$$i(t) = A(e^{-mt} - e^{-lt}) \tag{7.26}$$

$$A = \frac{1}{m} e^{-\ln/m/(m/l - 1)} \frac{I_{max}}{l - m} \tag{7.27}$$

$$\tau(t - \theta) = \begin{cases} 0 & t \leq \theta \\ 1 & t > \theta \end{cases} \tag{7.28}$$

R and L are the footing resistance and the tower inductance, I_{max} is the lightning current magnitude, d is the span $\rightarrow$ length,

α is the position of the point of impact on the span [$\alpha = 0$ for tower and $\alpha = 1$ for midspan], Z_{gw} is the ground wire characteristic impedance and T $(1 - \theta)$ is the Heariside function.

Results have been expressed in terms of magnitude and wave shape of induced over-voltage at the point of impact which include the top first, and end of a tower, adjacent towers, and the span between towers.

Tables 7.3 to 7.7 show ratios of V/I as well as other system parameters.

Figures 7.3 through 7.6 show plots for relative over-voltage with respect to a nondimensional time element or ratio. This ratio is that of the tower time constant L/R and T_g known as the front conventional duration of the incident wave (0.5 to 5.0 μs). The tower time constant L/R is in the range 0.5 to 6.6 μs.

Figure 7.7 shows correlation between the ratio of over-voltage tail duration and incident lightning stroke duration tail with respect to $L/R\ T_g$. This study reported a very useful conclusion: for a lightning stroke on the tower or ground wire, the back flash over-voltage may occur when tail duration is within the range of 2 to 15 μs.

Figures 7.8 through 7.12 show lightning over-voltage oscillograms secured by digital results. This study ends with the conclusion that reliable determination of over-voltage and its timing will lead to accurate predetermination of the rate of lines, rate of outage, and consequently to smooth coordination of insulation systems on the tower, ground wire, and at a point on a tower span.

7.IV SOLVED EXAMPLES

A. Given the inducing lightning over-voltage on a transmission line tower expressed by Equation (2.70), and given $Z_0 = 500 - z$, obtain expression for the lightning characteristic impedance. Use the lumped circuit model given by Equation (7.2).

TABLE 7.2
Input Data for Parametric Study

	0.5/80	0.53/26.2	1.0/80	1.03/117	2.22/219	2.5/80	5.0/80
Type of wave, $\mu s/\mu s$	0.5/80	0.53/26.2	1.0/80	1.03/117	2.22/219	2.5/80	5.0/80
Surge impedance, ω	260;450	330	260;450	330	330	260;450	260;450
Lightning channel impedance, ω	∞	330	∞	330	330	∞	∞
Tower footing resistance, ω	5;20	9;18;27	5;20	9;18;27	9;18;27	5;20	5;20
Tower inductance, μH	15;30	25	15;30	25	25	15;30	15;30
Span length, m	450	100;200;300	450	100;200;300	100;200;300	450	450

TABLE 7.3
Tower Lightning Stroke, U/I of the Tower

Z_c,	Ω	260				330			450			
R,	Ω	5		20		9	18	27	5		20	
L,	μH	15	30	15	30	23	24	25.8	15	30	15	30
	0.5	36.35	53.20	39.92	54.96	52.5	57.6	67.5	43.27	68.02	46.99	71.31
T_f,	1.0	24.47	38.13	28.02	40.21	50.5	54.0	58.5	28.53	45.63	31.39	48.19
μs	2.5	13.19	21.90	17.57	24.63	26.5	29.5	33.7	14.33	24.82	18.61	27.70
	5.0	7.62	13.16	15.87	17.50				7.96	14.25	16.85	18.57

TABLE 7.4
Tower Lightning Stroke, *U/I* Second Tower

Z_c,	Ω	260				330			450			
R,	Ω	5		20		9	18	27	5		20	
L,	μH	15	30	15	30	23	24	25.8	15	30	15	30
	0.5	12.01	26.23	14.47	28.49	14.0	16.5	21.2	10.53	23.69	12.41	26.22
T_f,	1.0	8.18	18.09	10.20	19.50	12.0	14.5	18.2	6.48	16.59	7.98	18.57
μs	2.5	4.11	8.82	5.05	9.97	6.5	8.0	10.0	3.07	8.29	4.06	9.21
	5.0	2.17	4.84	4.24	5.45				1.64	4.37	4.04	4.84

TABLE 7.5
Midspan Lightning Stroke, *U/I* Midspan

Z_c,	Ω	260				330			450			
R,	Ω	5		20		9	18	27	5		20	
L,	μH	15	30	15	30	23	24	25.8	15	30	15	30
	0.5	130.0	130.0	130.0	130.0	183.5	183.5	183.5	225.0	225.0	225.0	225.0
T_f,	1.0	129.1	129.1	129.1	129.1	172	172	172	223.5	223.5	223.5	223.5
μs	2.5	107.6	107.6	107.6	107.6	142.5	142.5	142.5	186.3	186.3	186.3	186.3
	5.0	75.5	76.1	75.6	76.1				129.8	130.8	129.9	130.9

TABLE 7.6
Midspan Lightning Stroke, U/I First Tower

		Z_c, Ω	260				330			450			
		R, Ω	5		20		9	18	27	5		20	
		L, μH	15	30	15	30	23	24	25.8	15	30	15	30
	0.5		34.84	51.55	37.98	54.04	37.0	38.4	43.5	42.38	64.54	46.32	67.05
T_f,	1.0		24.32	37.52	27.55	40.19	32.5	34.5	37.0	27.23	45.60	31.68	48.41
μs	2.5		13.09	21.72	17.49	24.62	18.5	21.0	24.25	14.12	24.60	18.55	27.59
	5.0		7.62	13.13	13.04	16.75				7.95	14.21	13.89	17.88

TABLE 7.7
Midspan Lightning Stroke, U/I Second Tower

		Z_c, Ω	260				330			450			
		R, Ω	5		20		9	18	27	5		20	
		L, μH	15	30	15	30	23	24	25.8	15	30	15	30
	0.5		14.9	29.29	17.39	30.84	9.20	10.5	14.4	12.96	29.80	15.0	32.14
T_f,	1.0		10.0	20.82	11.0	19.72	7.8	10.2	12.2	6.8	20.04	9.68	20.34
μs	2.5		5.27	11.55	5.68	10.47	4.25	5.37	6.72	4.87	11.33	5.34	10.60
	5.0		2.88	6.48	3.86	6.10				2.68	6.04	3.75	5.98

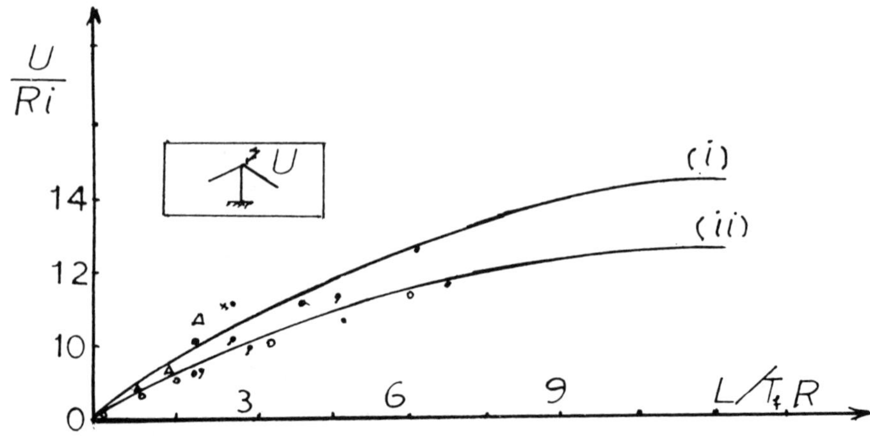

FIGURE 7.3   Lightning on a tower: relative overvoltage U/Ri as a function of the parameter L/T_fR. (i) $Z_c = 450\ \Omega$; (ii) $Z_c = 260\ \Omega$; ($\triangle$) $Z_c = 330\ \Omega$, $d = 100$ m; ($\bullet$) $Z_c = 450\ \Omega$, $d = 450$ m; ($\bullet$) $Z_c = 330\ \Omega$, $d = 200$ m; ($\otimes$) $Z_c = 260\ \Omega$, $d = 450$ m; ($\times$) $Z_c = 330\ \Omega$, $d = 300$ m.

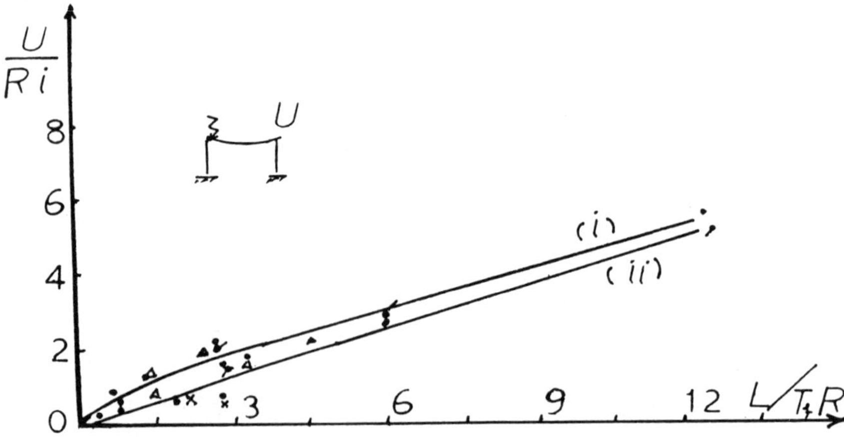

FIGURE 7.4   Lightning on an adjacent tower: relative overvoltage U/Ri as a function of the parameter L/T_fR. (i) $Z_c = 450\ \Omega$; (ii) $Z_c = 260\ \Omega$; ($\triangle$) $Z_c = 330\ \Omega$, $d = 100$ m; ($\bullet$) $Z_c = 450\ \Omega$, $d = 450$ m; ($\bullet$) $Z_c = 330\ \Omega$, $d = 200$ m; ($\otimes$) $Z_c = 260\ \Omega$, $d = 450$ m; ($\times$) $Z_c = 330\ \Omega$, $d = 300$ m.

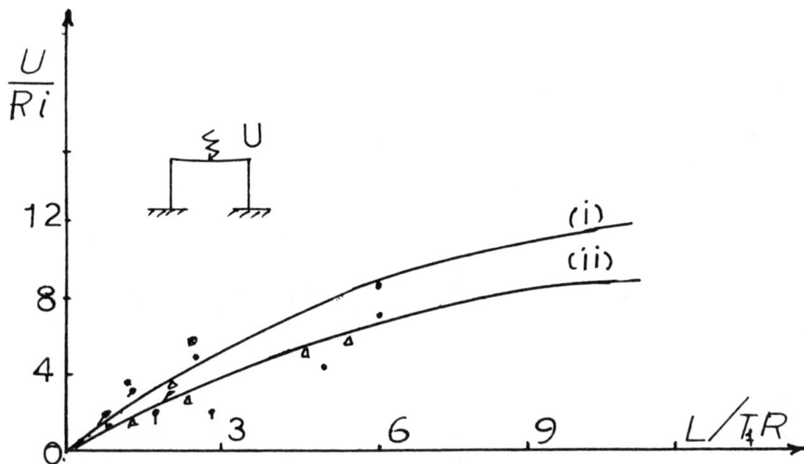

FIGURE 7.5   Lightning at midspan of the ground wire: relative overvoltage U/Ri on the adjacent towers, as a function of the parameter L/T_fR. (i) $Z_c = 450\ \Omega$; (ii) $Z_c = 260\ \Omega$; ($\triangle$) $Z_c = 330\ \Omega$, $d = 100$ m; ($\bullet$) $Z_c = 450\ \Omega$, $d = 450$ m; ($\bullet$) $Z_c = 330\ \Omega$, $d = 200$ m; ($\otimes$) $Z_c = 260\ \Omega$, $d = 450$ m; ($\times$) $Z_c = 330\ \Omega$, $d = 300$ m.

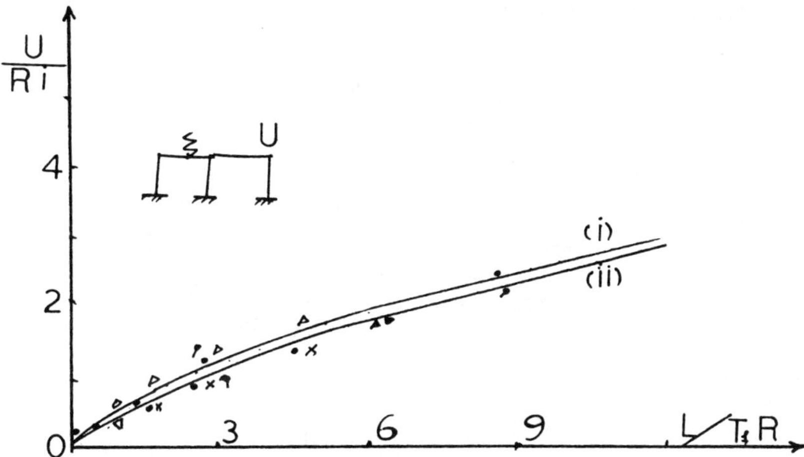

FIGURE 7.6   Lightning at midspan of the ground wire: relative overvoltage U/Ri on the second tower, as a function of the parameter L/T_fR. (i) $Z_c = 450\ \Omega$; (ii) $Z_c = 260\ \Omega$; ($\triangle$) $Z_c = 330\ \Omega$, $d = 100$ m; ($\bullet$) $Z_c = 450\ \Omega$, $d = 450$ m; ($\bullet$) $Z_c = 330\ \Omega$, $d = 200$ m; ($\otimes$) $Z_c = 260\ \Omega$, $d = 450$ m; ($\times$) $Z_c = 330\ \Omega$, $d = 300$ m.

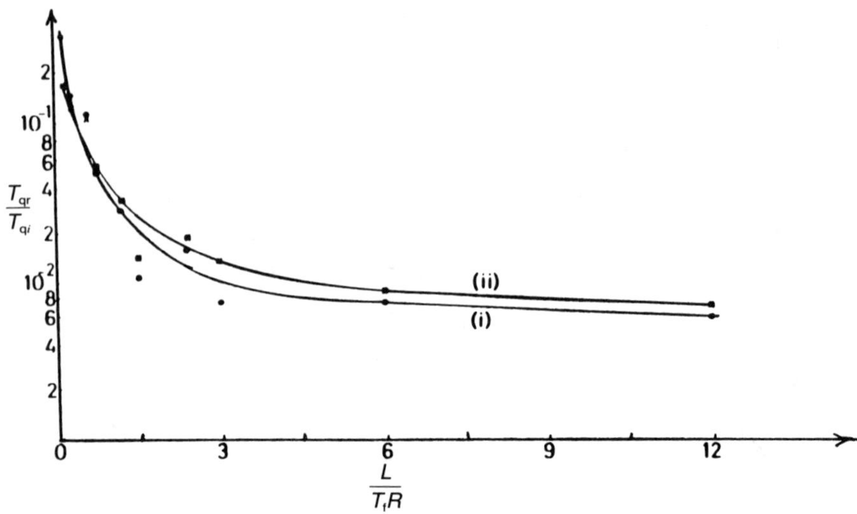

FIGURE 7.7   Lightning on a tower: ratio T_{qr}/T_{qi} as a function of the parameter L/T_fR. (i) $Z_c = 450$ Ω; (ii) $Z_c = 260\ \Omega$; $T_{qr} =$ time to half-value of the tower response; $T_{qi} =$ time to half-value of the incident lightning current.

Solution

   The inducing voltage is given by

$$U_i = [J_c(t) - J_v(t)]\sqrt{\frac{j}{\pi}}\frac{1}{R_{c,v}\hat{Y}}\sum_{n=0}^{\infty}\frac{j^n}{2^n n!}$$

$$\left[\frac{\rho^{n+\frac{1}{2}}\tau XP(-jn\rho)}{jn^2 R_{c,v}} - \frac{n+\frac{1}{2}}{jn^2 R_{c,v}}\int\rho^{n-\frac{1}{2}}\exp(-jn\rho)d\rho\right.$$

$$+\frac{\rho^{n-1}\exp(-jn\rho)}{j^n} - \frac{n-1}{j^n}\int\rho^{n-2}\exp(-jn\rho)d\rho$$

$$\left.+\frac{j2^{n-1}(n-1)!}{3\sqrt{2}}\rho^{3/2}\right] = e_g(t) \tag{A1}$$

$$l_g'(t) = J_g(t)A \tag{A2}$$

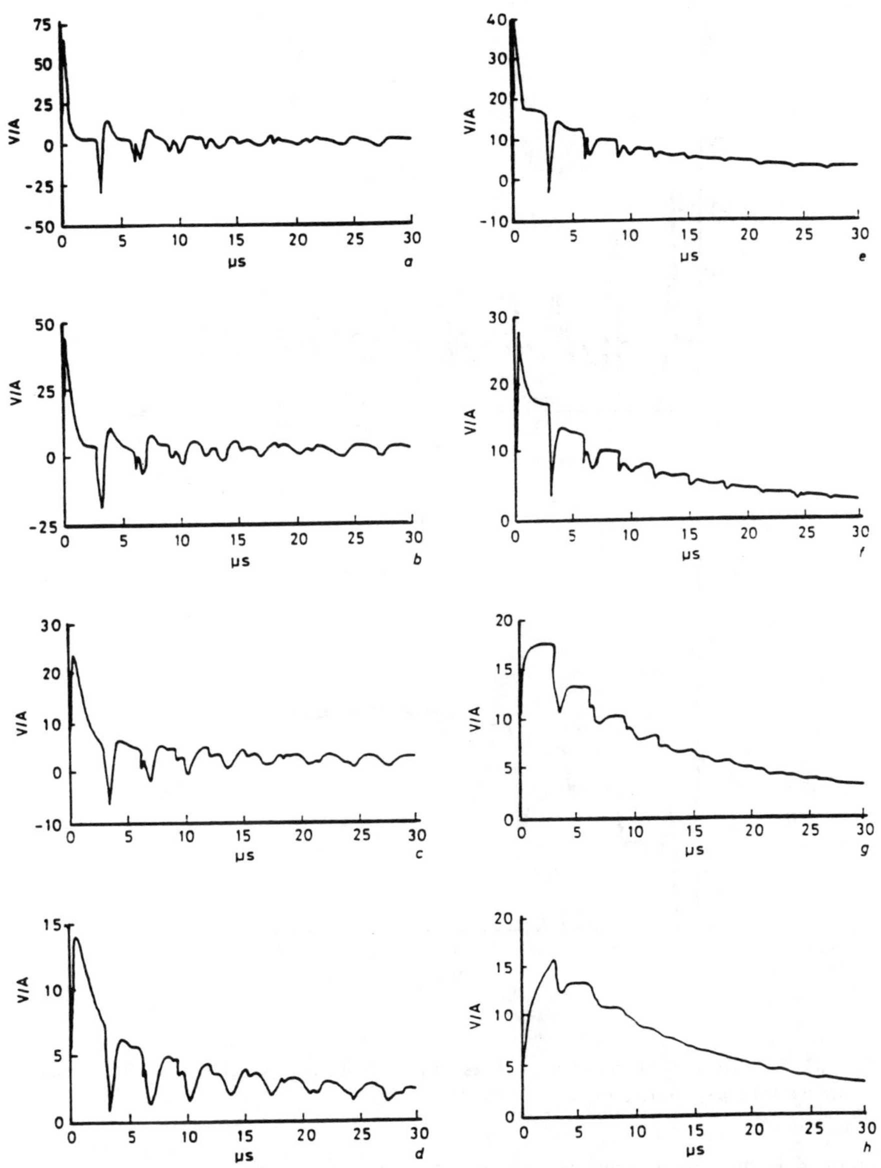

FIGURE 7.8   Tower lightning stroke, *U/I* tower—digital results: (a) 0.5 μs, 30 μH, 5 ω, Z_c = 450 Ω; (b) 1 μs, 30 μH, 5 Ω, Z_c = 450 Ω; (c) 2.5 μs, 30 μH, 5 Ω, Z_c = 450 Ω; (d) 5 μs, 30 μH, 5 Ω, Z_c = 450 Ω; (e) 0.5 μs, 15 μH, 20 Ω, Z_c = 260 Ω; (f) 1 μs, 15 μH, 20 Ω, Z_c = 260 Ω; (g) 2.5 μs, 15 μH, 20 Ω, Z_c = 260 Ω; (h) 5 μs, 15 μH, 20 Ω, Z_c = 260 Ω.

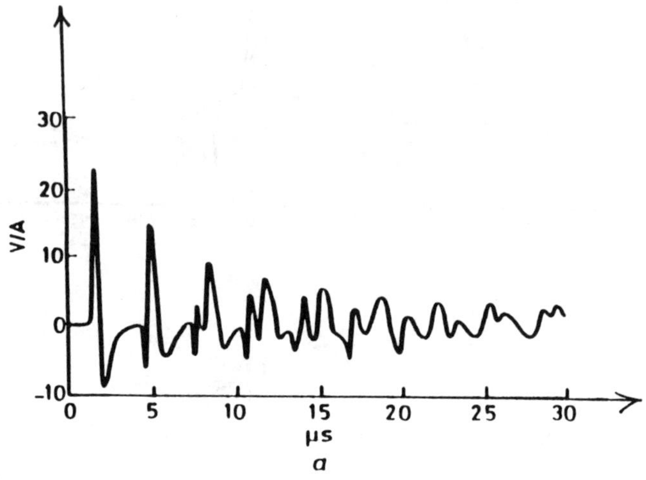

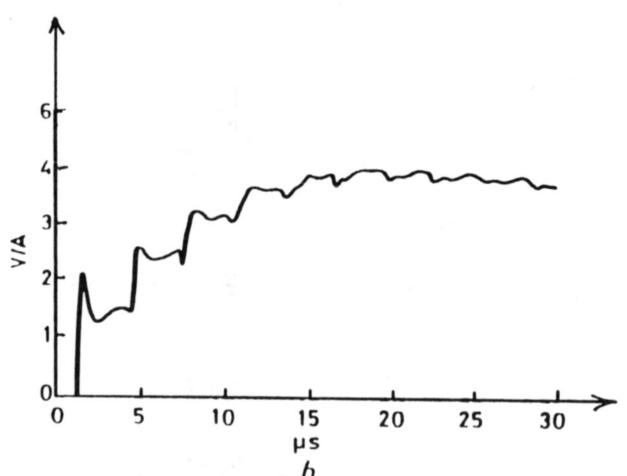

FIGURE 7.9   Tower lightning stroke, *U/I* first tower—digital results. (a) 0.5 μs, 30 μH, 5 Ω, Z_c = 450 Ω; (b) 5 μs, 15 μH, 20 Ω, Z_c = 450 Ω.

where *A* is the tower plane containing the arm and vertical axis.

$$J_g(t) = J_c(t) = A_c(t - t_1)U_{-1}(t) \quad \text{for } t > t_1$$

$$= A_c t \quad \text{for } 0 < t < t_1 \tag{A3}$$

$$J_g(t) = J_v(t) = A_v(t - t_1)U_{-1}(t) \quad \text{for } t > t_1$$

$$= A_v t \quad \text{for } 0 < t < t_1 \tag{A4}$$

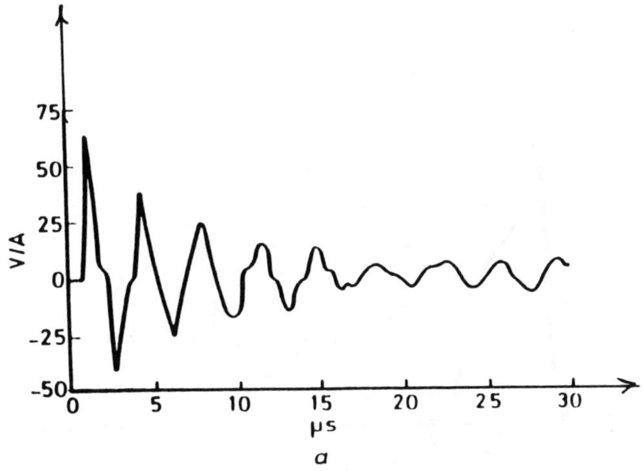

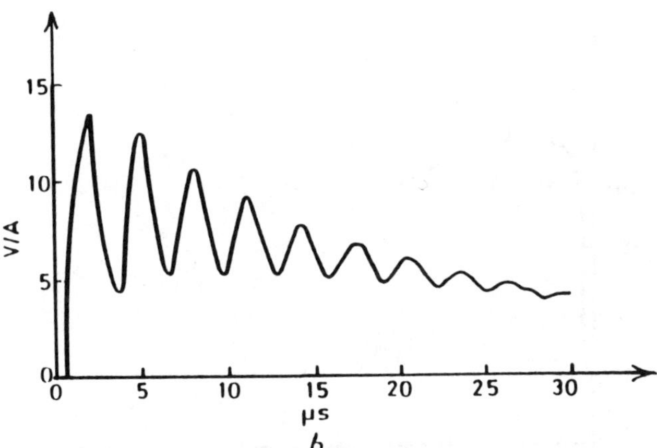

FIGURE 7.10 Midspan lightning stroke, U/I first tower—digital results. (a) 0.5 μs, 30 μH, 5 Ω, $Z_c = 450$ Ω; (b) 5 μs, 15 μH, 20 Ω, $Z_c = 450$ Ω.

Therefore, using Equation (7.2) in conjunction with Equations (A1), (A2), (A3), and (A4)

$$\frac{e_g(t)}{l'_g(t)} = \frac{Z_{ch}Z_0}{2Z_{ch} + Z_0} = K(t)$$

$$2K(t)Z_{ch} + Z_0K(t) = Z_{ch}Z_0 \tag{A5}$$

$$Z_{ch}[2K(t) - Z_0] = -Z_0K(t)$$

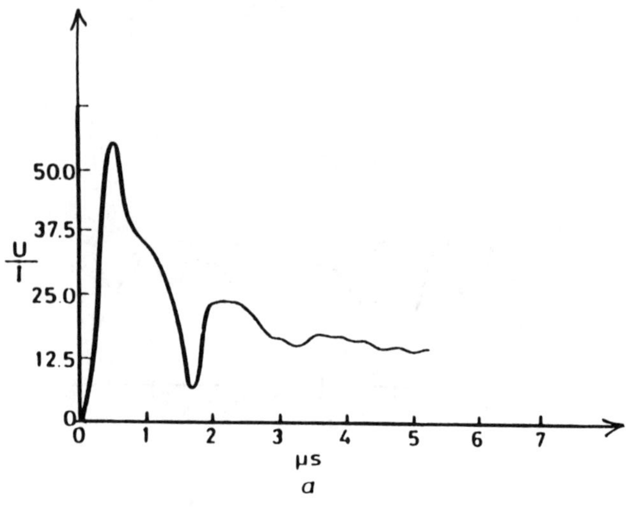

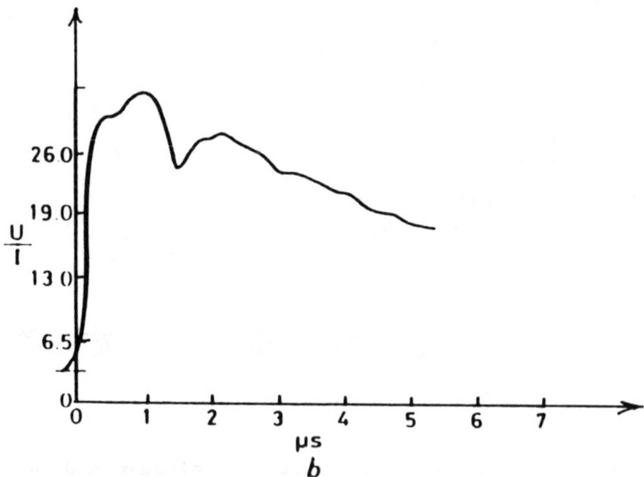

FIGURE 7.11   Tower lightning stroke, U/I tower—analogue results.

Hence,

$$Z_{ch} = -Z_0 K(t)/[2K(t) - Z_0] \tag{A6}$$

B. Using Equation (7.5), obtain an expression for the over-voltage $V_{ins}(t)$ across an insulation string at the arm of a transmission line tower, for $C_d \to \infty$.

Solution

$U_a(t)$ in Equation (7.5) represents the induced lightning voltage surge on a transmission line. Selecting the induced voltage developed by a lightning current

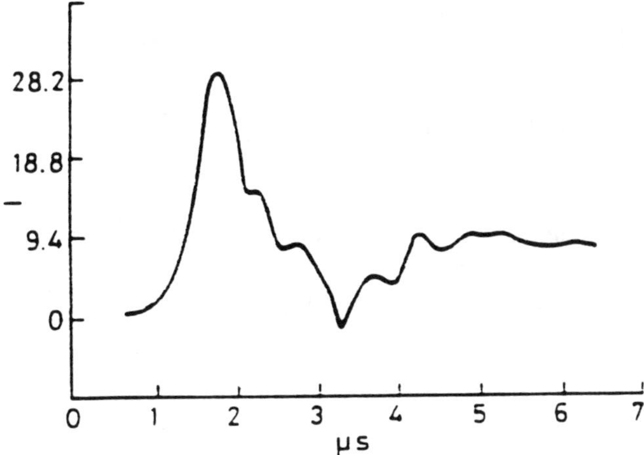

FIGURE 7.12   Midspan lightning stroke, U/I first tower—analogue results.

surge represented by a sharp rising linear front and a slow declining linear tail, $U_a(t)$ is given by Equation (2.70).

If $U_a(t)$ [the over-voltage induced on the line conductor]

= $U(t)$ [the over-voltage induced on the top of tower],

Therefore, from Equations (7.5) and (7.7):

$$U_{ins}(t) = 1 - K_g\left[1 - \left(1 - \frac{\sqrt{C_g}}{C_d}\right)\frac{U_0}{U_a(t)}\right]\sqrt{\frac{C_d}{C_g}} - U_a(t) \qquad (B1)$$

where $U_a(t)$ is expressed by Equation (2.70).

From Equation (B1),

$$U_{ins}(t) = 1 - \frac{U_0}{U_a(t)} - U_a(t) = 1 - \frac{U_0 - U_a^2(t)}{U_a(t)} \qquad (B2)$$

C. Consider a transmission line tower is subjected to a lightning over-voltage breakdown with current flow through the entire cylindrical tower continuum. If U_{is} is negligible and $U_t = U_{induced}$, obtain simplified expression for the insulator string voltage after a duration as to $U_{ac} = K_{ph}$ volts.

Solution

From Equation (7.5) with $U_t = U_{induced}$ becomes,

$$U_s = \frac{h_{arm}}{h_t} t \, U_{induced} - U_{induced} + U_{ac} \qquad (C1)$$

Since lightning surge breakdown is occurring right through the tower itself, Equation (3.38) represents solution for the induced tower over-voltage, whereby

$$U_{induced} = -j2_0\mu\delta_0(t - t_0)[Q_c - Q_v] \sum_{\substack{m_c=1 \\ m_v=1}}^{\infty} \sum_{\substack{n_c=1 \\ n_v=1}}^{\infty} \left(Y_{n,m}Z_{n,m} \frac{e^{-jX_{n,m}}}{X_{n,m}}\right) \quad (C2)$$

$U_{induced}$ in Equation (C2) above was derived in Chapter 3 on the basis of summation of magnetic moments.

Therefore, with $U_{induced}$ given by Equation (C2)

$$U_s(t) = U_{induced}\left[\frac{h_{arm}}{h_t} t - 1\right] + K_{ph} \quad (C3)$$

At $\quad t = t_0, \; \delta(t - t_0) = 1$

$$U_{induced} = -j2\mu(Q_c - Q_c)\sum_{c,v} \sum_{c,v} Y_{n,m}Z_{n,m} \frac{e^{-jX_{n,m}}}{X_{n,m}} \quad (C4)$$

7.V PROBLEMS

1. Lightning surge impacted an overhead transmission tower with an oscillatory induced over-voltage. Obtain an expression for the current surge through the tower using the basis of Equation (7.2).
2. Repeat problem 1 if the surge-induced voltage is an over-damped wave.
3. With the definition of tower self-inductance $L = dQ/di$, obtain solution for (L) when the voltage induced at the top of the tower $U(t)$ is the same as that induced at a transmission line (lightning stroke is a step time function with cochlear tail).
4. For the same situation described in problem 3, obtain solution for the dynamic capacitance C_d in accordance with the definition, $c_d = dQ(t)/dU(t)$.
5. From results and conditions of problems 3 and 4, obtain expression for the dynamic impedance of the tower Z_d in accordance with the definition given by Equation (7.3).
6. From results obtained in problem 5, secure solution for the tower instantaneous impedance $Z_i(t)$ in accordance with the definition given by Equation (7.4).
7. Consider lumped representation of a transmission line tower in terms of (L) and (R). Using solution for an inducing lightning voltage surge secured by summation of magnetic moments, obtain an expression for the tower-induced over-voltage $U(t)$.
8. Using results from problem 7 and with $U_a(t)$ in Equation (7.5) representing the voltage induced on a transmission line, solve for $K_{cd}(t)$. Consider $U_a(t)$ is due to an actual lightning surge with a sharp voltage and solve decaying linear rail.

9. Refer to Figure 7.2 for an actual lightning surge characterized as a sharp linear rise and slow linearly decreasing tail, and obtain an expression for the over-voltage induced at a span of ground wire.

10. From results secured in problem 9, obtain expression for the insulator string over-voltage U_{ins}.

11. Using Equations (7.23) to (7.28), refer to the calculation of the tower-induced over-voltage $U(t)$. Obtain expression for the tower system time varying surge impedance $Z(t)$.

12. Using $U(t)$ given by Equation (7.23) and the tower lumped representation in terms of L and R, obtain solution for tower current $I(t)$; one can assume zero initial and universal normal conditions.

13. From the result of problem 12 obtain expression for the tower time varying surge impedance $Z(t)$.

14. Referring to Figure 7.4, secure empirical mathematical representation for the relative over-voltage ratio with respect to L/T_gR. Deviation of the desired relationship could be based on taking the mean curve between case (i) and case (ii).

15. Repeat problem 14 regarding the two curves shown in Figure 7.6.

16. Repeat problem 14 regarding the two curves shown in Figure 7.7.

PRINCIPAL LIST OF SYMBOLS

Z_{ch} Surge impedance of lightning channel
Z_o Surge impedance of conductor
e_g Voltage generator
Q Conductor charge
k_{cd} Dynamic coupling
V_o Voltage of string insulation to ground
r_t Effective tower radius
$U(t)$ Induced voltage at tower
Z_t Tower surge impedance
U_{ac} AC conventional voltage
M Mutual inductance

REFERENCES

1. **Gary, C., Garabedian, R., Zhang, Z., Cristescu, D., and Enache, R.,** Breakdown characteristics of insulator strings stressed by short tail waves, *IEEE Proc.,* 133(8 Part A), 552, 1986.

2. **Whitehead, E. R.,** Tour d'horizon de la CIGRE concernant les performances des lignes de transport à très hautes tensions vis-à-vis des coups de foudrè, *Electra,* 33, 1974.

3. **Cristescu, D.,** Theoretical determining of protection areas against direct lightcning strokes, CIGRE, SC 33-79, WG: ol, Neptun-Romania, 1979.

4. **Akopian, A.,** Recherches de laboratoires sur les zones de protection des parafoudre à tiges multiples, CIGRE, Rep. No. 328, 1937.

5. **Gary, C., Le Roy, G., Hutzler, B., Lalot, J., and Duban Ton, C.,** *Les Propriétés Diélectriques de l'air et les Très Hautes Tensions,* Eyrolles, Paris, 1984.

6. **Rudenberg, R.,** *Elektrische Wanderwelle,* Springer-Verlag, Berlin, 1962.

7. **Kostemo, M. V.,** *Atmosferic Perenapriajenia, Grozozascita Visoko Voltnih Ustanovok,* GEI, Moscow, 1949.

8. **Gary, C., Timotin, A., and Cristescu, D.,** Prediction of surge propagation influenced by corona and skin effect, *IEEE Proc. A,* 130(5), 1983.

9. **Lundholm, R.,** Surtension produite lors d'un coup de foudre direct sur un pylone de ligne de transport d'énergie, CIGRE, Paris Rep. 333, 1958.

10. **Wagner, C. F. and Hileman, A. R.,** A new approach to the calculation of lightning performances of transmission lines, *IEEE Trans.,* PAS-778 (IIIB), 1959.

11. **Bewley, L. V.,** *Travelling Waves on Transmission Systems,* Dove Publications, New York, 1963.

12. **Caldwell, R. C. and Darveniza, M.,** Experimental and analytical studies of the effects of non-standard wave shapes on the impulse strength of external insulation, *IEEE Trans.,* PAS-92 (III), 1973.

13. **Gary, C., Cristescu, D., Enache, R., and Popa, B.,** High Voltage Insulation Behaviour at Short Wave Stresses, paper 44-04, presented at 4th I.S.H. Athens, 1983.

14. **Dinculescu, D., Cimador, A., and Fieus, R.,** On the lightening over-voltage shape on high voltage lines, *Rev. Roum. Sci. Tech. S.S.R.,* 20(3), 1970.

15. **Gary, C., Cimador, A., and Fieus, R.,** La foudre, étude de phénomène. Applications à la protection des lignes de transport, *Rev. Gen. Electr.,* 84(1), 1975.

16. **Enache, R.,** Aspects Regarding High Voltage Insulation Behaviour at Nonconventional Stresses, Doctor's degree thesis, Bucharest, 1985.

Chapter 8

CORONA EFFECTS

8.1 INTRODUCTION

Chapter 1 presented conventional transmission line equations for propagation of incident waves with consideration of longitudinal inductance and resistance as well as transverse conductance and capacitance. However, the conventional set of equations for analysis of wave propagation cannot be used for symmetrical operating conditions involving a return path through Earth, a voltage level beyond the corona threshold, and conditions at high frequencies. As an example, Fourier integral expansion of lightning overvoltage surge will contain broad spectrum of high frequencies in the MHE range.

MHE frequencies of voltage or current surge introduce the concept of skin effect in line conductors and Earth, while corona phenomenon alters conventional values of line dynamic capacitance as well as the dynamic conductance, changing the velocity of propagation for lightning surge wave front, and increasing the transversal line losses.

Modification for the conventional equations of wave propagation presented in Chapter 1, therefore, is essential in accounting for skin effect and corona, especially under the incidence of lightning over-voltage surges as well as switching impulses.

Analytical modification is centered on Maxwell's field equations listed below:

$$\overline{D} \times \overline{H} = \overline{J}_c + \frac{\partial \overline{D}}{\partial t} \tag{8.1}$$

$$\overline{D} \times \overline{E} = - \frac{\partial \overline{B}}{\partial t} \tag{8.2}$$

Equations (8.1) and (8.2) must be subjected to the general, nonlinear constituent relationships:

$$\overline{D} = \epsilon_1 E + \epsilon_2 \frac{\partial E}{\partial t} + \epsilon_3 \frac{\partial^2 E}{\partial t^2} + \dots \tag{8.3}$$

$$\overline{B} = \mu_1 H + \mu_2 \frac{\partial H}{\partial t} + \mu_3 \frac{\partial^2 H}{\partial t^2} + \ldots \qquad (8.4)$$

$\epsilon(W)$ and $\mu(W)$ could be complex.

Based on an earlier study conducted by Rádulet, Timotin, and Tugulea, a new set of wave propagation of a transmission line obtained by means of analytical approximation through a study conducted by C. Gary, A. Timotin, and D. Cristescu that includes Earth as a return path, using the concept of transient line parameters, has been developed that will take care of skin effect and corona. The new set of equations is in the form of integro-differential equations and convolutions that require nonclassical method of solution and eventually numerical approach to achieve reasonable, quick results.

8.II MODELING OF TRANSMISSION LINE PROPAGATION EQUATIONS

Modeling is based on the use of transient parameters and nonlinear capacitance, which implies that

$$q = q(U, dU/dt, d^2U/dt^2 \ldots) \qquad (8.5)$$

where q is the line charge in coulombs and U is the induced over-voltage which may exceed corona threshold.

The new set of propagation equations in terms of the transient line parameters developed by Rádulet, Timotin, and Tugulea is expressed as:

$$\frac{\partial i}{\partial X} = c_{dyn} \frac{\partial U}{\partial t} + G_0 U \qquad (8.6)$$

$$-\frac{\partial U}{\partial X} = l_0 \frac{\partial i}{\partial t} \int_0^t r(t - t') i(x, t') dt' \qquad (8.7)$$

where X is any point on the line from point of impact, $U(x,t)$, $i(x,t)$ is the voltage and current at any position and time at the transmission line.

The dynamic capacitance c_{dyn} is given as:

$$c_{dyn} = \frac{\partial q}{\partial U} = c_{dyn}\left(U, \frac{dU}{dt}, \frac{d^2U}{dt^2} \ldots\right) > c_0 \qquad (8.8)$$

where c_0 is the conventional transverse capacitance for the transmission line.

$$r(x,t) = -\frac{\partial U}{\partial X}\bigg| dt \; i(t) = U_{-1}(t) \qquad (8.9)$$

where $r(t)$ is known as the transient line resistance. Let T_i be known as the wire penetration time and T_s the Earth penetration time,

$$T_i = \mu_i \sigma_i a^2 \tag{8.10}$$

$$T_s = \mu_s \sigma_s h^2 = \mu_0 \sigma_s h^2 = v_i T_i \tag{8.11}$$

where h is the height of the line.

Introducing the relative time variables,

$$\theta_i = t/T_i \text{ and } \theta_s = t/T_s = \theta \tag{8.12}$$

Define $\rho_i(\theta_i)$ as the relative transient resistance of the line,

$$\rho_i(\theta_i) = \frac{r_i(t)}{r_0} = 1 + \sum_{n=1}^{\infty} \exp(-X_n^2 \cdot \theta_i) \tag{8.13}$$

and $\rho_s(\theta_s)$ as the relative transient resistance of Earth

$$\rho_s(\theta_s) = \frac{r_s(t)}{r_{os}} = \frac{1}{2\sqrt{\pi\theta_s}} + \frac{1}{4}\left[\exp\left(\frac{1}{\theta_s}\right) erfc\left(\frac{1}{\sqrt{\theta_s}}\right) - 1 \right] \tag{8.14}$$

ρ_i and ρ_s are represented in Figure 8.1, where X_n, $n = 1, 2, 3, \ldots$ are the zeros of the Bessel function $J_1(X)$ of the first kind and order, and $erfc(X)$ is the complementary error function.

$$r_{os} = \frac{1}{\pi h^2 \delta_s} = K_0 r_0 \tag{8.15}$$

Known as the reference line resistance for Earth, $r(t)$ is the line resistance

$$= r_i(t) + r_s(t) \tag{8.16}$$

and

$$\rho(\theta) = \frac{r(t)}{r_0} = \rho_i(v_i\theta) + K_0\rho_s(\theta) \tag{8.17}$$

r_0 is the line DC (direct current) resistance

$$U_i = \frac{T_s}{T_i} \text{ is the order of } 10^{-3}$$

$$K_0 = \frac{1}{\pi h^2 \delta_3 r_0} \text{ is the order of } 10^{-3}$$

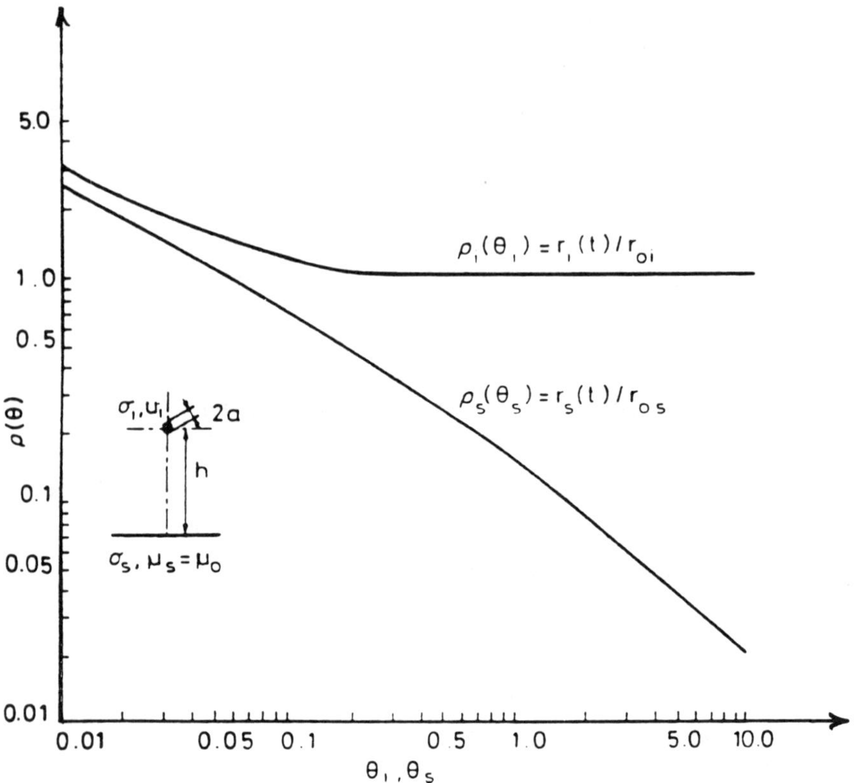

FIGURE 8.1   Relative transient resistances of the cylindrical wire ρ_i, and the homogenous earth ρ_s, as functions of the associated relative time variables θ_i and θ_s. $\tau_i = \mu_i\sigma a^2$; $\tau_s = \mu_s\sigma_s h^2$; $\theta_i = t/\tau_i$; $\theta_s = t/\tau_s$; $r_{oi} = r_o = 1/\pi\sigma_i\sigma^2$; $r_{os} = 1/\pi\sigma_s h^2$.

For computing purposes, let

$$\theta = t/T_s \text{ and } \xi = X/v_0 T_s \tag{8.18}$$

$U = U(\epsilon,\theta)$ and $a(\epsilon,\theta)$ as the normalized current defined by

$$a(\xi,\theta) = Z_0 i(x,t) \tag{8.19}$$

where

$$v_o = \frac{1}{\sqrt{l_0 C_0}} \text{ and } Z_0 = \sqrt{\frac{\rho_0}{C_0}}$$

Hence, the normalized line equations become

$$-\frac{\partial a}{\partial \xi} = K \frac{\partial U}{\partial \theta} + v_g U \tag{8.20}$$

$$-\frac{\partial U}{\partial \xi} = \frac{\partial a}{\partial \theta} = v_r \frac{\partial}{\partial \theta} \int_0^\theta \rho(\theta - \theta')a(\xi,\theta')d\theta' \tag{8.21}$$

The relative dynamic capacitance occurs as

$$K = c_{dyn}/c_0 = K\left(U, \frac{dU}{dt}, \frac{d^2U}{dt^2} \cdots \right) \tag{8.22}$$

v_r is the longitudinal loss coefficient and v_g is the transverse loss coefficient, wherein

$$v_r = r_0 T_s/\rho_0 \ll 1 \tag{8.23}$$

$$v_g = G_0 T_s/C_0 \ll 1 \tag{8.24}$$

8.III CORONA MODEL

Systematic laboratory studies of the impulsive corona discharge conducted by Gary, Timotin and Cristescu led to the following characterization of the voltage-charge cycle $q = g(v)$ as shown in Figure 8.2a, where U_{co} = corona inception voltage when $q = q_0 = c_0 U_{co}$; no corona appears and the capacitance is the geometrical value c_0. As the voltage increases, the dynamic capacitance c_{dyn} will increase from c_0 to c_{cor}, corresponding to a voltage level of U_{c1} where $q = q_1$. As the voltage U surges to a higher level on the order of three to four times U_{co}, the dynamic capacitance c_{dys} will approach the limit of c_{cor}. Then, when U reaches the limit U_{max} and begins to decrease monotonously, the dynamic capacitance c_{dyn} approaches the original limit of c_0.

Regarding the wave front where $U < U_{max}$, and $dU/dt > 0$, the two straight lines $q = c_0 U$ and $q = q_1 + c_{cor}(U - U_{c1})$ intersect at point P, where the voltage U has a value equal to the dynamic corona threshold U_{c1}.

Also, regarding the model shown in Figure 8.2a,

$$q = q(U) = c_0 U + (c_{cor} - c_0)(U_{c1} - U_{co})q(\eta) \tag{8.25}$$

where

$$\eta = (U - U_{co})/(U_{c1} - U_{co}),(0,1)\eta \tag{8.26}$$

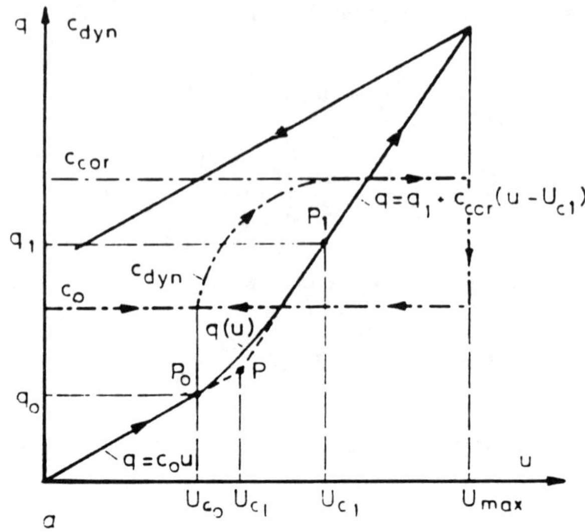

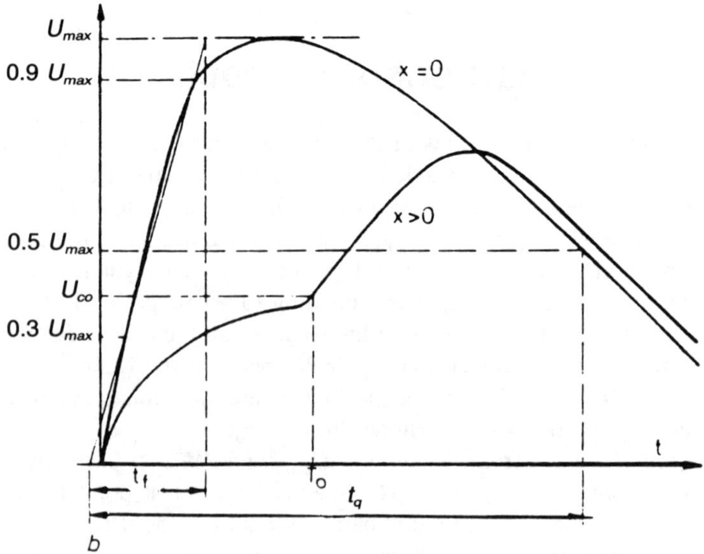

FIGURE 8.2   Simplified description of the corona model. (a) The (q,u) and $(C_{dyn},\, u)$ curves; (b) time constants of the input $(x = 0)$ and propagated $(x > 0)$ wave shapes.

and

$$q(0) = 0, q(1) = 1 \tag{8.27}$$

Also,

$$c_{dyn} = Kc_0 = \frac{d\theta}{dU} = [c_0 + c_{cor} - c_0]f(\eta) \tag{8.28}$$

where

$$f(\eta) = \frac{d\theta}{d\eta} = \eta^{\frac{1}{F_1 - 1}} \tag{8.29}$$

F_1 is shown implicitly in the following:

$$F_1 = \frac{U_{c1} - U_{co}}{U_{ct} - U_{co}} \tag{8.30}$$

In addition, the study conducted by C. Gary, A. Timotin, and D. Cristescu identified

$$c_{cor} \approx c_0\left[1 + 0.6ch\left(\frac{a}{24.1}\right)^{1.1}\right] \tag{8.31}$$

and

$$U_{ct} \approx U_{co} + \frac{P_0}{\sqrt{a}} t_f^{0.75} \left(\frac{U_{max}}{t_g}\right) \tag{8.32}$$

where a is the line radius in millimeters and t_f is the normalized visetime in μs.

The formula of U_{c1} is convenient only for the wave shape of laboratory tests as shown in Figure 8.2b, where the biexponential representation is given by

$$U(t) = U_{max}K\left[\exp\left(\frac{-t}{t_d}\right) - \exp\left(-\frac{t}{t_m}\right)\right] \tag{8.33}$$

with

$$K = \left(R^{\frac{R}{R-1}}\right)\Big/(R - 1) \tag{8.34}$$

$$R - t_d/t_m \tag{8.35}$$

where t_d is the tail time constant and t_m is the front time constant.

In the case of nonexponential form, we have:

$$U_{c1} \approx U_{co} + \frac{P}{\sqrt{a}}\left(\frac{U_{co}}{T_0}\right) \tag{8.36}$$

where

$$P = P_0 t_f^{0.75} \tag{8.37}$$

Also,

$$K = c_{dyn}/c_0$$

$$= 1 \text{ if } 0 < |U| \leq U_{co}, \frac{d|U|}{dt} > 0$$

$$= 1 + [K_{cor} - 1]f(\eta) \text{ if } U_{co} \leq |U| \leq U_{c1}, \frac{d|U|}{dt} > 0 \tag{8.38}$$

where

$$K_{cor} = \frac{c_{cor}}{c_0} \text{ if } U_{c1} \leq |U| \leq U_{max}, \frac{d|U|}{dt} > 0 \tag{8.39}$$

$$K_{ret} = \frac{c_{ret}}{c_0} \text{ if } |U| < U_{max}, \frac{d|U|}{dt} < 0 \tag{8.40}$$

Figures 8.3, 8.4, and 8.5 show characteristic behavior for the propagation of normalized impulse at distances up to 10 km. These curves show that as the Earth conductivity becomes less, more attenuation for the wave peak E_{max} results at a certain point; however the retardation is greater due to the increase of the transient line inductance, which is due to deeper penetration of the field into the Earth.

Figure 8.6 shows the time of the wave maximum $t_m = t_m(x)$ as a function of distance (x) on the line for various Earth qualities and input waveforms.

The Gary et al. study included useful information for the duration of the tail wave. Figure 8.7 shows distance dependence of the peak attenuation U_{max}/E_{max} for the reference Earth $(0.01\ S\ m^{-1})$ and for various tail duration. It was shown that the smaller the front rise time t_g, the greater the influence of the Earth on the peak voltage U_{max}. For truncation of impulse wave the study showed a direct effect on the attenuation of its peak. Figure 8.8 shows the time dependence of a normalized impulse wave to be 1.2/50 μs, abruptly truncated at 2 s at various distances from the origin. Peak attenuation is observed at 43% at $X = 3$ km, whereas it is 90% for untruncated wave.

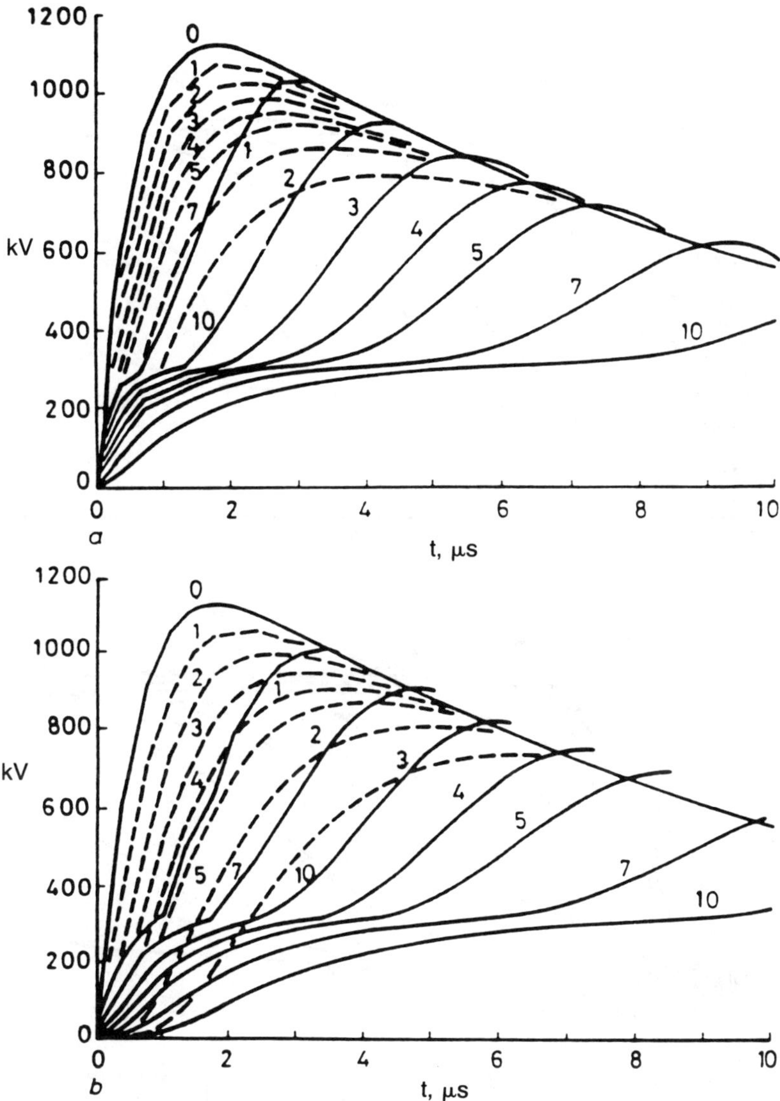

FIGURE 8.3 Propagation of normalized impulse with and without corona. Time dependence of $U(Xk, t)$ at indicated Xk (km). Dotted line, with corona; solid line, without corona. (a) $\sigma_s = 0.01$ $S\,m^{-1}$; (b) $\sigma_s = 0.001\ S\,m^{-1}$. $E_{max} = 1.120$ kV, 1.2/10 µs.

For skin effect, it is reported in the Gary and co-workers study that line conductor effect has little influence compared to Earth return path, where its transient resistance could be 1000 times greater than the corresponding wire resistance. The situation is different, however, for a ground conductor made of steel in which the penetration time is directly proportional to its permeability.

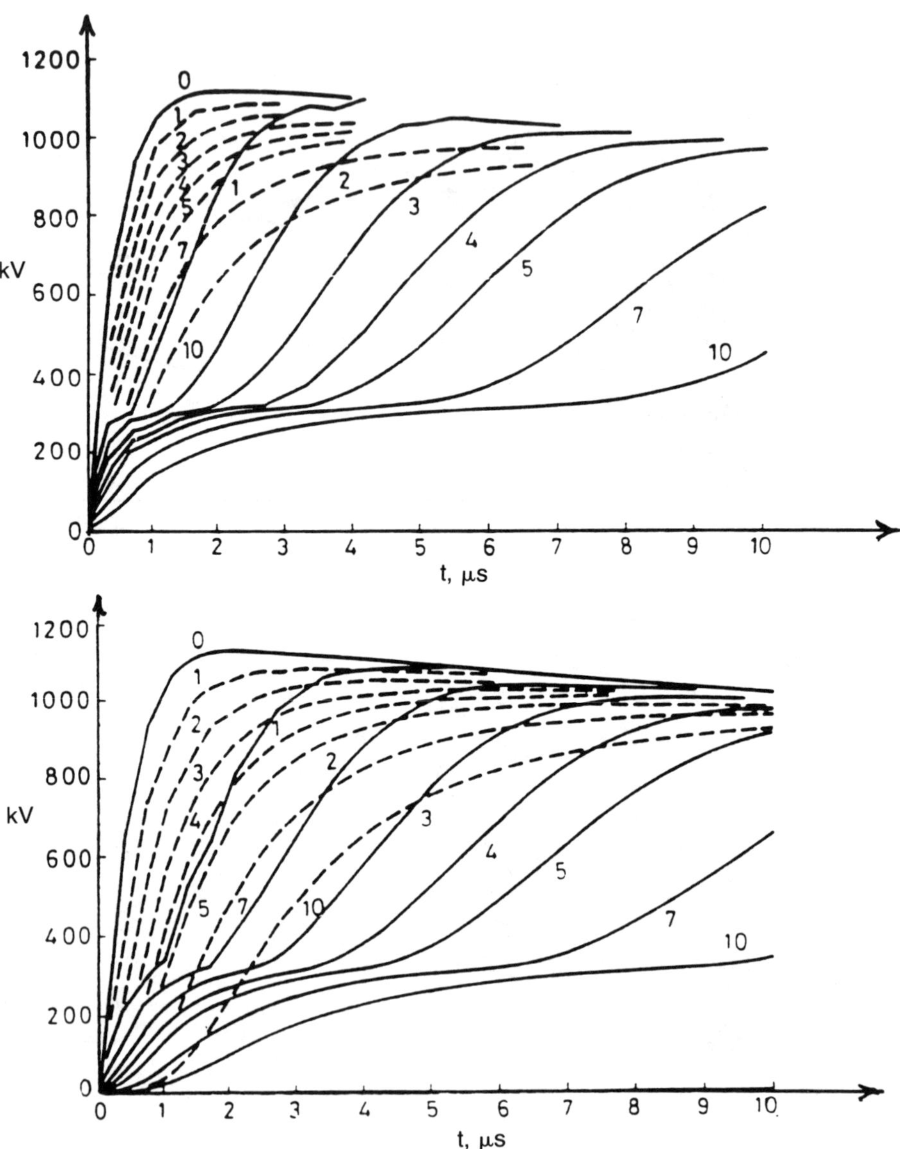

FIGURE 8.4 Propagation of normalized impulse with and without corona. Time dependence of $U(Xk, t)$ at indicated Xk (km). Dotted line, with corona; solid line, without corona. (a) $\sigma_q = 0.01$ S m^{-1}; (b) $\sigma_q = 0.001$ S m^{-1}. $E_{max} = 1.120$ kV, 1.2/50 μs.

This led to a reduction in the relative time values and an increase in the relative transient resistance values of the wire on the front wave. Figure 8.9 illustrates wave shapes at short distances from the point of impact, before reflection at the masts. It is observed that Earth electrical conductivity has little

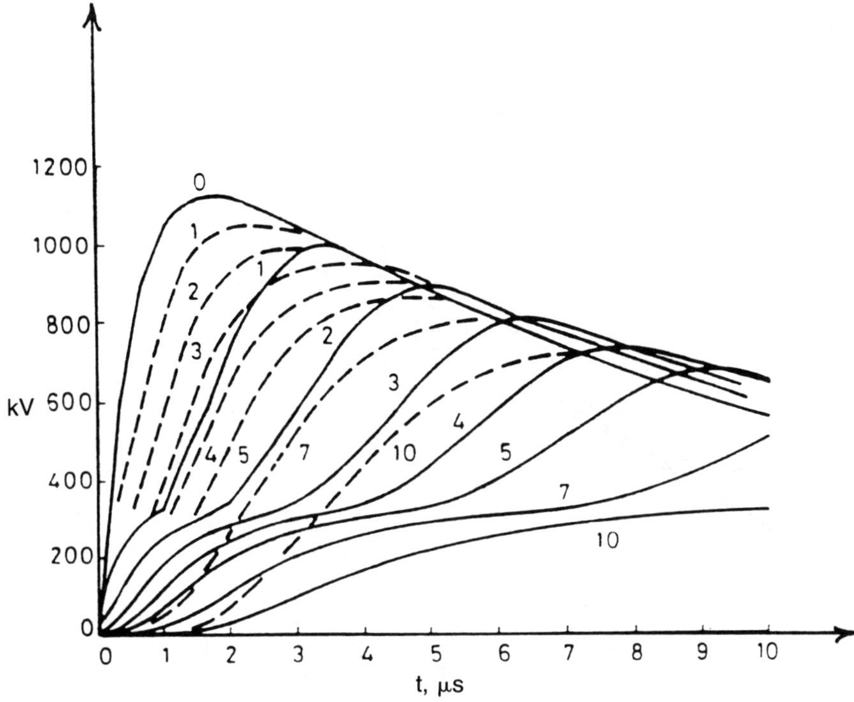

FIGURE 8.5   Propagation of normalized impulse with and without corona in the case of a very badly conducting ground. Time dependence of $U(Xk, t)$ at indicated Xk (km). Dotted line, with corona; solid line, without corona. $\sigma_s = 0.0001\ S\ m^{-1}$. $E_{max} = 1.120$ kV, 1.2/10 μs.

effect on the wave distortion. The study also indicated that skin effect losses can influence the wave peak attenuation via a short rise time. Figure 8.10 shows a staged attenuation of 0.2/6 μs wave at short distances.

8.IV CORONA AT HIGH DIRECT VOLTAGES

A. INTRODUCTION

For small gaps under an applied high DC voltage, Jhawar and Chalmers confirmed the generally accepted square-law dependence of corona current with respect to region voltage. The formula also can be applied to corona formation from trees during the passing of thunderstorms. The square-law formula

$$i \approx A[U - U_0]U \tag{8.41}$$

where U_0 is corona threshold voltage.

Work carried out by T. E. Allibone and J. C. Saunderson indicated that corona current from Earth-bound points of varying curvatures placed between two DC charged parallel plates confirmed the validity of the square-law rule for

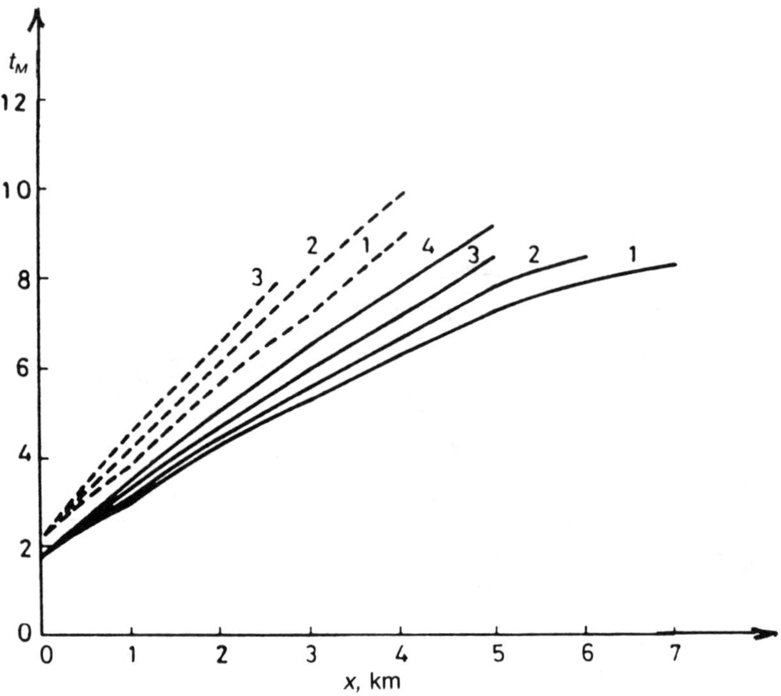

FIGURE 8.6   Distance dependence of the peak time t_M. 1.2/10 μs at $x = 0$; 1.2/50 at $x = 0$. σ_s = 0.01 S m^{-1}; 2 σ_s = 0.005 S m^{-1}; 3 σ_s = 0.001 S m^{-1}; σ_x = 0.0001 S m^{-1}. Positive polarity: $2a = D = 26.4$ mm.

a DC voltage up to 400 kV, where a corona developed at the current edges of the upper plane.

Knudsen and Iliceto also conducted work on corona using a DC voltage close to 1.4 MV using a point-plane geometry; however in the scope of current measurement at a very large plane electrode, they were not able to observe the well-known sharp peaks in current pulse that were observed by Trichel and others. Those peaks were identified as developing from negatively charged point electrodes in small gaps.

Allibone and Saunderson indicated by their work that by careful selection of the measuring parameters they have been able to record corona pulses that form at a negatively charged point several meters above a plane electrode. The current at the plane is the displacement convective current corresponding to Trichel corona pulses.

B. EXPERIMENTAL WORK

Investigations carried out by Allibone and Saunderson secured voltage measurement by a shielded resistance potentiometer, adjusted by a 2-m spark gap to 0.5% adjustment, with an applied DC voltage up to IMV on both polarities.

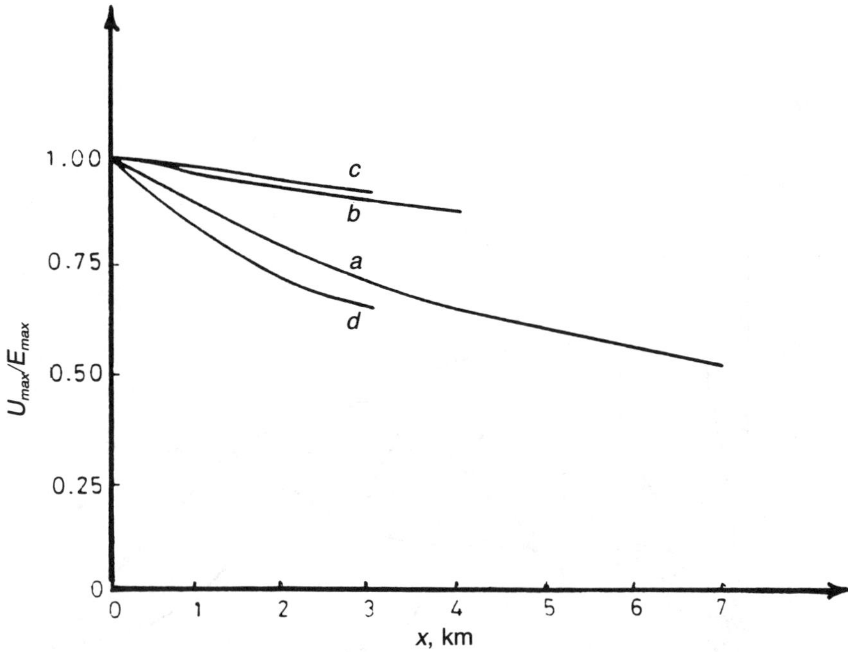

FIGURE 8.7 $U_{max} E_{max}$ as function of the distance x. (a) 1.2/10 μs at $x = 0$; (b) 1.2/50 μs at $x = 0$; (c) 1.2/80 μs at $x = 0$; (d) 0.2/6 μs at $x = 0$.

In measuring the corona between a sphere and a point, the earthen electrode was placed vertically below the suspended sphere and was connected to an oscilloscope using a 4-m length of Tektronix resistive probe having a band width of 4 MHE. The needle electrode had a radius of curvature of 0.05 mm; other electrodes were used such as a 1.9-cm diameter rod tapered to a cone with a 2-mm radius, a 1.9-cm rod-ended hemispherical, and a 1.2 × 1.2 cm square-sectioned rod cut square perpendicular to its length and left with sharp edges and corners.

Figure 8.11 shows corona currents between the needle and the 2-m sphere for an applied positive polarity voltage up to 950 kV. It was observed that no current streamers developed below 1000 kV. Figure 8.12 represents current-voltage characteristics for corona for other, less sharp electrodes, while Figure 8.13 illustrates the same curves for various plane-needle gaps.

C. OSCILLOGRAPHIC MEASUREMENTS

Trichell showed that corona pulsing currents from sharp electrodes charged negatively took the form of short bursts with time duration on the order of 1 μs. In a reliable observation of corona pulses, the needle was connected to ground a resistance (R) ranging from 1 to 10 km, as well as to the resistive probe leading to the oscilloscope. It is observed that the number of pulses released due to corona increases linearly with the increase of current value. Oscillograms of corona currents are shown in Figure 8.14.

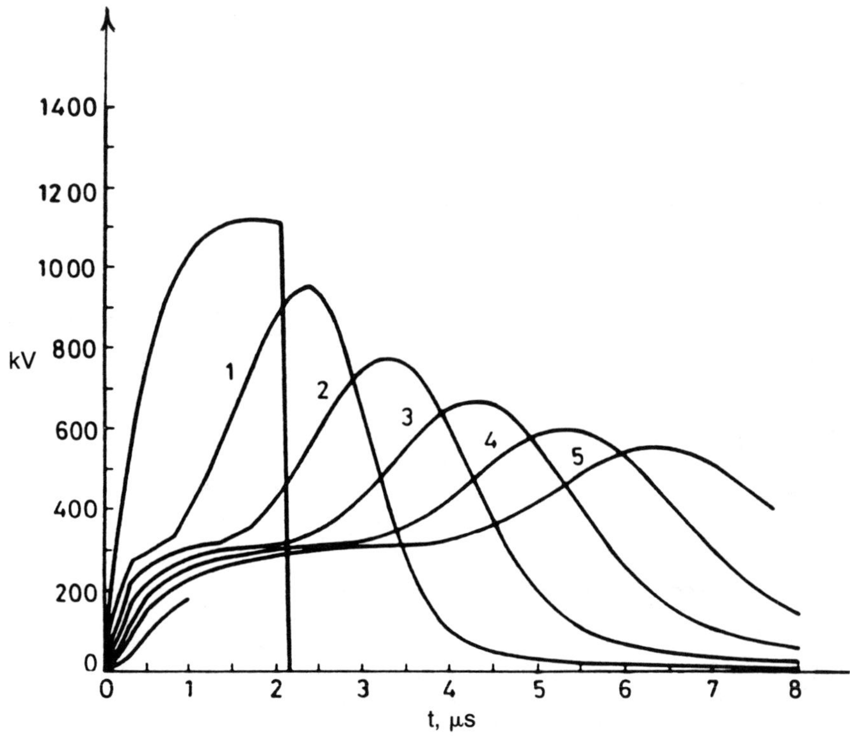

FIGURE 8.8   Propagation of a truncated 1.2/50 impulse. Time dependence of $U(Xk, t)$ at indicated 1.2/50 impulse.

Figure 8.15 shows measured frequency for the release of corona pulses with respect to total current. The pair of curves indicate corona current of 1 μamp of a frequency of 10^4 pulses per second and 40 μamp for a frequency of 10^6 pulses per second, while another pair of curves shows corona current of 1.9 μs of a frequency of 1.3×10^3 pulses per second at 200 kV; a corona current of 10 μamp had a frequency of 3×10^3 pulses per second at 350 kV.

The study carried out by Allibone and Saunderson concluded that the square-law formula for corona current with respect to stressing voltage was verified with reliable precision for a variety of rod-plane gaps with stressing voltage up to 1 MV of both polarities. Figure 8.16 shows correlations between corona current and plane-needle gap for both polarities. From Figure 8.16 it is observed that corona current varies inversely with respect to gap spacing, however as the spacing increases, the inverse power of the spacing (d) becomes 2.5. Measurement of corona current with respect to spacing was based on inverted geometry; the sphere point which gave the surprising result that sparkover voltage from 200 to 100 kV of this gap on positive polarity was 15% higher than that for the point-plane geometry on negative polarity.

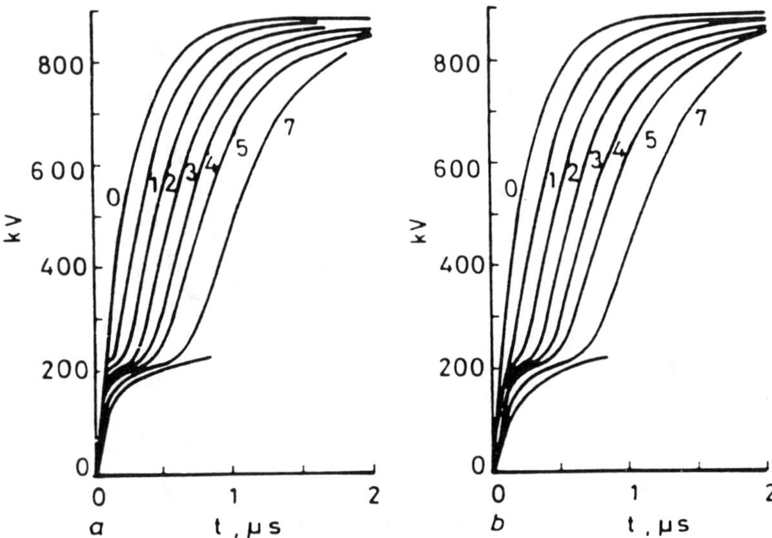

FIGURE 8.9   Propagation of a lightning wave, 0.8/40 μs, 880 kV, of positive polarity on a ground conductor ($h = 22$ m, $a = 8$ mm) of steel with $\mu_r = 200$, $\sigma_i = 7.10^6 \ \Omega^{-1} \ m^{-1}$, 150 mm², with corona effect. (a) $\sigma_s = 0.01 \ S \ m^{-1}$; (b) $\sigma_s = 0.001 \ S \ m^{-1}$.

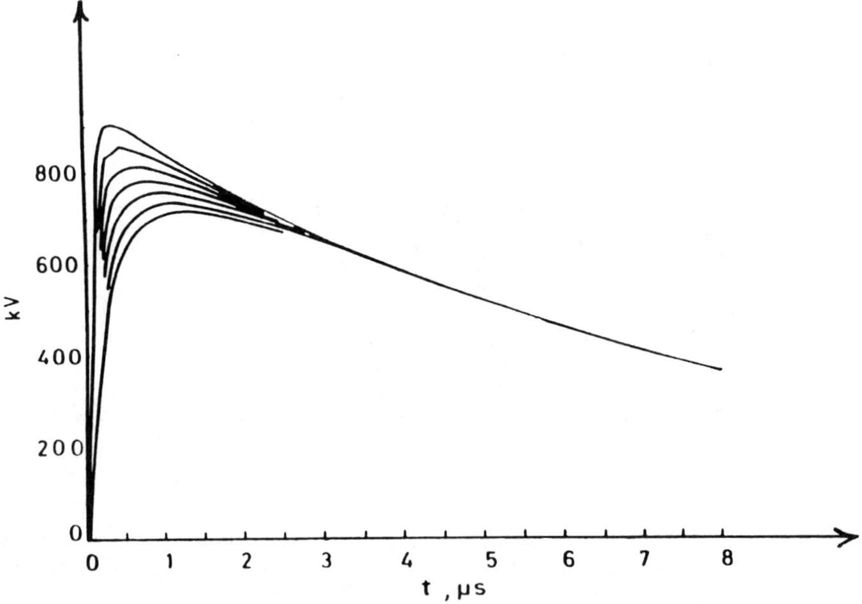

FIGURE 8.10   Progressive skin effect attenuation of the 0.2/6 μs wave at distances $x = 0.5$, 1, 1.5, 2, 2.5 and 3 km. $\Delta t = 83.3$ ns, $\sigma_s = 0.01 \ S \ m^{-1}$, $\tau_s = 1.81$ μs, 0.2/6 μs.

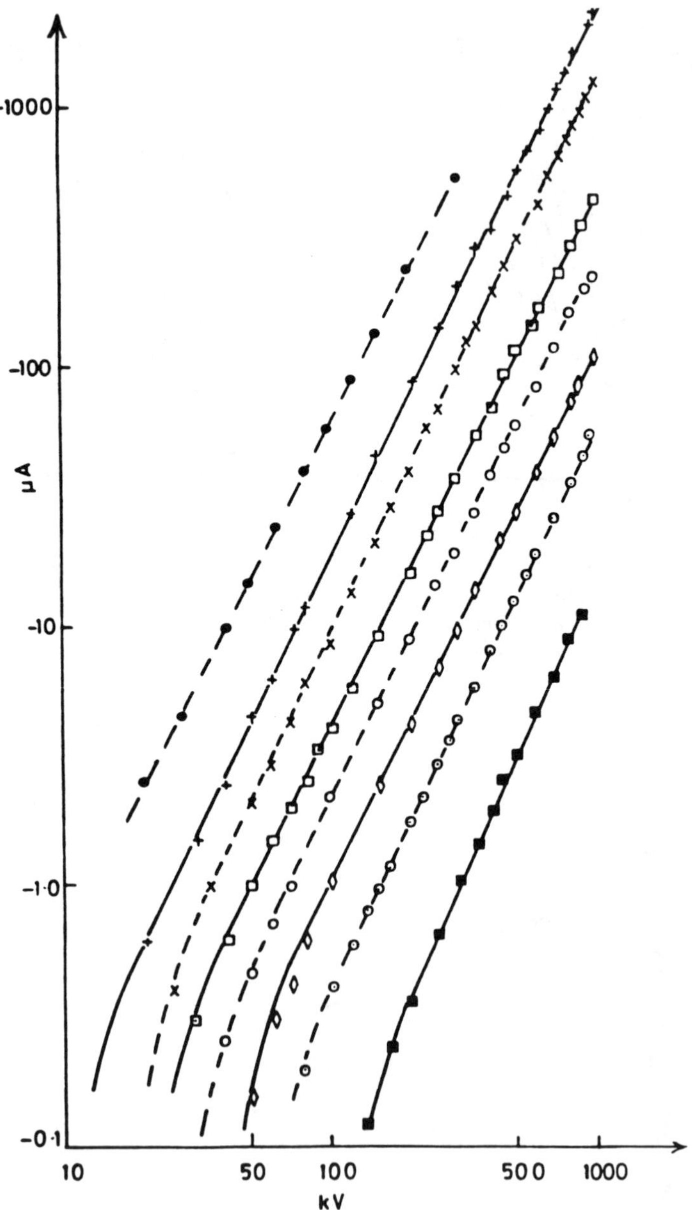

FIGURE 8.11   Corona current/voltage relationships for a 2 m sphere-needle gap of various spacings from 40 to 400 cm. (● -- ●), 40 cm; (+ — +), 70 cm; (× -- ×), 100 cm; (□ — □), 150 cm; (○ -- ○), 200 cm; (◇ — ◇), 250 cm; (⊙ -- ⊙), 300 cm; (■ — ■), 400 cm.

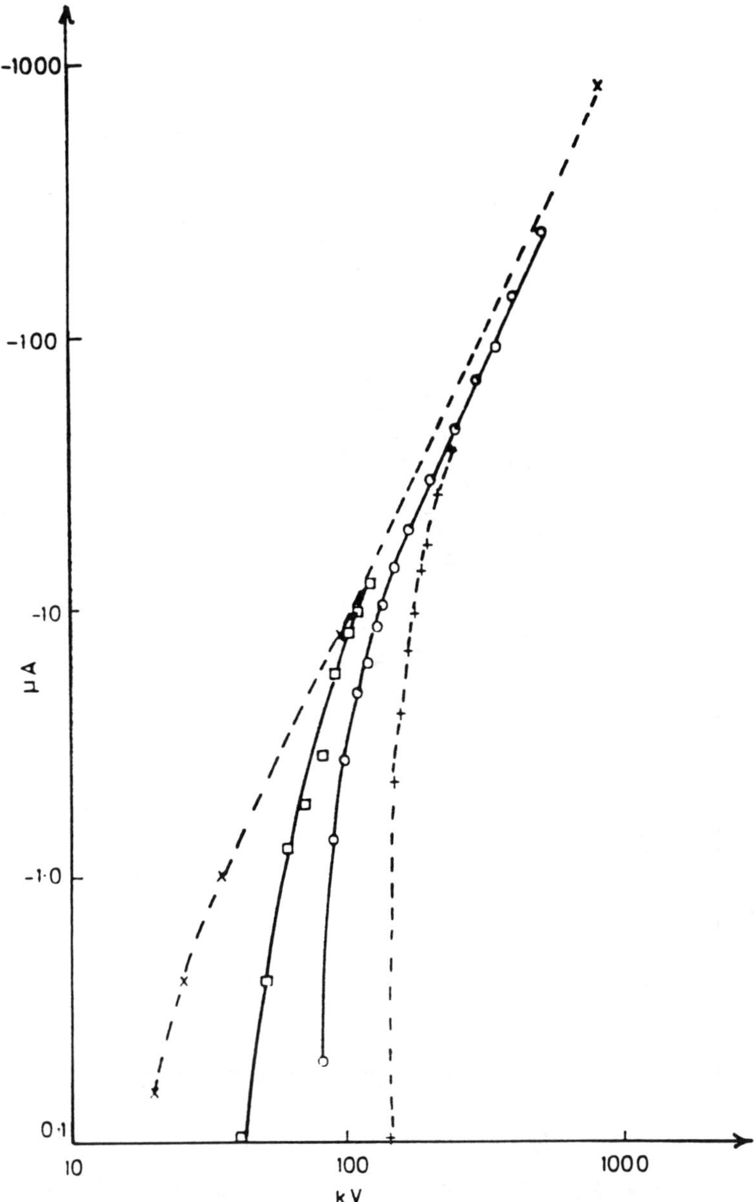

FIGURE 8.12   Corona current/voltage relationship for various electrodes with gap = 1 m. ($\times$ -- $\times$), needle $r = 0.05$ mm; ($\square$ — $\square$), square cut 1.2 cm $\times$ 1.2 cm; ($\bigcirc$ — $\bigcirc$) conical $r = 2$ mm; ($+ \text{---} +$), hemispherical $r = 0.95$ cm.

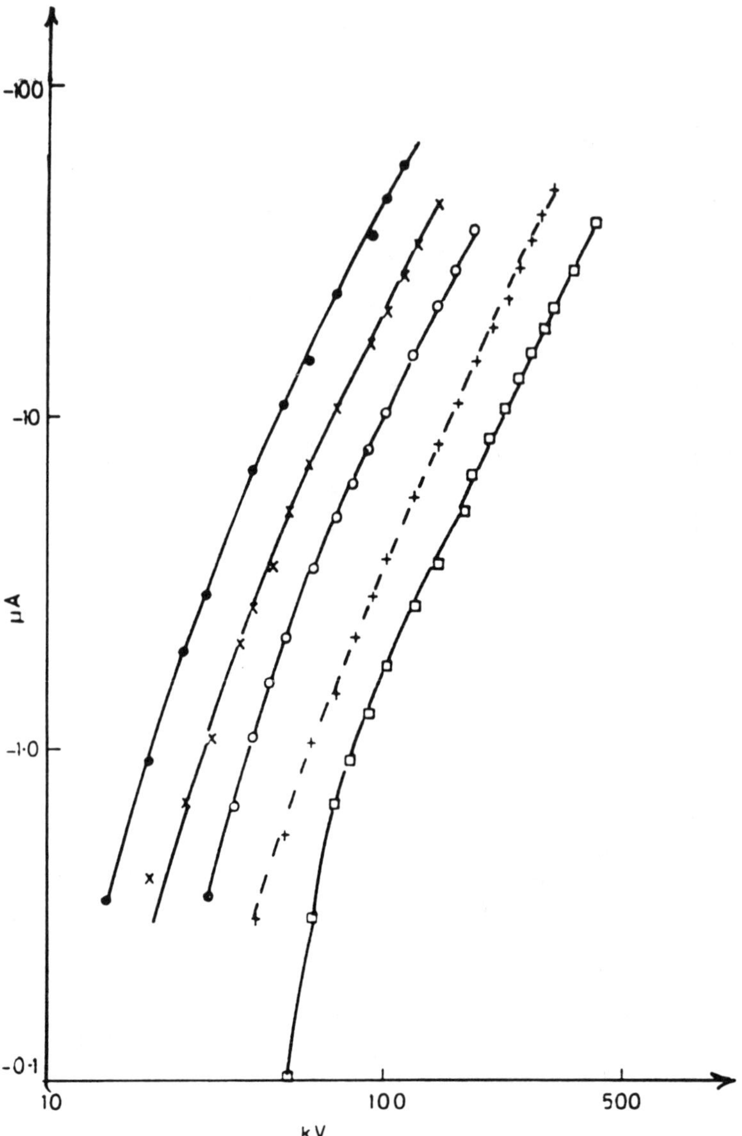

FIGURE 8.13   Corona current/voltage relationships for various plane-needle gaps. (●—●), 24 cm; (×—×), 36 cm; (○—○), 50 cm; (+---+), 75 cm; (□—□), 100 cm.

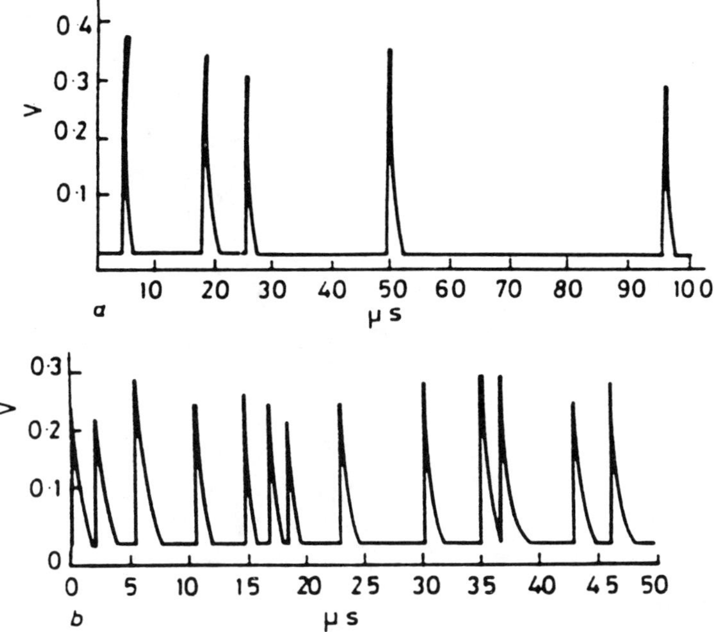

FIGURE 8.14   Oscillograms of corona pulses from negatively charged needle in a 2.5 m sphere-needle gap (R = kΩ). (a) Average of 50 pulses/ms; (b) average of 250 pulses/ms.

8.V HUMIDITY EFFECTS*

Indications are that voltage breakdown in a nonuniform field gap depends on the inception and formation of corona, which in turn depends on electrode profile, length of gap, and the rate of rise and duration of the voltage. It is also observed that the addition of 1 g of water per cubic meter of air will lead to an increase of voltage breakdown by 1%. A gap with a positively stressed electrode of small radius representing diverging field shows substantial dependence on humidity, while in a negatively charged electrode for converging field, the humidity effect was observed to be much less. The Allen study pointed to the fact that humidity has significant impact on sparkover voltage whether the applied voltage is steady, time varying, or a unidirectional train of impulses in a diverging field. Allen's study was directed on the rod-plane electrode system.

Breakdown strength in a rod-plane and a rod-rod gap system showed a decline as the time to crest T_{cr} increases and starts to increase again, with a minimum occurring in the range $100 < T_{cr} < 1000$ μs, illustrating the well-known pattern of U curves identified by Waters. The applied voltage could be a switching impulse or lightning pulse. In applying a switching impulse of relatively long

* The work in this Section is based on and extracted from a paper by N. L. Allen.[1]

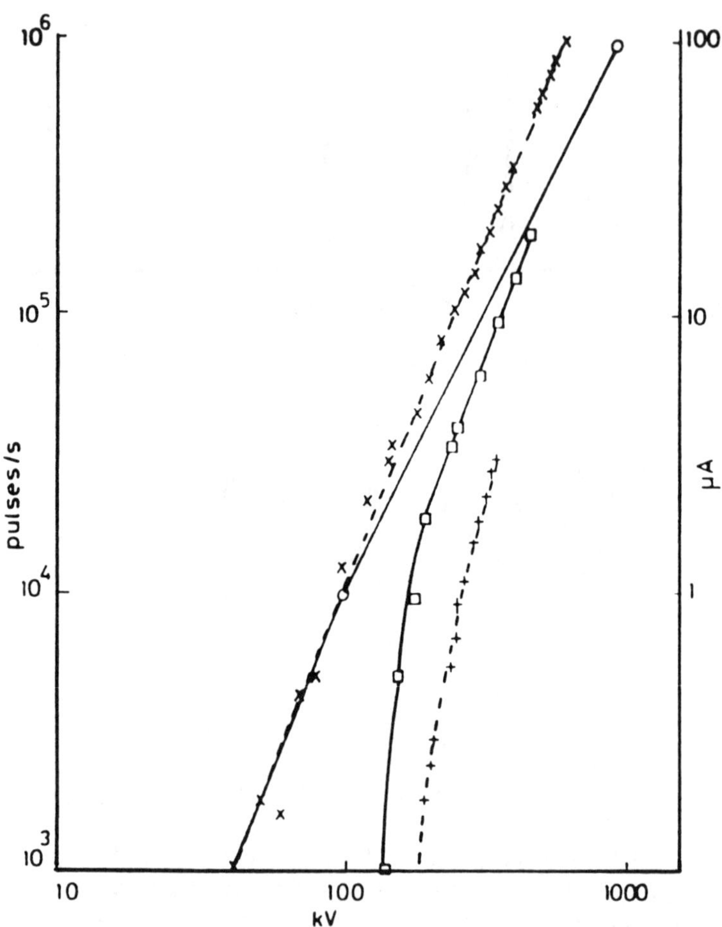

FIGURE 8.15   Corona currents and corona pulse frequencies. Sphere-needle gap of 2.5 m: (O—O), current from 1 to 100 μA; (×--×), pulses from 10^3 to 10^6 pulses/s. Sphere-(hemispherically-ended) rod gap of 2.5 m: (□—□), current from 0.1 to 2.0 μA; (+--+), pulses from 10^3 to 3×10^4 pulses/s.

duration, a significant role in the prebreakdown corona is the formation of the ionization leader. Gallimberi and Rea indicated in their work that lightning pulse characterized as 0.9/23 μs; the leader formation had been identified, but that the corona activity that precedes the final stage of breakdown is dominated by ionizing streamers.

A. ON BREAKDOWN

The effects of humidity on the 50% flash-over voltage breakdown form the subject of this subsection. U_{50} depends upon the gap length, pulse wave shape,

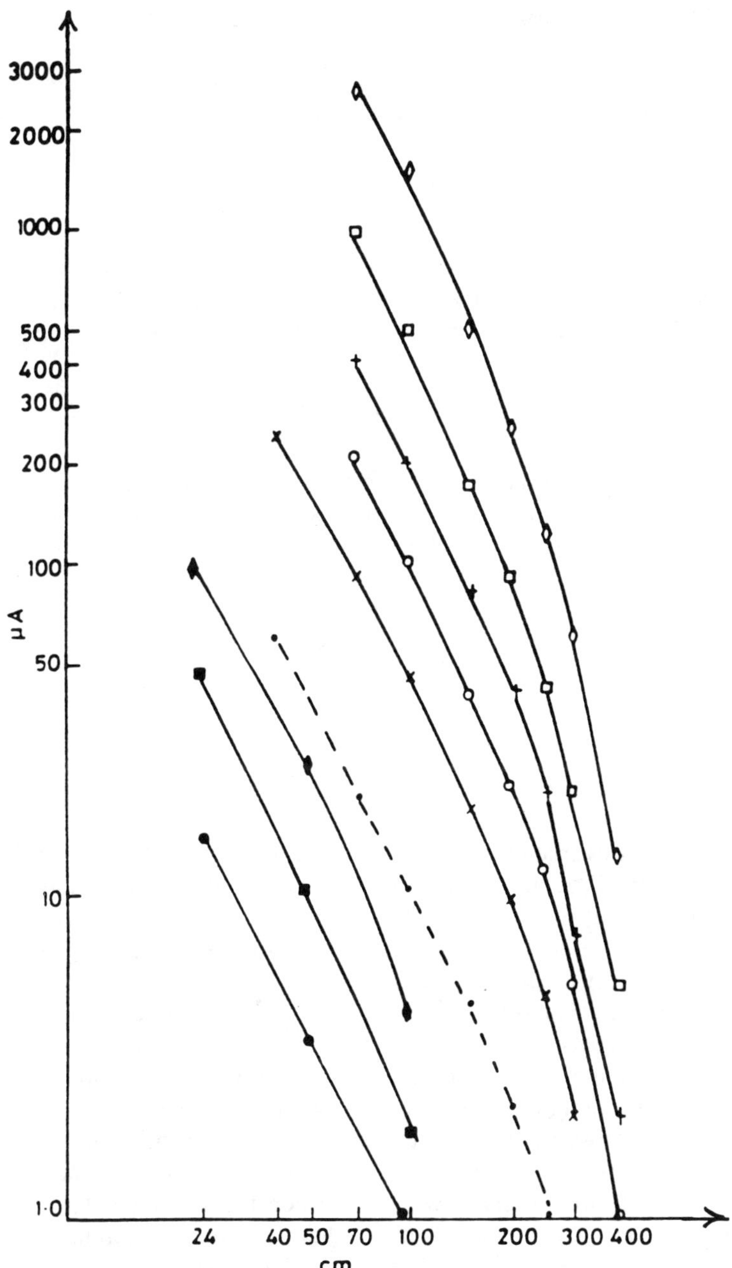

FIGURE 8.16   Corona current/gap spacing curves. Sphere-needle gap at: 100 kV (●), 200 kV (×), 300 kV (○), 400 kV (+), 600 kV (□), 1000 kV (◇). Plane-needle gap at: 600 kV (●), 100 kV (■), 150 kV (♦).

and humidity. The discussion will not touch on the effect of relative air density or profile of the rod electrode.

The study of N. L. Allen indicated that for a gap of a given length, relation between flash-over voltage and absolute humidity is given by

$$U_h = U_{std}\left[1 + \frac{K_h}{100}(h - h_{std})\right] \tag{8.42}$$

where U_h, U_{std} are the 50% flash-over voltages at humidity h and h_{std}, respectively; h_{std} is equal to 11 g/m^3 adopted by the IEC in the U.K.; K_h is a humidity coefficient, so if K_h is a constant, then U_h varies linearly with h. According to the IEC recommendation,

$$K_h = K_w \tag{8.43}$$

where K is a coefficient that varies almost linearly with h, and h has a value of unity when $h = 11$ g/m^3, and W is an exponent that varies with gap length. Over a limited humidity $[5<h<20$ g/m$^3]$

$$K \approx 1 + \frac{h - h_{std}}{100} \tag{8.44}$$

and

$$K_h \approx \left[1 + \frac{h - h_{std}}{100}\right]^w \tag{8.45}$$

Figure 8.17 shows variation between K_h with gap length and humidity. For gap length (d) meter, K_h is independent of electrode separation and it has an order of 1%/g/m^3. For all gaps and variations a range of humidities exists between 2.5 and 20 g/m^3. Also Figure 8.18 illustrates the pattern of 50% flash-over voltage with respect to absolute humidity and gap length for 2/40 μs.

Allen indicated that studies conducted since 1970 pointed to an important observation that K_h also depends on the time to crest T_{cn} of the straight lines for the flash-over voltage V_{50} against gap separation (d) at different humidities under the occurrence of lightning surge. On the other hand, for switching impulse, the set of curves rises less rapidly than linearly, and humidity correction factors vary among wave fronts. Consequently, Equation (8.42) is applicable mainly where lightning impulses are associated with K_h as a constant for all gap lengths. Allen referred to Aihara et al., who identified that a similarity relationship existed between impulse wave front and flash-over voltage for different gap spacings. Therefore, at a give humidity, a plot of V/M against T_{cn}/d generated a unique curve for all wave fronts.

$$M = 2d^{0.4} - \frac{4}{d + 3} \tag{8.46}$$

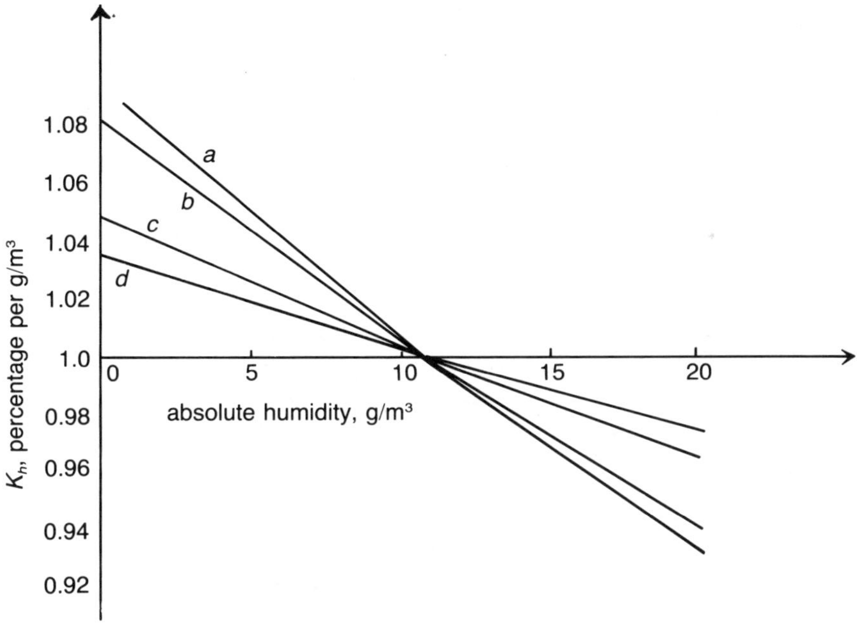

FIGURE 8.17   Values of K_h derived from the JEC recommended values of K and W for positive impulse voltages. (a) Gaps of 0.5 m and 1 m, $w = 1$; (b) gap of 2m, $w = 0.88$; (c) gap of 5 m, $w = 0.52$; (d) gap of 10 m, $w = 0.1$. The variation of w with gap adopted by the JEC.

A new humidity correction factor K_0 is found to be related to K_h by an empirical relationship

$$K_h = \alpha K_0 \tag{8.47}$$

and

$$\alpha = 0.56e^{-0.25d} + 0.44 \tag{8.48}$$

B. SPECIAL CONCERNS

1. Calculation of K_h is not always assured to correspond to a standard humidity.
2. Some doubt exists about precision measurement of humidity.
3. Some doubt exists as to whether measurement over-voltages or humidity have been corrected for standard air or density.
4. An increase in the number of shots to measure the U_{50} level indicates variations of K_h.
5. For gaps <1.5 m length, the electrode profile has significance under conditions of breakdown, while for long gaps, breakdown is determined by the predominately streamer nature of the preceding corona.

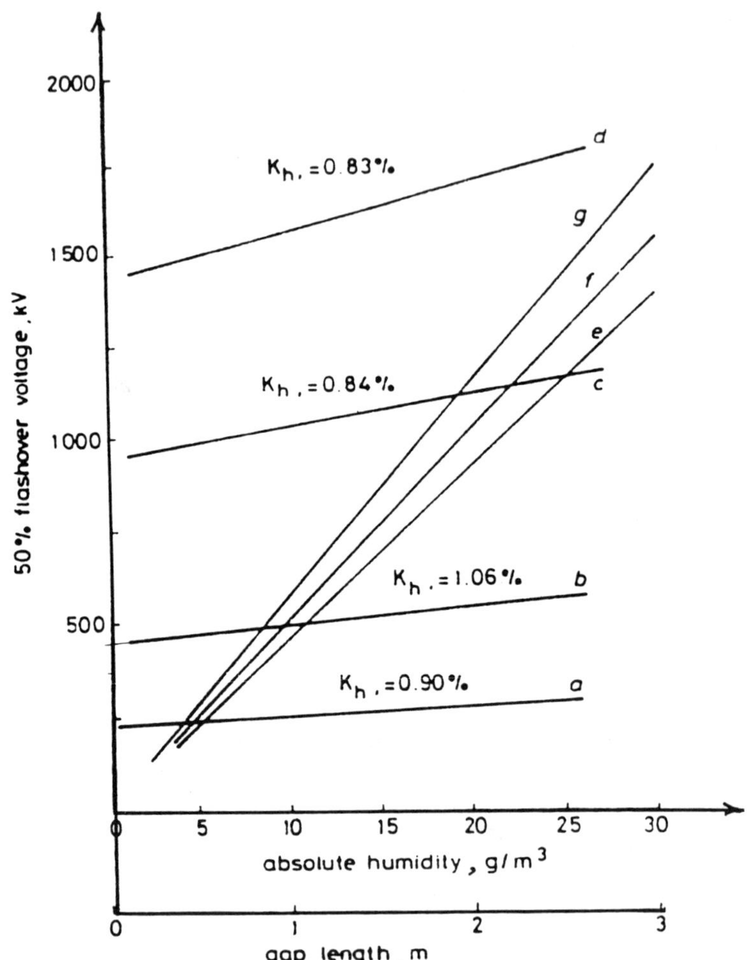

FIGURE 8.18   50% flashover voltages as a function of humidity for 2/40 µs impulses. (a) 0.5-m gap; (b) 1-m gap; (c) 2-m gap; (d) 3-m gap; (e) humidity = 0 g/m³; (f) humidity = 11 g/m³; (g) humidity = 25 g/m³.

6.   Tests carried out under ventilated and unventilated conditions led to differences in the densities of residual negative ions, which by detachment provide an initial release of electrons for the inception of corona and subsequent flashover. Therefore, ventilation conditions will have an indirect effect on the parameter K_h.

7.   Irradiation of electrode gaps after each experimental phase has a direct effect on spark flash-over and other parameters. The supply of ions due to proximity of an radioactive source will also enhance conditions for corona inception.

Pigini proposed for the case of lightning impulse a simple relation between the average potential gradient in the rod-plane gap of length (d) at V_{50} and the average gradient E_s^+, which is required for the propagation of a streamer.

$$A = (V_{50}/d)/E_s^+ \tag{8.49}$$

and

$$\frac{V_{50}(std)}{d} \bigg/ E_s^+(std) = \frac{V_{50}(\delta,h)}{d} \bigg/ E_s^+(\delta,h) = A \tag{8.50}$$

where $E_s^+(std) \rightarrow 500$ kV/m for a relative air density of $\theta = 1$ and standard humidity of 11 g/m^3.

Thus, if V_{50} is measured at any condition of air density θ and absolute humidity h, application of the humidity correction factor yields the value of $E_s^+(\theta,h)$, and hence the constant, so that the value of V_{50} (std) can be secured.

C. HUMIDITY U CURVE

The study conducted by Allen indicated that minimum level for V_{50} observed in the well-known U curves occurs at all humidities and is normally in the range $100 < T_{cr} < 1000$ μs; however, if for $T_{cr} < 100$ μs there is a rapid decrease in V_{50} with increasing T_{cr}, then the increase beyond the minimum is much slower.

To conclude this section, various important conclusions can be drawn:

1. The rate of increase for the potential gradient sought for the travel of a streamer in corona phenomenon is on the order of 1% for each gram per cubic meter in moisture content, and is independent of electrode gap length. Such a rate is about the same as the rate of change for V_{50} with respect to humidity in the case of lightning impulse.
2. The effect on voltage breakdown and corona is to be an involved correlation. The humidity correction factor is found to demonstrate a minimum at some instant toward time duration to T_{cr}. This is an indication of transition from streamer ions domination to leader ionization control, i.e., from a situation of relatively high potential gradient to a low potential gradient, especially in the case of a larger humidity correction factor for a relatively large T_{cr}, where the intrinsic humidity has direct dependence on the leader potential gradient.
3. Humidity effects in short electrode gaps of a few centimeters are found to be different than those in longer gaps, due to different field configurations between electrodes in the two situations.

8.VI CORONA AT SUSTAINED HIGH DIRECT VOLTAGES

Consider a coaxial cylindrical transmission line with inner central wire and the surrounding surface. When the wire is negatively charged serving the role of cathode, the positive ions formed nearby will bombard the cathode with intense energy, thereby releasing more electrons sufficient to establish a self-sustained discharge for corona. Those electrons will travel toward the cylindrical anode forming a path current flow external to the region of ionization. In the case of the central wire being positively charged, electrons in close proximity to the cathode move toward the anode, and as they attain a high field, an electron avalanche will be formed due to multi-ionization activities of electrons. Positive ions generated by electron avalanche will travel to the cylindrical cathode, and upon bombardment, electrons will be released, but with the possession of relatively small energies.

Upon reaching the critical potential between the cathode and anode, corona discharge commences whereby current flow will increase with voltage with some slight deviation between positive and negative corona.

Equation (3.60a) provided a solution for the induced electric field on a transmission line in rectangular coordinates. A method of converting those field components to cylindrical system is shown below:

$$\overline{E}_\rho = E_x \cos\phi + E_y \sin\phi$$

$$\overline{E}_\alpha = -E_x \sin\phi + E_y \cos\phi \tag{8.51}$$

Therefore,

$$\overline{E}_\rho = P\left[\frac{j2}{c} Q\mu_o V_1(t - t_o)\right] \sum_{m=1}^{\infty} \sum_{n=1}^{\infty} Y_{n,m} Z_{n,m}$$

$$\left[\frac{e^{-jX_{n,m}}}{X^2_{n,m}} + j\frac{e^{-jX_{n,m}}}{X_{n,m}}\right]\cos\alpha$$

$$+ P\left[\frac{-j2}{c} Q_o\mu V_1(t - t_o)\right] \sum_{m=1}^{\infty} \sum_{n=1}^{\infty} Z_{n,m} \frac{e^{-jX_{n,m}}}{X_{n,m}} \sin\alpha \tag{8.52}$$

$$\overline{E}_\alpha = -P\left[\frac{j2Q\mu_o}{c} V_1(t - t_o)\right] \sum_{m=1}^{\infty} \sum_{n=1}^{\infty}$$

$$\left[\frac{e^{-jX_{n,m}}}{X^2_{n,m}} + j\frac{e^{-jX_{n,m}}}{X_{n,m}}\right]\cos\alpha$$

$$+ P\left[\frac{-j2Q\mu_o}{c} V_1(t - t_o)\right]$$

$$\sum_{m=1}^{\infty} \sum_{n=1}^{\infty} Z_{n,m} \frac{e^{-jX_{n,m}}}{X_{n,m}} \cos\alpha \tag{8.53}$$

$$E_\tau = P\left[\frac{-j2Q\mu_2}{c} V_1(t - t_o)\right] \sum_{m=1}^{\infty} \sum_{n=1}^{\infty} Y_{n,m} \frac{e^{-jX_{n,m}}}{X_{n,m}} \tag{8.54}$$

Of course, the space coordinates indicated above are from the place of occurrence of the lightning conductive and convective strokes. Also, the parameter Q is a direct representation of the conductive and convective electric current stroke. Here, we can consider the outer cylindrical line surface as the cathode and the inner conductor wire as the anode.

The movement of ions from the cathode to the anode depends upon the field strength available in the region and their mobility,

$$i = 2\pi r\rho v_r \tag{8.55}$$

where v_r is the ion's velocity, ρ is the charge density, and γ is the radius of the cathode and measured from the central wire.

Taking ϵ_r as the prevailing electric field and K as the ion's mobility, therefore,

$$i = 2\pi\rho K E_r \tag{8.56}$$

Resorting to Poisson's equation for radial space variation only,

$$\frac{1}{r}\frac{\partial}{\partial r}(E_r r) = 4\pi\rho \tag{8.57}$$

where ρ is the charge density. Then, by Equation (8.45)

$$\frac{1}{r}\frac{\partial}{\partial r}(E_r r) = \frac{2i}{rKE_r} \tag{8.58}$$

Therefore,

$$(E_r r)^2 = \frac{2ir^2}{K} + C \tag{8.59}$$

As indicated in the book by J. D. Cobine, *Gaseous Conductors* (Dover Publications, 1941), the constant of integration C can be obtained by assuming that $\gamma \gg \gamma_i$, at which the prevailing electric field strength is equal to the sparking condition. In other words, for $r \gg r_i$

$$C = (r_i\, Esunri)^2 - \frac{2ir^2_i}{K} \tag{8.60}$$

And consequently

$$\overline{E}_r = \hat{a}_r \frac{1}{r} \left[\frac{2ir^2}{K} + (r_iE_{ri})^2 - \frac{2ir^{2i}}{K} \right]^{1/2} \tag{8.61}$$

However, ϵ_r is the field in the coaxial space from the central wire located at its own coordinate. Therefore, to account for the location of the place at which lightning stroke occurred, Equation (8.46) becomes:

$$E_{\rho-r} = \frac{1}{\rho - r} \left[\frac{2ir^2}{K} + (r_iE_{ri})^2 - \frac{2ir^{2i}}{K} \right]^{1/2} \tag{8.62}$$

E_i is the local field in the coaxial region at the sparking potential which is on the order of 32 kV/cm at atmospheric pressure and uniform field.

8.VII CORONA UNDER A SHARP RISE AND SLOW DECAYING TAIL OF LIGHTNING STROKE

The induced electric field on a point on a transmission line was given in Equation (4.43) through (4.46) where

$$\overline{E}_\rho = -\hat{a}_\rho \left[\frac{2C^2U_{max}\delta_o(t)[A_c - A_v]}{U_{i(X=L)}\delta'^o(t)} \right.$$
$$+ \frac{2C^2U_{max}\delta_o(t - t_1)[A_v - A_c]}{U_{i(X=L)}\delta'^o(t - t_1)}$$
$$+ \frac{2C^2U_{max}\delta'^o(t - t_1)[A_c - A_v]}{U_{i(X=L)}\delta''^o(t - t_1)}$$
$$\left. + \frac{2C^2U_{max}\delta_o(t - t_1)v_{-1}[A_v - A_c]}{U_{i(X=L)}\delta'^o(t - t_1)} \right] \tag{8.63}$$

$$\overline{E}_\alpha = -\hat{a}_\alpha \left[\frac{2C^2U_{max}\delta_o(t)[A_c - A_v]}{U_{i(X=L)}\delta'^o(t)} \right.$$
$$+ \frac{2C^2U_{max}\delta_o(t - t_1)[A_v - A_c]}{U_{i(X=L)}\delta'^o(t - t_1)}$$
$$+ \frac{2C^2U_{max}\delta'^o(t - t_1)[A_c - A_v]}{U_{i(X=L)}\delta''^o(t - t_1)}$$
$$\left. + \frac{2C^2U_{max}\delta_o(t - t_1)v_{-1}[A_v - A_c]}{U_{i(X=L)}\delta'^o(t - t_1)} \right] \tag{8.64}$$

and

$$\overline{E}_\tau = -\hat{a}_\tau \left[\frac{2C_2 U_{max}\delta_o(t)[A_c - A_v]}{U_{i(X=L)}\delta'^o(t)} \right.$$

$$+ \frac{2C^2 U_{max}\delta_o(t - t_1)[A_v - A_c]}{U_{i(X=L)}\delta'^o(t - t_1)}$$

$$+ \frac{2C^2 U_{max}\delta'^o(t - t_1)[A_c - A_v]}{U_{i(X=L)}\delta''^o(t - t_1)}$$

$$\left. + \frac{2C^2 U_{max}\delta_o(t - t_1)v_{-1}[A_v - A_c]}{U_{i(X=L)}\delta'^o(t - t_1)} \right] \quad (8.65)$$

where U_i $(X = L)$ is the inducing voltage at $X = L$ given by Equation (2.96).

The solution for the induced voltage at a point on a transmission line is given by the set of equations (2.100) through (2.104).

As indicated in Section IV the electric field at the initiation of corona could be written by:

$$E_{\rho-r} = \frac{1}{\rho - r} \left[\frac{2l'_{c,v}(t)r^{2i}}{K} + (r_i E_{ri})^2 - \frac{2l'_{c,v}(t)r^{2i}}{K} \right]^{1/2} \quad (8.66)$$

where $l_c\,v(t)$ is represented as indicated in Chapter 2:

$$J_c = A_c t \quad 0 < t < t_1$$

$$= A_c(t - t_1) \quad t > t_1$$

$$J_v = A_v t \quad 0 < t < t_1$$

$$= A_v(t - t_1) \quad t > t_1 \quad (8.67)$$

Condition of field breakdown at the transmission line could be secured from the field equations of (8.55) whereby at $t = t_1$ (where t_1 is the time of rise of lightning urge) breakdown will commence due to lightning surge.

Corona may commence at these conditional singularity times:

$$\delta'^o(t) = \delta'^o(t - t_1) = \delta''^o(t - t_1) \quad (8.68)$$

Representation for the $\overline{E}_{\alpha-\varrho}$ and $E_{\tau-z}$ could be obtained by solving Poisson's equation in three dimensions:

$$\frac{1}{r}\frac{\partial}{\partial r}(\overline{E}_r r) + \frac{1}{r^2}\frac{\partial \overline{E}_\varrho}{\partial Q} + \frac{\partial \overline{E}_z}{\partial Z} = 4\pi\rho \quad (8.69)$$

Again, ρ is the charge density in interspace of the coaxial transmission line. Solutions for $\overline{E}_Q$ and $\overline{E}_Z$ in conjunction with that of $\overline{E}_r$ are left as problems for the student. The reader may also determine the appropriate boundary conditions based on treating the line conductor as perfect and the interspatial region as perfect dielectric.

The vectorial radial corona field expressed by Equation (8.16) would match the sparking potential that has the direct linkage between Equation (8.70) from the Cobine book *Gaseous Conductors* and Equation (2.66), the equation for the sparking voltage:

$$U_s = BPd/\ln \frac{APd}{\ln(1/r)} \tag{8.70}$$

where P is the pressure in millimeters of mercury, d is the separation between the advancing electron avalanche and the transmission line, and A and B are constants. However, this author must indicate that Equation (8.60) is valid when the ionization process is dominated by secondary electrons, as Cobine expressed in his book.

For convenience to the reader, Equation (2.66) is rewritten below:

$$U_{induced}(x,t) = \frac{2C^2U_{max}}{U_{inducing}(X = L)U_{-1}(t - t_o)} \, U_{inducing}(x,t)U_{-1}(t - t_o) \tag{8.71}$$

and

$$U_{inducing}(x,t) = U_c(x,t) + U_v(x,t) \tag{8.72}$$

Or, as Cobine expressed, the sparking potential at atmospheric pressure and in uniform field is given by

$$U_s = 30d + 1.35 \text{ kV/cm} \tag{8.73}$$

where d is in centimeters. Therefore, from Equations (8.19) and (8.21):

$$30d + 1.35 = \frac{2C^2U_{max}}{U_{inducing}(X = L)v_{-1}(t - t_o)} \, U_{inducing}(x,t) \tag{8.74}$$

The average distance at which corona will commence could be expressed by:

$$d = \left[\frac{2C^2U_{max}U_{inducing}(x,t)}{30U_{inducing}(X = L)U_{-1}(t - t_o)} - 0.045 \right] \text{cm} \tag{8.75}$$

Then the average uniform corona field can be expressed by

$$E_{ave} = \frac{2C^2 U_{max} U_{inducing}(x,t)}{U_{inducing}(X = L)U_{-1}(t - t_o)d} \text{ V/m} \tag{8.76}$$

A similar expression to Equation (8.24) could be obtained when the corona sparking voltage is a sustained high direct voltage.

8.VIII CORONA AT LINE TOWER

The Gary et al. solution for the induced voltage on a transmission line tower in Chapter 7 is given by Equation (7.23), while the incident current impulse is given by Equation (7.26).

One condition for sparking and, of course, the commencement of corona could be observed when the impulse parameter $m = \rho$, which could be seen from Equation (7.27). Another case that may initiate sparking is for the parameter $m = 0$. A third possibility that also depends on the current parameter is for $m = 0$ and a very large l. However, another, more informative criterion for corona commencement at the tower is when the voltage induced given by Equation (7.23) approaches in terms of order of magnitude the approximate representation of sparking voltage given by

$$U_s = (30d + 1.35)\text{kV/cm} \tag{8.77}$$

that case every parameter and variable in Equation (7.27) will be involved.

8.IX CORONA IN GASEOUS CONTINUUM

A. CORONA IN He GAS

Spontaneous field breakdown in helium gas containing He^+ and He^{2+} is reviewed in Chapter 5 wherein breakdown could occur under field and concentration criteria and due to combined effects of field and concentration. Those presentations were contained in Section I.C of Chapter 5. In addition Section I.D presented the concept of magnetically induced electric field and current densities, and sequentially, condition of microscopic voltage breakdown has been presented in the set of Equations (6.31) to (6.33) whereby the gaseous continuum was treated as anisotropic gaseous mixture.

B. CORONA IN N_2 GAS

Section 5.III, discusses breakdown characteristics with respect to unitary gases such as N_2, H_2, CO_2 and air, and specifically on the response of N_2 with 1 to 20% CCl_2F_2 by pressure under lightning impact and switching impulses. It was indicated that breakdown leading to corona occurs at a higher level for negative polarity pulses under both lightning and switching. Detailed breakdown voltage level with respect to rod/plane gap for nitrogen is shown in Figure 5.2 and for a mixture of N_2/CCl_2F_2 is shown in Figures 5.3 through 5.10.

C. CORONA IN SF_6 GAS

Chapter 5 presents a fairly detailed discussion on the ionization behavior under intensified fields. This involves the roles of electrons, positive ions, photons, and secondary processes of ionizations. SF_6 is being used increasingly as an effector insulator in cables and transformers, and demonstrates corona response whenever the prevailing field exceeds the effective ionization potential of SF_6 at the existing pressure and temperature.

8.X SOLVED EXAMPLES

A. State the conventional set of transmission line wave propagation equations and the new transient line parameters equations. Show quantitative differences between the two sets.

Solution

The conventional set is given below:

$$-\frac{\partial i}{\partial X} = C_0 \frac{\partial U}{\partial t} + G_0 U \tag{A1}$$

$$-\frac{\partial V}{\partial X} = l_0 \frac{\partial i}{\partial t} + r_0 i \tag{A2}$$

The new transient parameters set is

$$\frac{\partial i}{\partial X} = C_{dyn} \frac{\partial U}{\partial t} + G_0 U \tag{A3}$$

$$-\frac{\partial U}{\partial X} = l_0 \frac{\partial i}{\partial t} + \frac{\partial}{\partial t} \int_0^t r(t - t') i(x,t') dt' \tag{A4}$$

It is seen by looking at Equations (A1) and (A3) that c_{dyn} has replaced the geometrical capacitance c_0, where

$$c_{dyn} = c_{dyn}\left(U, \frac{dU}{dt}, \frac{d^2U}{dt^2} \cdots \frac{d^nU}{dt^n} \right) \tag{A5}$$

Also, from Equations (A2) and (A4), U_0 is replaced by the transient line resistance $r(t)$, where

$$r(x,t) = \frac{\partial}{\partial t} \int_0^t r(t - t')i(x,t')dt'$$

$$= - \left.\frac{\partial V}{\partial X}\right| i(t) = V_{-1}(t) \tag{A6}$$

The behavior of relative transient resistances for a cylindrical wire and homogeneous Earth is shown in Figure 8.1, while changes in c_{dyn} can be observed in Figure 8.2a. c_{dyn} depends on c_0 beyond the threshold corona limit, and up to a level is set by the charge $q(t) = q_1 + c_{cor}[U - U_{c1}]$ followed by monotonously to c_0.

B. Using Equations (8.20) and (8.21), derive second order differential equations describing the transient normalized voltage and current to include corona effect.

Solution

Differentiate Equation (8.20) with respect to θ and Equation (8.2) with respect to ξ.

$$-\frac{\partial^2 a}{\partial \xi \partial \theta} = k\frac{\partial^2 U}{\partial \theta^2} + v_g \frac{\partial U}{\partial \theta} \tag{B1}$$

$$-\frac{\partial^2 V}{\partial \xi^2} = \frac{\partial^2 a}{\partial \theta \partial \xi} + v_r \frac{\partial^2}{\partial \theta \partial \xi} \int_0^\theta \rho(\theta - \theta')a(\xi,\theta')d\theta' \tag{B2}$$

Therefore, from Equations (B1) and (B2),

$$-\frac{\partial^2 U}{\partial \xi^2} + K\frac{\partial^2 U}{\partial \theta^2} = -v_g\frac{\partial U}{\partial \theta} + v_r\frac{\partial^2}{\partial \theta \partial \xi} \int_0^\theta \rho(\theta - \theta')a(\xi,\theta)d\theta' \tag{B3}$$

Then differentiate Equation (8.20) with respect to ξ and Equation (8.21) with respect to θ:

$$-\frac{\partial^2 a}{\partial \xi^2} = K\frac{\partial^2 U}{\partial \theta \partial \xi} + v_g\frac{\partial U}{\partial \xi} \tag{B4}$$

$$-\frac{\partial^2 U}{\partial \xi \partial \theta} = \frac{\partial^2 a}{\partial \theta^2} + v\frac{\partial^2}{\partial \theta^2} \int_0^\theta \rho(\theta - \theta')a(\xi,\theta')d\theta' \tag{B5}$$

From Equations (B4) and (B5) we obtain

$$-\frac{\partial^2 a}{\partial \xi^2} + K \frac{\partial^2 a}{\partial \theta^2} = v_g \frac{\partial U}{\partial \xi} - kv_r \frac{\partial^2}{\partial \theta^2} \int_0^\theta \rho(\theta - \theta')a(\xi,\theta')d\theta' \qquad \text{(B6)}$$

C. Figure 8.13 shows a set of curves correlating corona current with respect to stress voltage for a plane-needle gap. Set up a table and plot voltage v_s spacing when corona current is -1.0 and -10 μamp.

Solution

From Figure 8.13, draw a horizontal line at 1.0 μamp and then at 10 μamp. The data for voltage v_s spacing is written below:

At i = 1.0 μamp

Voltage (kV)	33	50	69	83	99
Spacing (cm)	24	36	50	75	100

At i = 10 μamp

Voltage (kV)	66	83	99	127	143
Spacing (cm)	24	36	50	75	100

D. Using Figure 8.18, derive an equation correlating 50% lightning impulse flash-over voltage with respect to humidity (h), h_{std} and $V_{50\text{-}std}$, at a gap length of 1 m.

Solution

At $d = 1$ m, $h = h_{std} = 11$ g/m³, $V_{50\text{-}std} = 500$ kV and $K_h = 1.06\%$. For line b, Figure 8.18,

$$V_{50} = k_1 h + k_2 \qquad \text{(D1)}$$

K_1 and K_2 are constants. Therefore, at $h = 0$, $V_{50} = 450$ kV.

$$450 = k_2 \qquad \text{(D2)}$$

and at $h = h_{std}$, $V_{50} = 500$ kV

$$500 = k_1 h_{std} + k_2 \qquad \text{(D3)}$$

From Equations (D2) and (D3)

$$K_1 = 50/h_{std}$$

Therefore,

$$V_{50} = \frac{50}{h_{std}} h + 450 \tag{D4}$$

but $V_{50\text{-}std} = 500$ kV, and Equation (D4) becomes

$${}^{\cdot} V_{50} + 50 = \frac{50}{h_{std}} h + V_{50\text{-}std} \tag{D5}$$

Equation (D5) is valid for $K_h = 1.06\%$ and $d = 1$ m.

Equation (8.27), which correlates (U_h) with respect to (h), is valid when K_h is constant for all values of gap lengths (d).

8.XI PROBLEMS

1. Two planes carrying electric surface charge density $\overline{D}$ coulomb/m^2. If the dielectric continuum is simple but characterized in terms of dynamic capacitance expressed by Equation (8.8), derive generalized expression for the energy stored in this condenser.

$$\text{Let } Q = Q(V,V',V'',\ldots V^n) \text{ where}$$

$$V' = \partial V/\partial t, \ V'' = \partial^2 V/\partial t^2$$

$$\text{and } V^n = \partial^n V/\partial t^n$$

2. Repeat the solution of problem 1 for a dielectric medium characterized as linear.
3. Repeat solution of problem 1 for a dielectric medium characterized as anisotropic, making ϵ a rank two tensor.
4. Obtain expression for the time varying power in problems 1, 2, and 3.
5. If one plane in the condenser system described in problem 1 is movable, then using the concept of energy density stored, obtain expression for the electric force in Newtons.
6. Obtain expression for the transient resistance $r(x, t)$ based on the data and results in problems 1, 2, and 3.
7. Given:

$$\rho(\theta) = \rho_0 U_1(t), \ K = K_0 \delta_0(t)$$

$$a(\epsilon,\theta) = A_0 \delta(t - t_0)$$

$$V_r = V_g = 10^{-3}$$

Solve for $U(\xi,\theta)$ between a line conductor and Earth return.

8. Obtain empirical equation for $\rho = \rho_i + \rho_s$ from Figure 8.1, then using $\rho(xi,\theta)$ repeat the solution of problem 7.

9. Referring to Figure 8.2a, obtain empirical phased solution for $q(U)$, describing various phases before corona, at corona, and beyond.

10. Repeat problem 9 for the dynamic capacitance c_{dyn}.

11. Referring to the lower group of curves shown in Figure 8.4, obtain expression for the normalized impulse with and without corona with respect to point (x) at $t = 5$ μs.

12. Repeat problem 11 for Figure 8.5 at $t = 1$ and 5 μs.

13. Referring to Figure 8.10, derive an empirical equation relating peak impulse voltage against distance (x) and time (t). Then establish a new relationship for $U(x, t)$.

14. A gap of 1 m separating a needle of 0.1 m from a metallic plane of infinite extent is shown below. If a source of electric charge exists between the needle and surface plane it is expressed by:

$$\rho(x,y,z,t) = \left(\frac{K_1 + K_2 Z}{x^2 + y^2}\right) e^{-\frac{t}{T_0}} \text{ Coulombs/m}^3$$

Using Poisson's equation, solve for the electric potential in the region between the needle and surface plane as a function of (x, y, z, t). The needle could be assumed as a perfect line conductor, while the plane surface is a perfect conductor.

15. Repeat problem 14, if the line conductor is perfect, while the plane conductor is imperfect having a finite scaler θ.

16. Repeat problem 14, if the needle is replaced by a cylindrical rod of perfect conductor and its tip is facing the surface conductor is a hemisphere with a certain radius.

17. Repeat problem 14, if the needle position is horizontal, parallel to the plane conductor, at a height of d_2 meters, as shown below. The needle could be assumed as a perfect line conductor.

18. Repeat problem 17, if the needle is replaced by a cylindrical rod conductor of length (L) and radius r. Assume the two ends of the needle as flat circular surfaces.

19. Repeat problem 14, if the gap between the bottom needle and the lower plane (d) is variable according to the expression: $d(Z) = (d_0 \, d_1 Z)$.

20. Repeat problem 19, if the needle line conductor is perfect while the plane surface conductor has a finite electrical conductivity (σ).

21. Repeat problem 16, if $d(Z) = d_0 + d_1 e^{aZ}$.

22. Repeat problem 17, if $d(Z) = d_0 + d_1 Z$.

23. Repeat problem 18, if $d(Z) = d_0 + d_1/a + Z$.

24. Using Figure 8.11, tabulate and plot the stressing gap voltage U_g in kilovolts v_s separation between the sphere and needle gap, then derive an empirical formula for $U_g \, v_s(Z)$. Take corona current at 10 μamp.

25. Repeat problem 24 for corona current at 1, 100, and 1000 μamp.

26. In a piecewise manner derive a functional relationship for corona pulsing frequency and current with respect to gap stressing voltage for the needle-sphere gap according to Figure 8.15.

27. Referring to Figure 8.16, tabulate and plot stressing voltage with respect to spacing between sphere and needle gap when corona current is 10, 50, 100, and 200 μamp.

28. Establish a mathematical relationship correlating the 50% lightning flash-over voltage against absolute humidity (h), $V_{50\text{-}std}$ and h_{std} at $K_n = 0.90\%$ for $d = 2$ m and $d = 3$ m. Use Figure 8.18.

29. Repeat problem 28 for $d = 0.5$ m and $K_n = 1.06\%$.

30. Repeat problem 28 for $K_n = 0.84\%$.

31. Referring to Figure 8.18, express a relationship between V_{50} against absolute humidity (K_n) for $d = 1$, 2, and 3 m.

32. Referring to Figures 8.19 and 8.20, express the 50% of switching impulse flash-over voltage against T_{cr} for $d = 1$ m.

33. Repeat problem 32 for $d = 2$ and 4 m.

34. Using Figure 8.20, express the humidity correction factor K_h against electrode gap separation (d) for $T_{cr} = 1$, 10, and 100 μs.

35. Using Figures 8.17, 8.18, and 8.20 with Equations (8.31) and (8.33), show a plot of V_{50}/M against T_{cr}/d.

36. With reference to Equation (7.23), identify all possibilities for $U(X,t)$ to attain the level of sparking voltage to initiate corona at the top of transmission tower.

37. With reference to Equation (8.17), solve Poisson's equation to express E_r, E_ϕ, and E_z for the coaxial cable with the inner conductor or infinitesmal wire.

38. From solutions of E_r, E_ϕ, and E_z obtained in problem 37, establish solutions for a corona commencing field in three dimensions.

REFERENCES

1. **Allen, N. L.,** Breakdown and humidity in the rod-plane gap, *IEEE Proc.*, 133(Part A, No. 8), 562, 1986.

2. **Allibone, T. E. and Saunderson, J. C.,** Observation of corona phenomena at very high direct voltages, *IEEE Proc.*, 133 (Part A, No. 8), 569, 1986.

3. **Gary, C., Timotin, A., and Cristescu, D.,** Prediction of surge propagation influenced by corona and skin effect, *IEEE Proc.*, 130 (Part A, No. 5), 264, 1983.

4. **Harada, T., Aihara, Y., and Aoshima, Y.,** Influence of humidity on lightning and switching impulse flashover voltages, *IEEE Trans. Power Apparatus Systems,* PAS 90 (No. 4), 1433, 1971.

5. **Harrington, R. F.,** *Time-Harmonic Electromagnetic Fields,* McGraw-Hill, New York, 1961.

6. **Rádulet, R., Timotin, A., and Tugulea, A.,** General theory of the transient parameters of electric lossy lines in presence of the ground, *Stud. Cercet. Energ. Electroteh.,* 16(3), 417–449, 1966.

7. **Cobine, J. D.,** *Gaseous Conductors,* Dover Publications, New York, 1941–1958.

Chapter 9

FREQUENCY SPECTRUM OF SURGE IMPEDANCES DUE TO LIGHTNING

Quantitative evaluation is carried out in this chapter for the effect of lightning surges on the surface impedances of power system installations such as transmission lines, transformers, line towers, etc. Since time duration up to the peak of lightning surge front is on the order of 1 s and for most of the surge tail it is on the order of 100 s, an increase in the characteristic impedances for transmission lines and transformers (even the ungrounded) is confined to only a very short time, about 100 s. Consequently, analysis regarding the concept of skin effect is substituted by examination for the behavior of the surface impedances across the frequency spectrum as reliable basis for skin effects calculations.

9.I SURFACE IMPEDANCE OF TRANSMISSION LINES

With reference to Section 3.VII, Equation (3.60a) present the solution for the induced electric field on an overhead transmission line, where

$$\overline{E} = \hat{a}_x \gamma(t) g_1(X) + \hat{a}_y \gamma(t) g_2(X) + \hat{a}_z \gamma(t) g_2(X) \tag{9.1}$$

where $\gamma(t)$, $g_1(X)$, and $g_2(X)$ relate to the common time function and the associative functions of X indicated in Equation (3.60a). Also,

$$\gamma(t) = \left[\frac{j2}{c} \mu Q_o v_1(t - t_o) \right] \left[\frac{\delta^2(t)}{c^2} \cos ct + \frac{3\delta(t)}{c^2} \sin ct \right.$$

$$\left. - \frac{2\delta'(t)}{c^3} \sin ct - \frac{1}{c^3} \int_0^\tau \sin ct \delta''(t - \tau)^{d\tau} \right] \tag{9.2}$$

where $Q_o = J_{c,v}(t)$.

In Equation (3.72), the corresponding induced magnetic field vector H_2 was given. Subsequently, the induced current density vector $\overline{J}$ was given by Equation (3.75). Expressions for the intrinsic surge impedance were given by Equations

(3.100) and (3.101), while $Z_x = 0$, Z_y and Z_z in the frequency domain could be secured by obtaining the Fourier integral for both surface impedances:

$$Z_y = \frac{P(t)}{\int P(t)dt} G_1(X_{n,m}) \tag{9.3}$$

$$Z_z = \frac{P(t)}{\int P(t)dt} G_2(X_{n,m}) \tag{9.4}$$

To obtain $Z_y(W)$ and $Z_z(W)$, a Fourier integral ought to be taken for the function:

$$F(t) = \left[P(t) \bigg/ \int P(t)dt \right]$$

$$S(t) = \int P(t)dt = \frac{2}{c^2} - \frac{1}{c^3} \sin ct + \frac{1}{c^2} \delta^2(t) \sin ct$$

and the Fourier transform of $S(t) = S(\omega)$:

$$S(\omega) = \frac{2}{c^2} \left[\delta(\omega) + \frac{1}{j\omega} \right] + \frac{j\pi}{c^3} [\delta(\omega - c) - \delta(\omega + c)]$$

$$+ \frac{[\delta(c - \omega) - \delta(c + \omega)]j\pi}{c^2} [2 - \delta(o)] \tag{9.5}$$

Next,

$$P(\omega) = \int_{-\infty}^{\infty} P(t)e^{-j\omega t}\, dt \tag{9.6}$$

However, since $S(\omega) = $ Fourier transform of $\int P(t)\, dt$. Therefore,

$$P(\omega) = j\omega S(\omega) \tag{9.7}$$

The author has to point out that taking the Fourier transform of $P(t)$ and the $\int P(t)\, dt$ separately is valid since the surface impedance expression came from the separate division of $\overline{E}/\overline{H}$, where $P(t)$ is associated with the $\overline{E}$ vector and $\int P(t)\, dt$ is associated with the $\overline{H}$ vector. Therefore,

$$Z_y(\omega) = j\omega G_1(X_{n,m})$$

$$Z_z(\omega) = j\omega G_2(X_{n,m}) \tag{9.8}$$

Equation (9.8) illustrates that throughout time duration of lightning surge to its frontal peak, transmission surface impedance will be extremely high such that induced current will reside almost at the surface of the overhead transmission line. The author must point out also that $Z(\omega)$ indicated in Equation (9.8) is due to a lightning return stroke current density taken as a step function as an instant peak and then followed by an infinite time decaying tail.

Solution for $Z_y(\omega)$ and $Z_z(\omega)$ for an actual lightning return stroke as shown by Equation (2.68) is to be left as a problem for students to solve.

9.II SURGE IMPEDANCE OF MULTILAYERS TRANSFORMER

In space unidimensional spread of any layer for the multilayer winding transformer, the surface surge impedance $Z(x,t)$ has been secured as in Equation (5.49), where

$$Z(x,t) = L \frac{A_2\delta(t) + A_3(A_c - A_v)[t - v(t - t_1)]}{A_2v(t) + A_3(A_c - A_v)\left[\frac{1}{2}t^2 - tv(t - t_1)\right]} \tag{9.9}$$

As it has been observed for the frequency spectrum for the transmission line surface surge impedance, the corresponding form of $Z(\omega,x)$ according to Equation (9.9) can be expressed by

$$Z(x,\omega) = jwL \tag{9.10}$$

where L is the transformer lumped self-inductance.

In Section 5.V, vectorial presentation for the transformer surface impedances in three dimensions and in the steady-state were established according to $Z_\alpha\rho$ given by Equation (5.72), $Z_{\tau\rho}$ given by Equation (5.75), and $Z_{\tau\alpha}$ given by Equation (5.83).

9.III SURGE IMPEDANCE OF TRANSMISSION LINE TOWER

Expression for the voltage induced on tower from work carried out by Gary et al.[2] given in Equation (7.23), whereby expression for the tower surge impedance is given by:

$$Z(t) = \frac{Z_{g\omega}}{2} - \frac{AK_1}{i(t)} \left\{ \sum_{i=2}^{n} \tau(\theta_1)K_1^{i-2}\left(e^{-k_1\theta_1}K_1^{i-2} \right.\right.$$

$$\sum_{k=1}^{i-1}\left[\frac{1}{(k_2 - 1)^k} - \frac{1}{(k_2 - m)^k} \right] \frac{\theta_1^{i-k-1}}{(i - k - 1)!}$$

$$\left. + \frac{e^{-m\theta_1}}{(k_2 - m)^{i-1}} - \frac{e^{-i\theta_1}}{(k_2 - 1)^{i-1}} \right) + \sum_{j=2}^{n'} \tau(\theta_2) k_1^{j-2}$$

$$\left(e^{-k_2\theta_2} \sum_{r=1}^{j-1} \left[\frac{1}{(k_2 - 1)^r} - \frac{1}{(k_2 - m)^r} \right] \frac{\theta_2^{j-r-1}}{(j - r - 1)!} \right.$$

$$\left. + \frac{e^{-m\theta_2}}{(k_2 - m)^{j-1}} - \frac{e^{-i\theta_2}}{(k_2 - 1)^{j-1}} \right) \right\} \tag{9.11}$$

$$= \frac{1}{2} Z_g - \frac{H}{i(t)} \tag{9.12}$$

where H is the entire expression multiplied by $1/i(t)$ in Equation (9.11), for

$$i(t) = A[e^{-mt} - e^{-lt}] \tag{9.13}$$

$$Z(t) = \frac{1}{2} Z_{g\omega} - \frac{H}{A(e^{-mt} - e^{-lt})} \tag{9.14}$$

$$\approx \frac{1}{2} Z_{g\omega} - \frac{H}{A} [e^{mt} + e^{(2m-1)t} + e^{(3m-2t)t}$$

$$+ e^{(4m-3t)t} + \ldots + e^{[nm-(n-1)t]t} \tag{9.15}$$

The Fourier integral of Equation (9.15) will identify $Z(\omega)$; therefore,

$$Z(\omega) = \frac{Z_{g\omega}}{2} \left[\delta(\omega) + \frac{1}{j\omega} \right] - \frac{H}{A} \left[\frac{1}{j\omega - m} \right.$$

$$+ \frac{1}{j\omega + (1 - 2m)} + \frac{1}{j\omega + (2t - 3m)} + \ldots$$

$$\left. + \frac{i}{j\omega + [(n - 1)t - nm]} \right] \tag{9.16}$$

The observation from Equation (9.16) is that at a very high value of ω, say $\omega \rightarrow \infty$

$$Z(\omega) \rightarrow \frac{1}{2} Z_{g\omega} \theta(\omega) \big|_{\omega \rightarrow \infty} \tag{9.17}$$

Considering the extreme case where $i(t) = F \delta(t)$, $Z(\omega)$ becomes:

$$Z(\omega) = \frac{1}{2} Z_{g\omega} \left[\delta(\omega) + \frac{1}{j\omega} \right] - \frac{H}{A} \frac{1}{I(\omega)} \tag{9.18}$$

The Fourier integral of

$$\frac{1}{\delta(t)} = \int_{-\infty}^{\infty} \frac{e^{-j\omega t}}{\delta(t)} \, dt$$

$$= \int_{-\infty}^{\infty} [\delta(t)]^{-1} e^{-j\omega t} \, dt$$

$$= - \int_{-\infty}^{\infty} \delta(t) e^{-j\omega t} \, dt = -1$$

Therefore,

$$Z(\omega) = \frac{1}{2} Z_{g\omega} \left[\delta(\omega) + \frac{1}{j\omega} \right] + \frac{HF}{A} \tag{9.19}$$

9.IV SOLVED EXAMPLES

A. Express the surface impedance of a multilayer transformer in three dimensions at $\omega \to \infty$.

Solution

Equation (9.9) gives $Z(x,t)$ which is independent of x, therefore $Z(y,t) = Z(x,t) = Z(z,t)$

$$\overline{Z} = \overline{Z}_x + \overline{Z}_y + \overline{Z}_z \tag{A1}$$

$$= \hat{a}_x Z(x,t) + \hat{a}_y Z(y,t) + \hat{a}_z(z,t) \tag{A2}$$

Then,

$$\overline{Z}(\omega) = [\hat{a}_x j\omega + \hat{a}_y j\omega + \hat{a}_z j\omega]L \tag{A3}$$

$$|Z(\omega)| = j\omega\sqrt{j\omega} \tag{A4}$$

$$\lim_{w \to \infty} |Z(\omega)| = \to \infty \tag{A5}$$

B. Obtain expression for the surge impedance of transmission line tower where $i(t)$ is a sneaking representation of a step function.

Solution

$$Z(t) = \frac{1}{2} Z_{g\omega} = \frac{H}{i(t)} \tag{B1}$$

A sneak representation for a step function is given by:

$$i(t) = I_{max}e^{-at} \tag{B2}$$

and

$$I(\omega) = \frac{I_{max}}{j\omega + a} \tag{B3}$$

Therefore,

$$Z(\omega) = \frac{1}{2} Z_{g\omega}\left[\frac{1}{j\omega} + \pi\delta(\omega)\right] + \frac{I_{max}}{j\omega + a} \tag{B4}$$

9.V PROBLEMS

1. Obtain solution for the frequency spectrum of the surge surface impedance of the overhead transmission line when the conductive current in the return stroke is expressed by Equation (2.68).

2. Repeat problem 9.1, then the conductive current in the return stroke is given by:

$$J_c(t) = A[e^{-m_1 t} - e^{-m_2 t}]$$

3. Obtain expression for the frequency spectrum of the three-dimensional surge surface impedance of a cylindrical transformer when the conductive current in the return stroke is that given by Equation (2.68).

4. Repeat problem 9.3 when the conductive current in the return stroke is given by:

$$J_c(t) = I_{max} \sin wte^{-at}$$

5. Obtain expression for the frequency spectrum of the line tower surface impedance when the conductive current in the return stroke is expressed by:

$$J_c(t) = I_{max} \cos wte^{-at}$$

6. Repeat problem 9.5 when the conductive current in the return lightning stroke $J_c(t)$ is a step function with absolute value of I_{max}.

7. At a point on a transmission line, establish an expression for the surface surge impedance for a region containing the line over a long length terminated to an ungrounded transformer. The overall surge impedance sought is in the frequency domain and in a direction to ground-assume-conductive lightning stroke as a step function.

8. At a point at the top of a transmission line tower for a power system containing the transmission line and to an ungrounded transformer, obtain an expression for the overall surface surge impedance in the frequency domain with respect to ground-assume-conductive current lightning stroke as a step function.

9. Repeat problem 7 for the overall surface surge impedance for a point in space within a plane parallel to ground surface.

10. Repeat problem 8 for the overall surface surge impedance for a point in space within a plane parallel to ground surface.

11. Repeat problem 7 when the conductive current lightning stroke is an actual time-dependent function, according to Equation (2.68).

12. Repeat problem 8 when the conductive current lightning stroke is an actual time-dependent function, according to Equation (2.68).

13. Repeat problem 7 when the conductive current lightning stroke is a time-dependent function expressed by:

$$J_c(t) = I_{\max}[e^{-m_1 t} - e^{-m_2 t}]$$

14. Repeat problem 7 when the conductive current lightning stroke is a time-dependent function expressed by:

$$J_c(t) = I_{\max} \sin wt e^{-m_1 t}$$

15. Repeat problem 8 when the conductive current lightning stroke is a time-dependent function expressed by:

$$J_c(t) = I_{\max}[e^{-m_1 t} - e^{-m_2 t}]$$

16. Repeat problem 8 when the conductive current lightning stroke is a time-dependent function expressed by:

$$J_c(t) = I_{\max} \sin wt e^{-m_1 t}$$

17. Repeat problem 14 when the conductive current lightning stroke is a periodic time function.

18. Repeat problem 16 when the time-dependent function for the conductive current lightning stroke is periodic.

19. Repeat problem 7 when the conductive current lightning stroke is a truncated step function at $t = T$ μs.

20. Repeat problem 8 when the conductive current lightning stroke is a truncated step function at $t = t$ μs.

PRINCIPAL LIST OF SYMBOLS

γ, g_1, and g_2	Associative functions
Z_y	The y oriented surface surge impedance
$P(t)$	Special time function given in Chapter 3
$Z(w)$	Surge impedance in the w domain
a, m_1, m_2	Constants

REFERENCES

1. **Denno, K.,** *Power System Design and Applications for Alternative Energy Sources,* Prentice-Hall, Englewood Cliffs, NJ, 1989.
2. **Gary, C., Garabedian, R., Zhang, Z., Cristescu, D., and Enache, R.,** Breakdown characteristics of insulator strings stressed by short tail waves, *IEEE Proc.,* 133(Part A, no. 8), 552, 1986.
3. **Harrington, R. F.,** *Time-Harmonic Electro-Magnetic Fields,* McGraw-Hill, New York, 1961.
4. **Papoulis, A.,** *Circuits and Systems & Modern Approach,* Holt, Rinehart, and Winston, New York, 1980.
5. **Rüdenberg, R.,** *Electrical Shock Waves in Power Systems,* Harvard University Press, Cambridge, MA, 1968.
6. **Ramo, S., Whinnery, J. R., and Van Duzer,** *The Fields and Waves in Communication Electronics,* 2nd ed., John Wiley & Sons, New York, 1965.

Chapter 10

TESTING EQUIPMENT AND LIGHTNING FLASH COUNTERS

10.I SIMULATION OF H.V. TESTING CIRCUIT

In this section, a simulation system is described that enables computations on high impulse voltage measuring systems to be carried out effectively. For a given excitation, the system response as well as the response time parameters can be computed. The measuring system is composed of a chain of concentrated and distributed parameters, each represented by a quadripole. Total response at the output will be secured by sequential summation of the response at each quadripole beginning at the entry of the first unit. The input voltage is given either by its analytical or digital form which is usually recorded by means of a fast A/D transient digitizer. On the other hand, given the output response that has been distorted by the measuring system, the input function could be retracted and secured by deconvolution procedures based on the same simulation system. The simulation system described in this section is the work of Profs. P. N. Nikolopoulos and F. V. Topalis, and is described in *IEEE Proceedings*. This system can also be used in computer-aided design of high voltage systems and for the correction of errors introduced by these systems.

The simulation shown in Figure 10.1 describes generation, measuring, and other concentrated distributed components. Response at the output usually is distorted due to stray elements such as capacitances, inductances, and resistances, as well as effects of reflections and retractions of traveling waves. Such distortion increases with the system dimensions. As the surge level of the tested high voltage increases, dimensions of the testing simulation system also increase, compounding the state of distortion. The simulation system of Nikolopoulos and Topalis requires that only geometric and electric characteristics of the system be known.

Simulation system computation is to be carried out by taking into account all system parameters, including all processes of differentials and integrals of voltages and currents. The system model is divided into elements, each represented by two concentrated parts and one distributed part. The concentrated parts of the nth element represent its resistive, capacitive, and inductive components. The equivalent circuit of the element is generally a series complex impedance

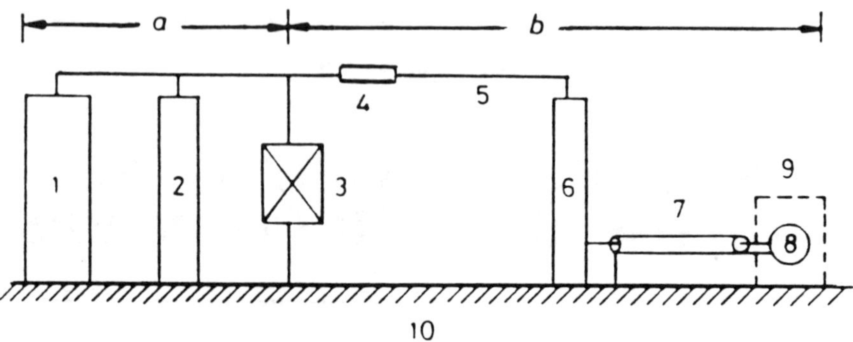

FIGURE 10.1   Typical impulse-voltage test arrangement. (a) Generation of impulse voltage; (b) measurement of impulse voltage. 1, Impulse-voltage generator; 2, front capacitor; 3, test object; 4, damping resistor; 5, high-voltage lead; 6, voltage divider; 7, measuring cable; 8, oscilloscope or digital recorder; 9, screened chamber; 10, grounding system.

$Z_i(s)$ or a complex admittance $Y_i(s)$ or a combination of both. The distributed part represents the stray inductance $L_{s,n}$ and stray capacitance $C_{s,n}$ for the nth element, which is simulated by a wave impedance, Z_{1n}. This wave impedance is considered to be constant along each element and is calculated by means of the total stray capacitance of the nth element to ground:

$$Z_{ln} = \frac{L_{s,n}}{C_{s,n}} = \frac{1}{v_n C_{s,n}} \tag{10.1}$$

where v_n is the velocity of voltage wave propagation in the nth element.

A typical example of the representation of the high voltage arm of the divider is given in Figure 10.2, where the resistive and capacitive elements are equally distributed at each element. The quantities (C_n) and (R/N) corresponding to an element are subdivided into parts of $C.2n$ and $R/2n$, placed at both ends of the element. These parts are concentrated, whereas the distributed part is the wave impedance Z_{1n} substituting $L_{s,n}$ and $C_{s,n}$.

A. MEASURING CIRCUIT

Design elements for the circuit in Figure 10.1 indicated by Nikolopoulos and Topalis consists of a 1.6-MV impulse-voltage generator and a 1.2-MV damped capacitive-voltage divider with a capacitance C_H = 1156.5 PF, and damping resistor R_H = 96 at the top of the divider. The 4 m high divider is connected to the impulse generator through a 4.2-m lead of 2.5 mm² cross-section. A 20-m coaxial cable, with a 75 matching resistance at its input, connects the low voltage arm of the divider with an 8-bit fast A/D converter (transient digitizer) placed in a screened chamber.

For step-response measurements, the impulse voltage generator was replaced by a step with a pulse amplitude of up to 50 V and a rise time of <1 ns.

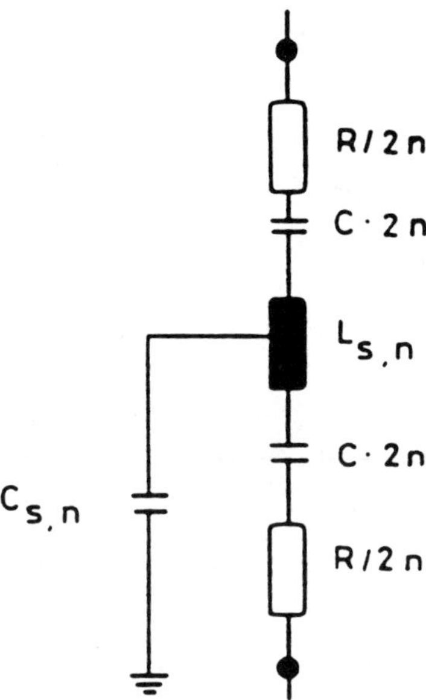

FIGURE 10.2 Representation of the nth element of a damped capacitive divider. R = high-voltage arm resistance; C = high-voltage arm capacitance; $L_{s,n}$ = stray inductance; $C_{s,n}$ = stray capacitance.

For the generation of low voltage impulses, a surge generator was used which could produce full or chopped impulses of up to 500 V. The addition of a vertical lead was inevitable in low voltage tests. The response of the system as well as its response time parameters were measured, and the recorded output voltage was constructed by means of the same equivalent circuit.

B. RECONSTRUCTION OF RESPONSE CHARACTERISTICS

Reconstruction was found necessary for various surge voltages due to stray inductive and lead effects. Figure 10.3 shows reconstruction for a high impulse lightning surge of 1.2/50 μs, and results were quite satisfactory for all the recorded voltages of different levels. In Figure 10.3 the front time of the recorded output of the oscillatory system was found to be 1.01 μs and that of the calculated input equals 1.13 μs. The corresponding front times of the damped system were 1.46 and 1.37 μs, respectively.

Figure 10.4 shows reconstruction of a lightning impulse voltage chopped in the front. The measured peak values of the recorded output voltages were found to be 54.4 and 54.2 kV, respectively, and the corresponding peak values of the calculated inputs were 56.4 and 56.6 kV, respectively.

10.II LIGHTNING FLASH COUNTER (LFC) AND CALIBRATION CIRCUIT

This section discusses a low power LFC using active circuit elements working within a narrow band filter centered at 10 kHz resulting in better discrimination

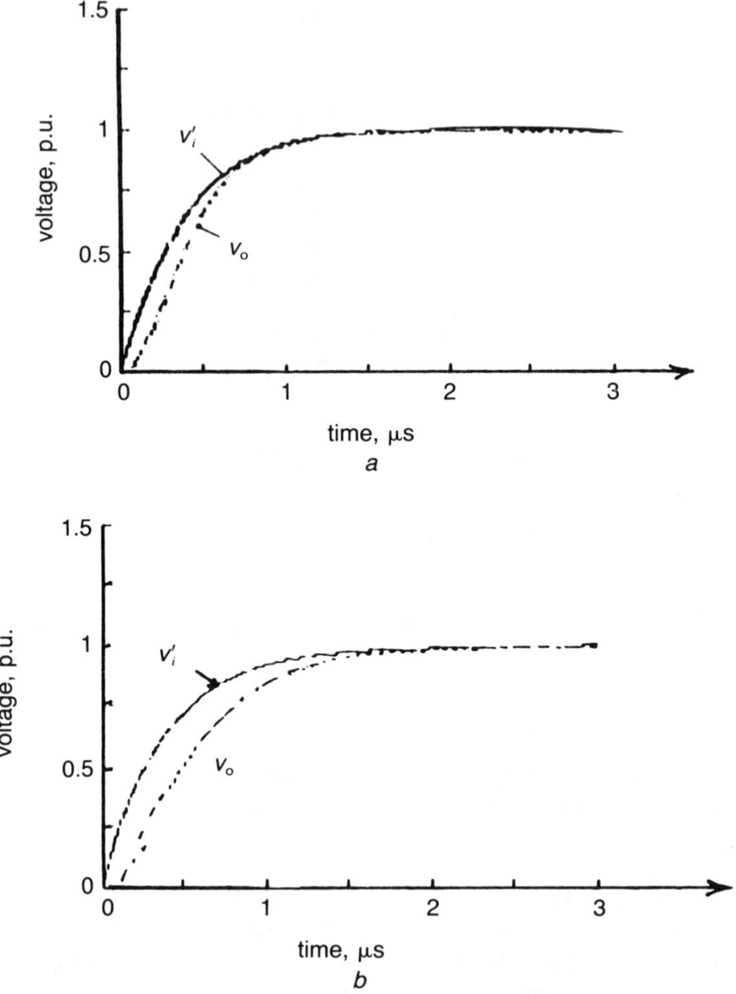

FIGURE 10.3   Reconstruction of standard 1.2/50 full lightning-impulse voltages. v_o = recorded output; v_i' = calculated input. a = oscillating system; b = damped system.

between cloud and ground discharges. Liable operational circuits to generate a double exponential pulse with variable rise time coupled to a testing circuit that ensures reliability and stability aspects for the circuit battery and calibration for various LFC networks have been devised. The design of the CIGRE-LFC network shows more reliable discrimination for the pulse signal than for the sine wave signal and for ground from cloud discharges than can be identified by conventional LFC networks. The work in this section is based upon that of E. O. Oladiran, E. Pisler, and S. Israelsson (*IEEE Proc.*, 135(Part A, No. 1), 22, 1988).

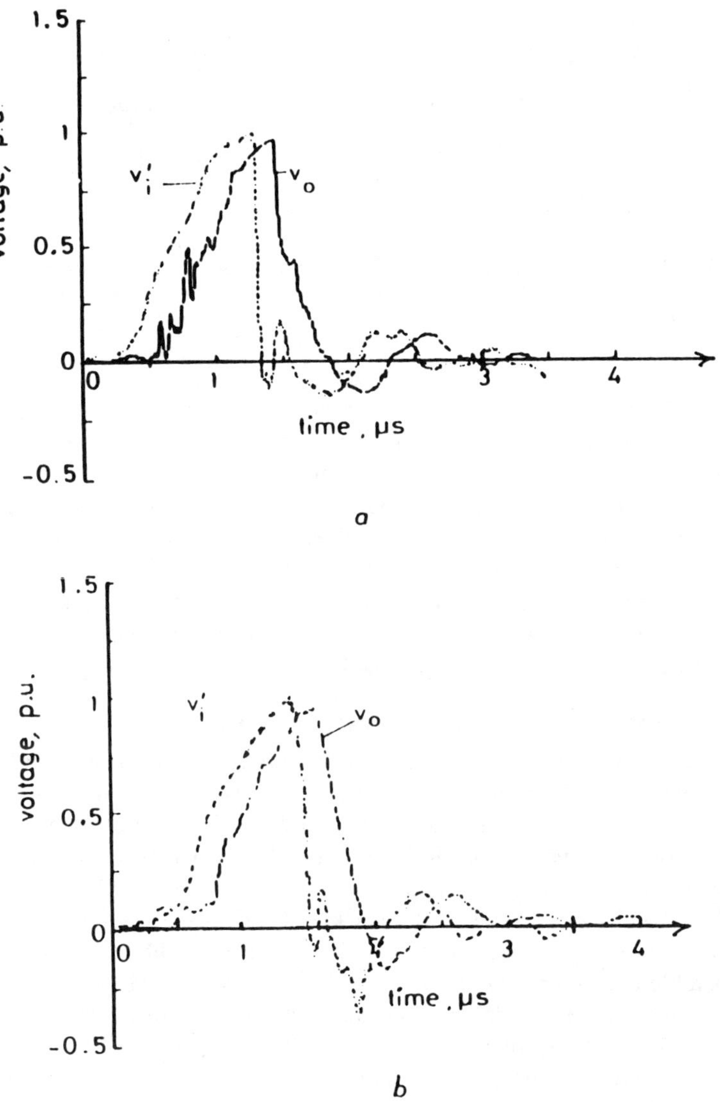

FIGURE 10.4 Reconstruction of lightning-impulse voltages chopped in the front, v_o = recorded output; v_i' = calculated input. a = oscillating system; b = damped system.

A. PROPERTIES OF AN IDEAL LIGHTNING FLASH COUNTER NETWORK

An ideal LFC network is characterized by the consumption of a minimum amount of power during the quiescent state, such that consumption of power will occur only at counting time. Quiescent current is usually on the order of 10 to 20 µA with regular battery voltage of 9 to 12 V and an Ah capacity of around 0.50.

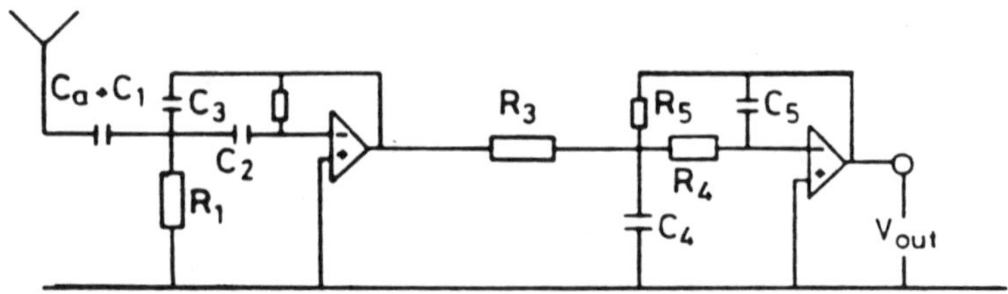

FIGURE 10.5   Narrowband active filter for use in lightning flash counters. $C_a = 58$ μF; $C_1 = 60$ pF; $C_2 = 243$ pF; $C_3 = 119$ pF; $C_4 = 246$ pF; $C_5 = 40$ pF; $R_1 = 78$ kΩ; $R_2 = 320$ kΩ; $R_3 = 110.5$ kΩ; $R_4 = 67$ kΩ; $R_5 = 209$ kΩ. $C_1 = $ cable capacitance, and the components are of 1% tolerance.

For effective discrimination between cloud and ground charges, researchers have suggested that reduction is the range of band filter and the resonant frequency should be shifted to a higher level. Active circuit design has not been much emphasized due to the magnitude of change in voltage during lightning and the current required to keep the active elements in the quiescent operative state, although the current technology has made the latter point less of a problem. In the *IEEE Proc.* article Oladiran et al. presented experimental and theoretical behavior for a micropower characterized by a highly stable, active network circuit to replace the present wideband RC-LFC networks.

B. NARROW-BAND NETWORK FOR GROUND FLASH COUNTER

Since LFCs do not reliably distinguish between cloud flashes and positive ground discharges, an improved network for LFC was proposed by Oladiran and co-workers, shown in Figure 10.5, which used such active elements as the LS 776 operational amplifier and RC elements. This network is characterized by quiescent current $= 15$ μA; frequency ≈ 9 kHz, Δf_o for 3 dB attenuation: 25 $\rightarrow$ 5 kHz band instead of the 50 $\rightarrow$ 2 kHz band of the RSA-10 network; the gain at the resonant frequency is 0 to 96, which is about -0.4 dB. This is a considerable improvement over the RSA-10 network which has a gain of -29.1 dB at resonance. Sinusoidal response characteristics of this filter are shown in Figure 10.6 and the calculated responses are shown in Figures 10.7A and 10.7B, both for relative and absolute attenuation.

C. PULSE CALIBRATION OF THE LIGHTNING FLASH COUNTERS

The distinguishing characteristic element of various types of lightning flashes is the duration of time rise of the pulse front. Negative surges are characterized by a fast rise time, while positive flashes are characterized by a slow rise and hence a longer duration. Lightning surges are, as indicated repeatedly, not periodic and the closest representation is the double exponential wave form

$$v(t) = A[e^{-\alpha_1 t} - e^{-\alpha_2 t}]$$ (10.2)

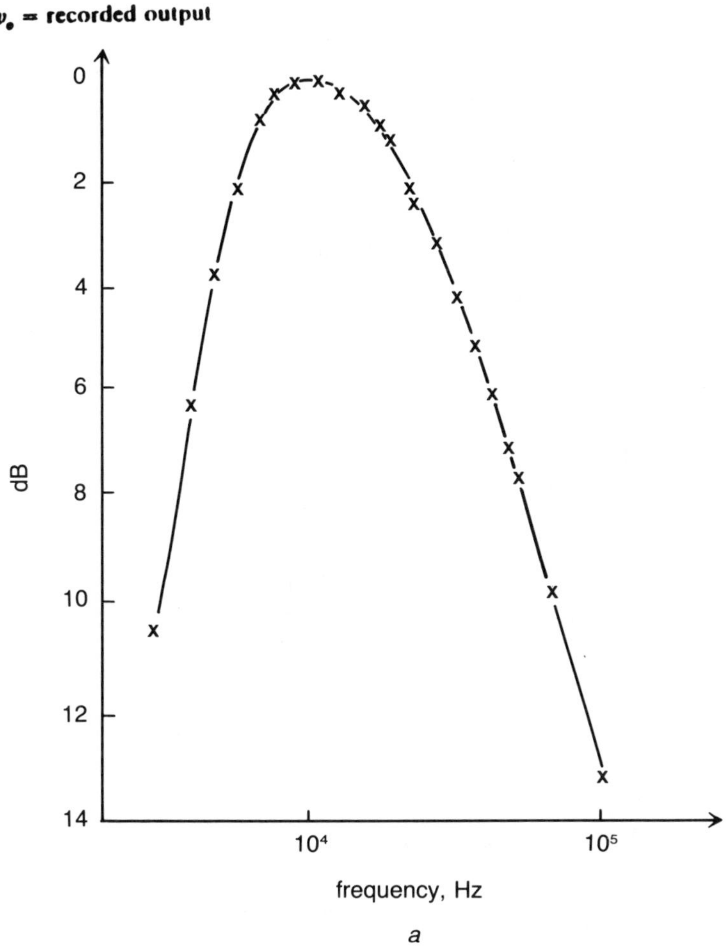

FIGURE 10.6   Experimental frequency characteristics of the active filter. (a) With gain − 1; (b) With gain − 2.

To determine the response of any LFC network to changing rise time, different rise times were subjected to corresponding peak voltages, as shown in Figures 10.8 and 10.9. Table 10.1 shows the peak responses and the corresponding rise times T_m for various LFC counters and the damping factors.

Figure 10.9 was obtained according to Oladiran et al. by normalizing the simulated output with respect to the observed maximum (see Table 10.1). It was also observed that although passive filters exhibit large damping, they demonstrated a reliable measure of discrimination in their responses to surges of different rise times. The graph in Figure 10.9 shows a reliable measure for the behavior of LFC that can be obtained by the use of voltage pulse, rather than sinusoidal voltages, a calibration which should be used to fix the threshold point for triggering.

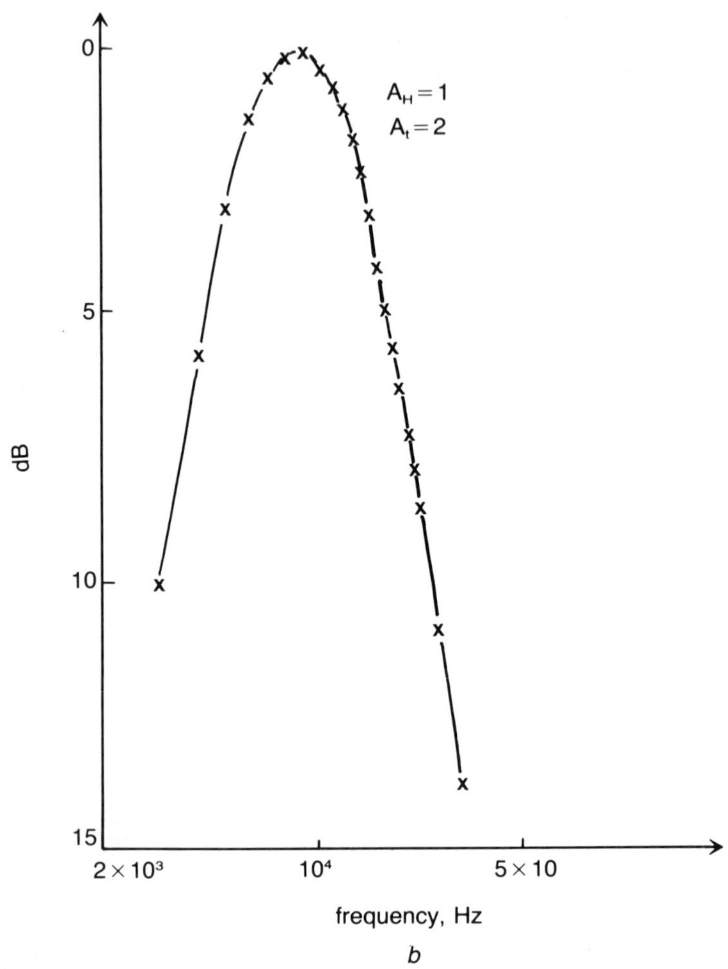

FIGURE 10.6b

D. RESPONSE CHECKING OF THE COUNTER

The main concern regarding reliable operation of the LFC is that its continuity be assured after a long lapse. One measure to ensure reliability of the LFC is the installment of a shunting element by which the supply voltage is entirely applied across the counter by pressing a switch. It is believed that if the mechanical counter is activated, the battery is still in good condition. However, two points must be satisfied to ensure compliance:

1.   Upon a flash by the counter, the remaining circuit components must retain integrated continuity.

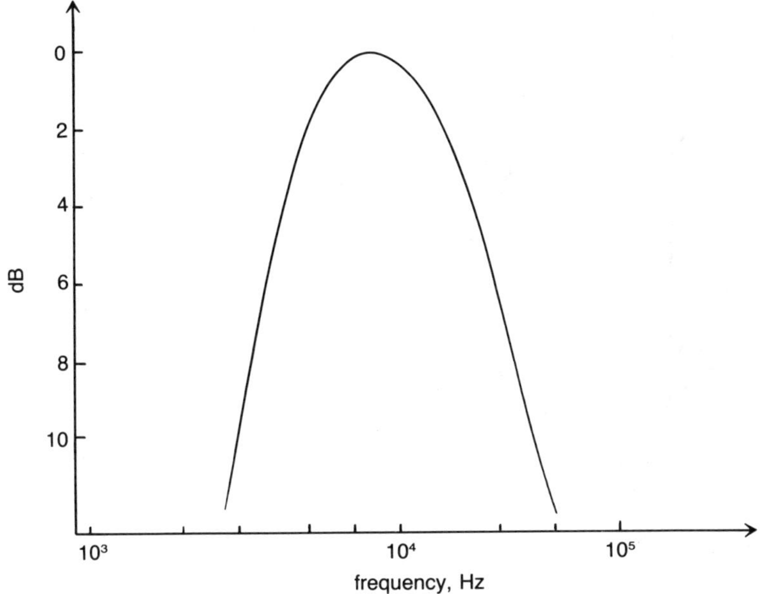

FIGURE 10.7A   Calculated frequency characteristics of active filter.

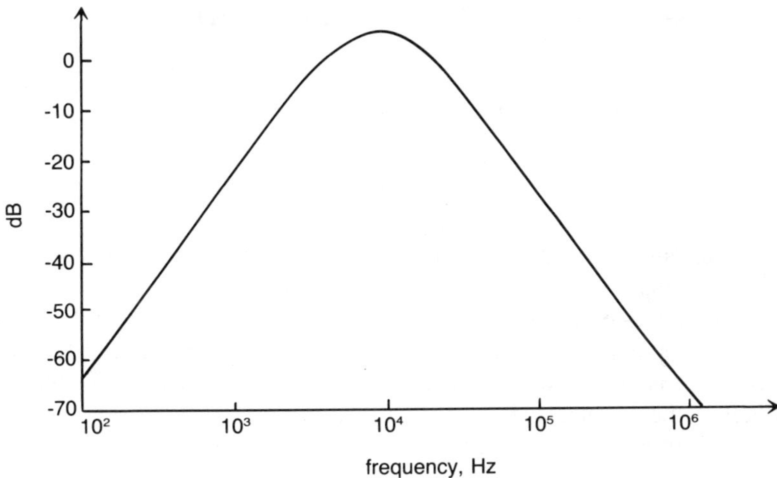

FIGURE 10.7B   Calculated attenuation characteristics of the 10 kHz active filter.

2. To ensure that the counter is current as well as voltage dependent, this character will block trigger a large current pulse upon the imposition of relatively small voltage across the LFC. When the battery became weak, the relay could be activated by a lightning flash while the mechanical counter would not.

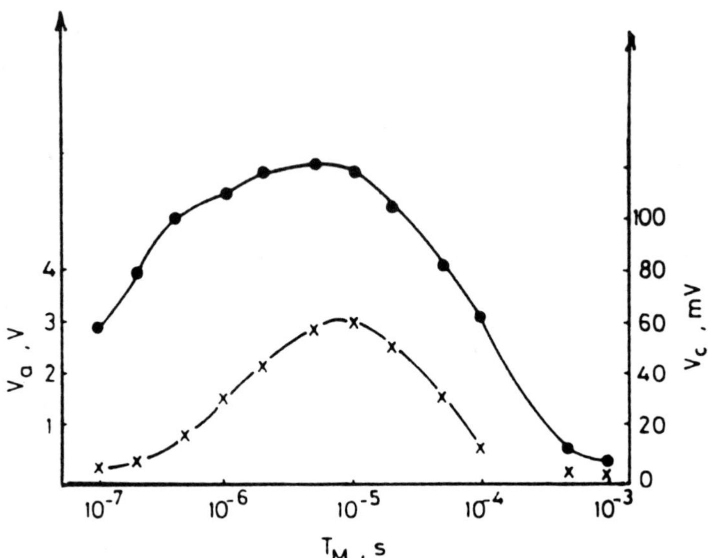

FIGURE 10.8   Output voltages of CIGRÉ and active filter when double exponential pulses of amplitude of 3 V were applied at the inputs. Note the change of scale: ×, active; ●, CIGRÉ.

TABLE 10.1
Damping Factors of the Various Filters for Double-Exponential Impulse

	RSA-10	V_{04}	V_{10}	V_{210}	Active	Active
T_M	5 μs	5 μs	5 μs	5 μs	10 μs	10 μs
A_1	0.039	0.028	0.053	0.052	0.052	1.0

Oladiran and associates have proposed a calibration system whereby the pulses are applied to the system and the count rate of the counter checked using a stable multivibrator as shown in Figure 10.10. The current will not flow until switch K is depressed, S_1 and S_2 are the contacts for the relay, and the system is completely isolated from the LFC until a calibration check is made. Thus, operation of the counter and any shift in the threshold value for triggering are assured reliability in system calibration.

10.III THEORETICAL ASPECTS

With reference to Figure 10.5, the transfer function of the two-stage active filter (see Oladiran et al.) is given by:

$$G(s) = \frac{HAS^2}{(BS^2 + CS + 1)(DS^2 + ES + 1)} \tag{10.3}$$

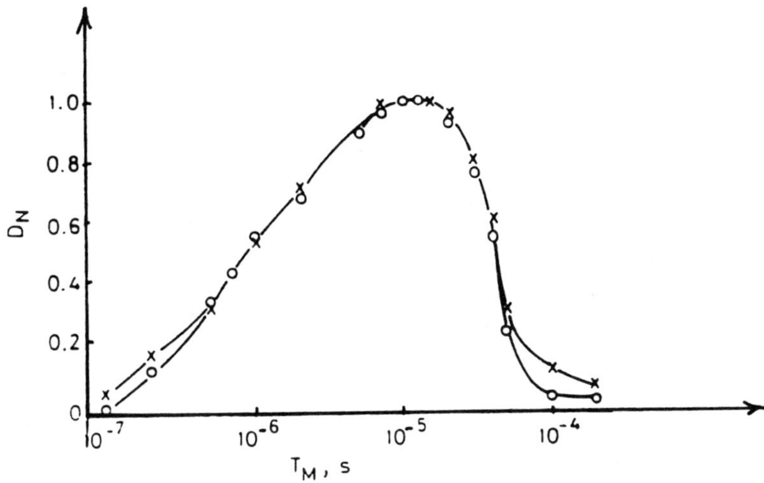

FIGURE 10.9A   Relative damping by an active filter of double exponential pulses of various rise times. ×, experimental values; ○, calculated values.

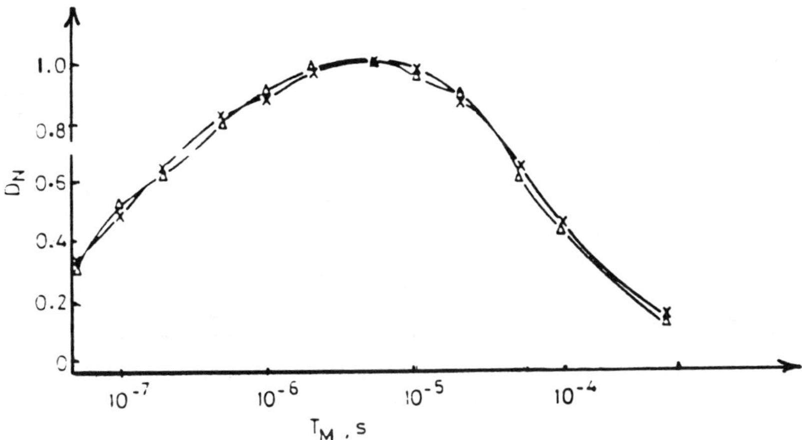

FIGURE 10.9B   Relative damping by CIGRÉ filter of double exponential pulses of various rise times. ×, experimental values; △, calculated values.

For a double exponential input, the response $h(t)$ is given by:

$$h(t) = 2e^{\alpha_1 t}[P_{D1} \cos W_1 t - Q_{D1} \sin W_1 t]$$

$$+ 2e^{\alpha_2 t}[P_{D2} \cos W_2 t - Q_{D2} \sin W_2 t] + K_3 e^{-\alpha t} - K_4 e^{-\beta t} \quad (10.4)$$

P_{D1}, P_{D2}, Q_{D1}, Q_{D2}, K_3, and K_4 are functions of a_1, a_2, W_1, W_2, α, and β.

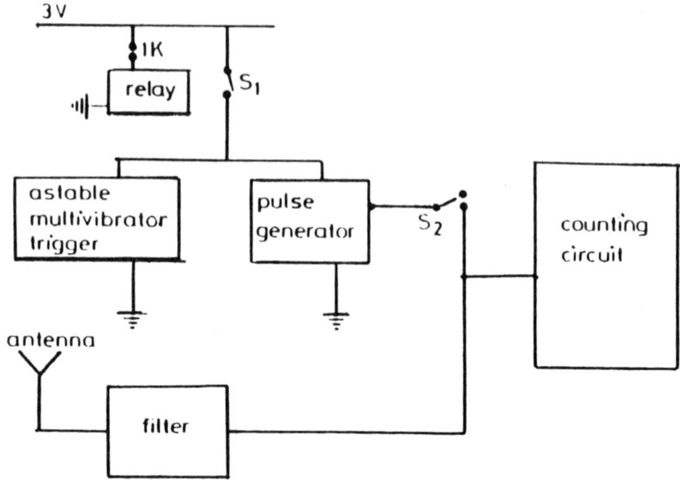

FIGURE 10.10   Schematic diagram for *in situ* calibration check of the LFC.

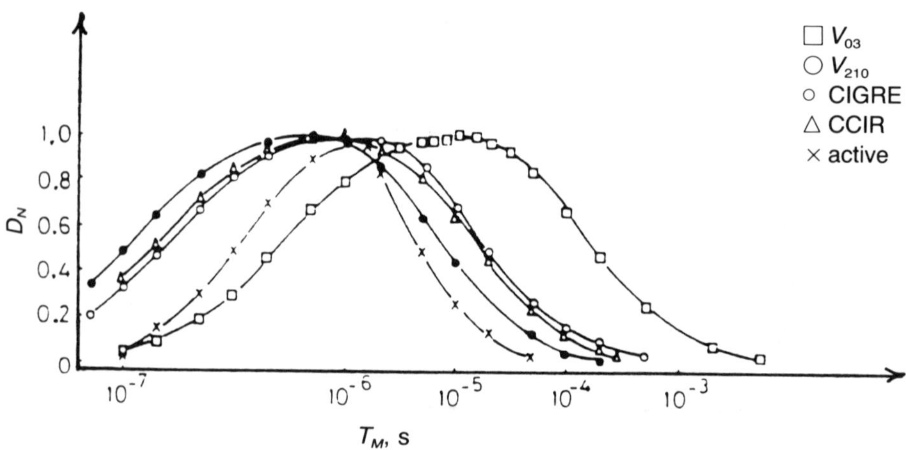

FIGURE 10.11   Calibration of counters using double-exponential pulses.

Figure 10.11 compares the response of passive filters to active filters, while Figure 10.12 compares three passive filters of different sinusoidal resonant frequencies of 10, 25, and 50 kHz.

Based on the above discussion and analysis presented in Section II Oladiran et al. made the following observations:

1.   The response of filters due to field changes generated by lightning flashes depends on the rise time of field variations.

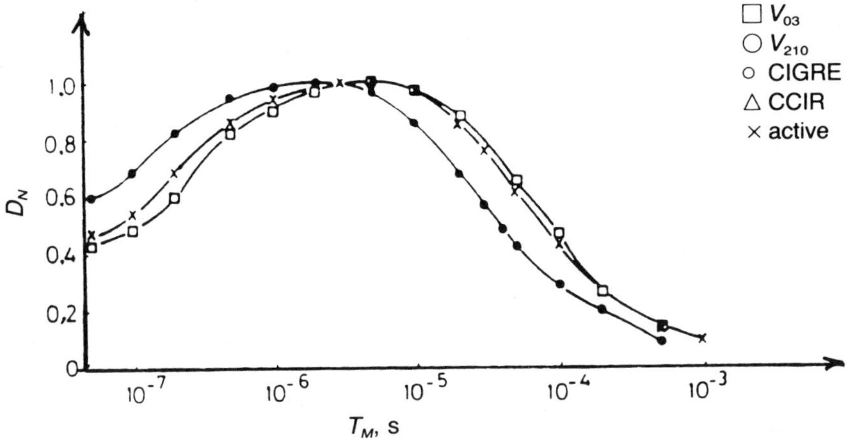

FIGURE 10.12   Response of modified CIGRÉ filter to double-exponential pulses (C_1 = 20 kHz, C_2 = 50 kHz) compared to response of 10 kHz CIGRÉ networks.

2.   Response is also dependent upon the resonant or center frequency as well as its bandwidth. For a narrow range of pulse rise time, clear discrimination can be secured.

3.   Figure 10.11 indicates that shifting the center or resonant frequency to higher values for the purpose of securing maximum sinusoidal response does not produce reliable discrimination by the LFC network unless combined with a very narrow band network response.

4.   The rise time response of the 500 Hz LFC network in the range 0.1 to 20 μs is the same as the response of the 10 kHz LFC network in the range 20 to 100 μs. Therefore, discrimination of flashes by the two networks is only dependent on the threshold value for triggering voltage set for each network.

10.IV FIELD EVALUATION OF LIGHTNING EARTH FLASH COUNTERS (LEFC)[1]

Material to be presented in this section is entirely extracted from a paper authored by R. B. Anderson, H. R. Van Niekerk, H. Kroninger and D. V. Meal, published in *IEE Proceedings*, Vol. 131, Pt. A, No. 2, pp. 118-124, March 1984.

Statistical research on LEFCs has emphasized reliable calibration and effective discrimination from cloud flashes. The 10 kHz counter has shown remarkable reliability in responding to earth flashes in subtropical areas where cloud flashes dominate. Interest is high for measurement of flash density as well as the isokeraunic level based on thunderstorm vs. nonthunderstorm days. This measurement is taken so that an approximate equivalence can be arrived at and an acceptable estimate of the earth's flash density can be obtained in cases where only the keraunic level is available. Research conducted in South Africa is at an advantage because informational data for earth flash density are known and can be applied to lightning protection projects.

Malan[15a] designed the first earth flash counter based upon the ratio of 5:100 kHz radiation that he had previously measured. However, research started in about 1965 by the Council for Scientific and Industrial Research Organization (CSIRO) in South Africa found that the ratio of 5:1 for the level of radiation could not be applied at the same time for the two components occurring during earth flash. Since the 5:10 kHz radiation was dominant for earth flashes, it was decided that experimental counters be designed to respond to this frequency band, and this later proved to more accurately identify discrimination against cloud flashes. Transistorized counters were designed for these frequencies and tested in the field with the immediate promise of high-level discrimination against cloud flashes.

A. LIGHTNING MEASURING SYSTEM

The effective range of LFC is defined as the radius within which the actual number of flashes occurring over a long period of time is equal to the number registered by the counter. Such a radius could be applied in the case of earth or cloud flashes. On the assumption of uniform distribution of flashes over the recording area, the effective counter range R_g is given by:

$$R_g = \left[2 \int_0^\infty P(r)r\,dr \right]^{1/2} \text{ km} \tag{10.5}$$

However, if the variable (r) is replaced by nondimensional $\dfrac{r}{R_g}$, the unit for R_g in Equation (10.5) becomes per unit (p.u.), where $P(r)$ is the probability that the counter will count a flash to earth at a range (r) in kilometers from the counter. Figure 10.13 illustrates plots for P_r, with respect to the ratio of the distance to an earth flash to the counter effective range $\left(\dfrac{r}{R_g}\right)$ for the three counters.

The CIGRE 500 Hz and 10 kHz LFCs are described in Reference 3. The RSA-5 was an original transistorized 500 Hz counter fitted with a vertical aerial but without altering the sensitivity to suit.

Recordings for $P(r)$ function with respect to (r) were carried out over a period of 12 to 13 years. The results for some individual counters are shown in Table 10.2 and Figure 10.13.

B. DISCRIMINATION AGAINST CLOUD FLASHES

Let N_g equal earth flash density given by:

$$N_g = KY_g/\pi R_g^2 \tag{10.6}$$

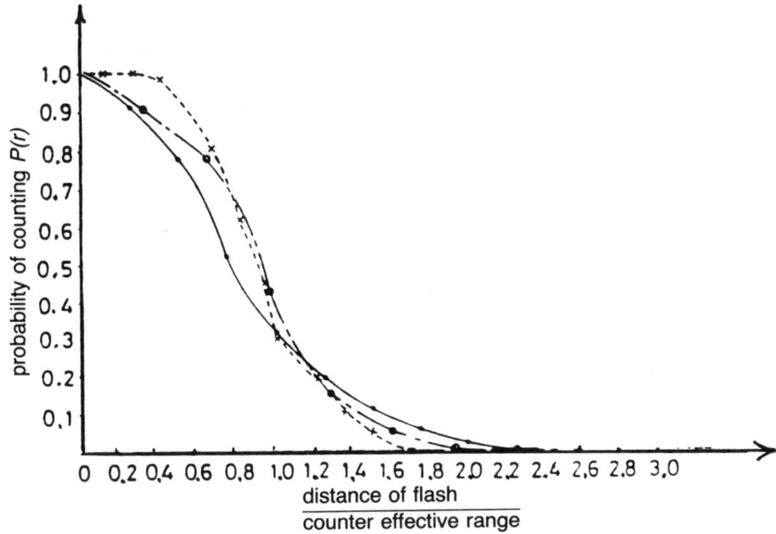

FIGURE 10.13   Probability of counting as a function of the ratio of the distance to an earth flash to the counter effective range. (●——●), RSA 10 (10 kHz); (×---×), CIGRÉ 500 Hz; (☉——☉), RSA 5.

where K = total registration of a counter, Y_g = correction factor <1, accounting for a proportion of earth flashes recorded by the counter, and R_g = effective range of counter to earth flashes.

In the literature Y_g has been described as a function of N_c, N_g, R_g, and R_c, where R_c is the effective range of counter to cloud flashes. It was also indicated that in order to calculate Y_g, it would be necessary to know not only the ratio of $R_c{:}R_g$, but also the respective densities of N_c and N_g as well as their ratio. A reliable measure of R_c could not be easily secured because the flash itself extends over a large distance that frequently exceeds 10 km, which is also on the same order as the expected value of the effective range. Similar questionable accuracy is voiced regarding the ratio of $N_c{:}N_g$, except in a situation where all flashes have been observed with reasonable reliability. This could occur only when R_c <10 km.

Because of this reliability problem, attention has been directed to measuring Y_g directly, as shown below:

$$Y_g = \frac{EF}{EF + CF} \tag{10.7}$$

where EF = number of identified earth flashes that operated the counter over a long period, and CF = number of identified cloud flashes that operated the counter observed over the same period.

TABLE 10.2
Effective Range R_g of Counters for Earth Flashes

Season	CIGRE (500 Hz, R_g, km)	CIGRE (10 kHz); RSA-10 (R_g, km)	RSA-5 (500 Hz), R_g, km)
1971/72	35.3	—	—
1972/73	36.8	—	16.8
1973/74	36.9	19.6	15.9
1974/75	—	17.4	15.8
1975/76	—	17.8	15.3
1976/77	—	23.4	15.8
1977/78	—	21.8	16.8
1978/79	37.9	20.7	14.0
1979/80	—	20.5	16.0
1980/81	—	19.9	14.8
Weighted mean	36.8	19.9	15.5
Total no. of observations	4972	6797	7097

TABLE 10.3
Observed Earth Flash Correction Factor (Y_g) for the CIGRE 10 kHz Counter

Season	No. of storms	No. of identified flashes			Y_g = proportion of earth flashes
		CF Cloud	EF Earth	Total	
1972/73	4	45	514	559	0.92
1973/74	4	17	702	719	0.98
1974/75	1	83	517	600	0.86
1975/76	1	13	276	289	0.96
1976/77	2	6	377	383	0.99
1977/78	7	4	531	535	0.99
1978/79	8	96	931	1027	0.91
1979/80	6	79	638	717	0.89
1980/81	4	2	79	81	0.97
Total	37	345	4565	4910	0.93

Anderson et al. reported that cloud flashes that operated the counter were easily identified because they occurred at distances close to the counter. On the other hand, only the earth flashes actually observed were recorded, but some that occurred outside the range of the recording equipment and were not seen were omitted even though they operated the counter. Therefore, it can be said that had the latter flashes also been observed they would have increased the value of the correction factor Y_g toward unity. The results, accumulated over 9 years, are given in Table 10.3.

In Table 10.3 only 345 cloud flashes in 4910 observations were observed to operate the counter, giving an overall weighted correction factor of Y_g =

TABLE 10.4
Resultant Cloud and Earth Flash Density and Their Ratio for Pretoria

	Calculated Flash Density			
Season	RSA-5 counter K	N_g CIGRE 10 kHz	N_c (R_c = 7.9 km)	Apparent ratio $N_c{:}N_g$
1972/73	10,492	9.2	21.4	2.3
1973/74	7,496	5.8	15.3	2.6
1974/75	9,899	6.4	20.2	3.2
1975/76	6,866	5.5	14.0	2.5
1976/77	7,825	7.0	16.0	2.5
1977/78	5,788	5.5	11.8	2.1
1978/79	7,091	6.5	14.8	2.2
1979/80	7,637	5.9	15.6	2.6
1980/81	7,406	6.3	15.1	2.4
Total	7,833	6.4	16.0	2.5

0.93. Accuracy was cited because the cloud flashes that operated the counter were within range of the counter, while earth flashes that were too far away to be observed but nevertheless operated the counter were excluded; their addition to the records would have increased Y_g toward unity. Therefore, the conclusion was that the 10 kHz counter could be used reliably for earth flashes within an accuracy of 10% if cloud flashes were ignored. Better accuracy could be secured if a corrected Y_g was applied to the result, as it was in the conditions in South Africa, with R_g = 19.9 km and Y_g = 0.93:

$$N_g = K/1340 \text{ earth flashes/km}^2 \tag{10.8}$$

Also, using the value of earth flash density from the 10 kHz counter as a base, the resultant cloud flash density for Pretoria was calculated as indicated in Table 10.4. The mean ratio of $N_c{:}N_g$ of 2.5 appears low for a subtropical area, but it is consistent with the annual measurement of the earth flash range and corrected Y_g.

From Equation (10.6), the ratio of the registration of two different counters operating in the same area is given by:

$$\frac{K_1}{K_2} = \left(\frac{R_1}{R_2}\right)^2 \bigg/ \frac{Y_1}{Y_2} \tag{10.9}$$

C. APPLICATIONS OF COUNTERS

In situations where only earth flashes occur, Figure 10.14 is a representation of the corresponding characteristic function. In cases where cloud flashes are partially counted, a unique curve could be established, provided that both $R_c{:}R_g$ and $N_c{:}N_g$ are known. However, it was shown by calculation that the differences

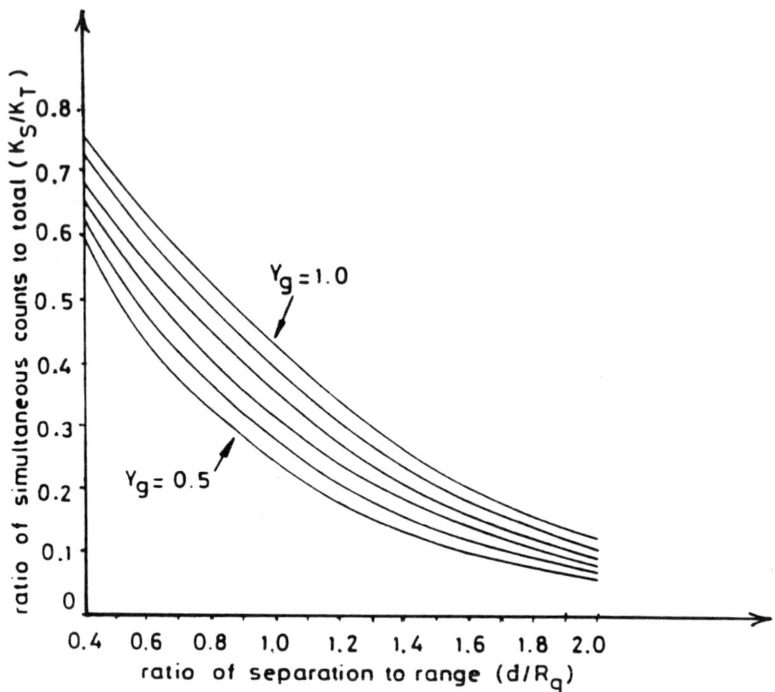

FIGURE 10.14   Ratio of simultaneous counts to total counts as a function of the ratio of station separation distance to the effective earth-flash range.

in values of d/R_g that could be obtained for values of Y_g varying between 0.9 and unity would be <10% if the counters were placed a distance (d) apart and are approximately equal to the expected value of the effective range R_g, i.e., $d/R_g = 1$. Therefore, if the number of simultaneous counts can be secured by relaying the number of counts from one counter to another, through either land-line or radio dispatch, and it occurs over a long period of time, Figure 10.14 can be used to estimate the ratio of d/R_g and thus check the value of R_g. This is especially possible for the 10 kHz counter, wherein the expected value of R_g = 20 km, and the expected value of Y_g lies between 0.9 and unity for more temperate parts of the globe. For tropical areas, Y_g for the 10 kHz counter lies between 0.8 and 0.9.

10.V THEORETICAL CRITERION FOR LIGHTNING FLASH COUNTERS

Current research in the literature is focused on the design of reliable networks to measure LFCs as effectively as possible, seeking to differentiate between cloud and ground flashes as well as between positive and negative polarities of the flashes.

In Chapters 1, 2, and 3, the discussion centered on (1) calculating the inducing electric field generated by a sustained lightning impulse, a pulse characterized with a sharp rising front and decaying tail and (2) the work of Sobocki, which presented functional forms of the inducing voltage for a pulse with superimposed oscillations, a critically damped front, and an overdamped front. In this section, this author felt it important to outline analytical criteria to determine a base or expectation for the LFC.

A. SUSTAINED DIRECT CURRENT STEP PULSE

Solutions for the inducing electric field strength due to a sustained lightning step were established in the set of Equations (2.41) through (2.44). Consider the gaseous atmospheric continuum in which the thunderstorm's range is anisotropic, implying that the medium electrical conductivity is a rank two tensor. Therefore,

$$\begin{bmatrix} \overline{J}_\rho \\ \overline{J}_\alpha \\ \overline{J}_\tau \end{bmatrix} = \begin{bmatrix} \sigma_{11} & \sigma_{12} & \sigma_{13} \\ \sigma_{21} & \sigma_{22} & \sigma_{23} \\ \sigma_{31} & \sigma_{32} & \sigma_{33} \end{bmatrix} \begin{bmatrix} \overline{E}_\rho \\ \overline{E}_\alpha \\ \overline{E}_\tau \end{bmatrix} \tag{10.10}$$

or

$$J_\rho = \sigma_{11}E_\rho + \sigma_{12}E_\alpha + \sigma_{13}E_\tau \tag{10.11}$$

$$\overline{J}_\alpha = \sigma_{21}\overline{E}_\rho + \sigma_{22}\overline{E}_\alpha + \sigma_{23}\overline{E}_\tau \tag{10.12}$$

$$\overline{J}_\tau = \sigma_{31}\overline{E}_\rho + \sigma_{32}\overline{E}_\alpha + \sigma_{33}\overline{E}_\tau \tag{10.13}$$

The occurrence of flash is identified by having any or all J_ρ, J_α, and J_τ to approach the value of breakdown, i.e., infinity.

Inspection of Equations (2.4) to (2.43) identifies the following conditions for a flash:

1. $J_\rho \to \infty$ which could happen when $E_\rho \to \infty$, $E_\alpha \to \infty$ and/or or $E_\tau \to \infty$

$$E_\rho \to \infty, E_\alpha \to \infty \quad \text{and/or} \quad E_\tau \to \infty$$

$$R_c = \rho, R_c \to 0, \hat{y} \to 0 \quad \text{and} \quad \rho = R_c \tag{10.14}$$

$$\hat{Y}_{\rho_1} = \sigma_{11} + j\omega\epsilon_{11} = 0$$

$$\hat{Y}_{\rho_2} = \sigma_{12} + j\omega\epsilon_{12} = 0$$

$$\hat{Y}_{\rho_3} = \sigma_{13} + j\omega\epsilon_{13} = 0 \tag{10.15}$$

or in terms of the flashing frequency:

$$\omega = -j\,\frac{\sigma_{11}}{\epsilon_{11}}$$

$$\omega = -j\,\frac{\sigma_{12}}{\epsilon_{12}}$$

$$\omega = -j\,\frac{\sigma_{13}}{\epsilon_{13}} \tag{10.16}$$

Equations (10.16) indicate the frequency at which a flash may occur, and the commencing of the flash is clearly associated with the anisotropic characteristics of the atmospheric medium.

2. $J_\alpha \rightarrow \infty$ which could happen when

$$\overline{E}_\rho = \overline{E}_\alpha = \overline{E}_\tau \rightarrow \infty$$

$$R_c = 0, \text{ and } R_c = \rho \tag{10.17}$$

and $\hat{Y}_\alpha = 0$, implying:

$$\hat{Y}_{\alpha_1} = \sigma_{21} + j\omega\epsilon_{21} = 0$$

$$\hat{Y}_{\alpha_2} = \sigma_{22} + j\omega\epsilon_{22} = 0$$

$$\hat{Y}_{\alpha_3} = \sigma_{23} + j\omega\epsilon_{23} = 0 \tag{10.18}$$

or in terms of σ:

$$\sigma_{21} = -j\omega\epsilon_{21}$$

$$\sigma_{22} = -j\omega\epsilon_{22}$$

$$\sigma_{23} = -j\omega\epsilon_{23} \tag{10.19}$$

3. $J_\tau \rightarrow \infty$ which could happen when

$$\overline{E}_\rho = \overline{E}_\alpha = \overline{E}_\tau \rightarrow \infty$$

$$R_c = 0, \quad \text{and} \quad R_c = \rho$$

and, of course, when $\hat{Y}_\tau \to 0$ which implies

$$\hat{Y}_{\tau_1} = \sigma_{31} + j\omega\epsilon_{31} = 0$$

$$\hat{Y}_{\tau_2} = \sigma_{32} + j\omega\epsilon_{32} = 0$$

$$\hat{Y}_{\tau_3} = \sigma_{33} + j\omega\epsilon_{33} = 0 \qquad (10.20)$$

or if we desire condition (10.20) in terms of W, the flashing frequency:

$$\omega = -j\,\frac{\sigma_{31}}{\epsilon_{31}}$$

$$= -j\,\frac{\sigma_{32}}{\epsilon_{32}}$$

$$= -j\,\frac{\sigma_{33}}{\epsilon_{33}} \qquad (10.21)$$

B. ACTUAL LIGHTNING PULSE

This refers to a lightning pulse where the front has a sharp linear rise and a linear decay. Solutions for the inducing electric field were presented in Equations (2.89). Again, commencement of flashing and its multiplicity could occur when the vectorial current density in an anisotropic continuum becomes theoretically ∞, implying satisfaction of the set of Equations (10.11), (10.12), and (10.13), i.e., $J_\rho = J_\alpha = J_\tau \to \infty$ with E_ρ, E_α, and E_τ given by Equation (2.89).

1. $\bar{J}_\rho \to \infty$ could occur when:

$$R_{c,v} = 0 = \hat{Y}$$

implying that $\bar{E}_\rho = \bar{E}_\alpha = \bar{E}_\tau = \infty$. $\bar{E}_\rho \to \infty$ could be achieved according to the same criteria listed by Equations (10.15). However, inspection of Equation (2.89) describing E_ρ indicates that an infinite value for E_ρ could occur when:

$$A_c \to \infty \quad \text{at} \quad t = t_1$$
$$R_c \to 0$$

Flashing frequencies are similar to those expressed by Equations (10.16).

2. $\bar{J}_\alpha \to \infty$, implying:

$$A_c \to \infty$$

$$\bar{E}_\rho = \bar{E}_\alpha = \bar{E}_\tau \to \infty$$

$$R_c = 0, \quad \text{and} \quad R_c = \rho$$

Flashing frequency from a set similar to Equations (10.9),

$$\omega = -j\,\frac{\sigma_{21}}{\epsilon_{21}}$$

$$= -j\,\frac{\sigma_{22}}{\epsilon_{22}}$$

$$= -j\,\frac{\sigma_{23}}{\epsilon_{23}} \tag{10.22}$$

3. $\bar{J}_\tau \rightarrow \infty$, implying:

$$A_c \rightarrow \infty$$

$$R_c = 0, \quad \text{and} \quad R_c = \rho$$

and, of course,

$$\bar{E}_\rho = \bar{E}_\alpha = \bar{E}_\tau \rightarrow \infty$$

implying that a set of flashing frequencies could be expressed in the same manner as those in Equation (10.22).

C. FRONT WITH OSCILLATIONS

According to Equation (2.113), a lightning pulse front with superimposed oscillations identifies α_{co} and b as parameters characterizing the wave front that was expressed by Sobocki. This author feels that α_{co} is a function of position with respect to the location of lightning stroke initiation and the point at which breakdown occurs. It is also this author's opinion that α_{co} is a function of the space admitivity $\hat{Y}$, which is equal to $\bar{\bar{\sigma}} + \overline{\overline{j\omega E}}$. Therefore,

$$\alpha_{co} = g_1(\rho,\alpha,\tau,\hat{Y}) \tag{10.23}$$

According to J. D. Cobine, expression for the sparking or flashing voltage is given by:

$$U_s = -Bpd/\ln\frac{\ln(1/r)}{Apd} \tag{10.24}$$

where P is the pressure in millimeters of mercury, r is the ratio of positive ionization coefficient to the electron ionization coefficient, and d is the separation between effective cathode and anode in centimeters.

Turning to Equation (2.113), at the moment of sparking or flashing, we may equate the sparking voltage given by (2.113) at $t = T_m$, which is equal to $n\pi/b\alpha_{co}$, to the sparking voltage U_s given by Equation (10.24). Therefore,

$$\frac{U_m}{K_c}\left[1 - e^{-\alpha_{co}T_m}(\cos b\alpha_{co}T_m + \frac{1}{b}\sin b\alpha_{co}T_m)\right]$$

$$= -Bpd/\ln\frac{\ln(1/r)}{Apd} \tag{10.25}$$

A and B are constants for gases such as air (as gaseous mixture), argon, helium, mercury, or under flashing in air or at atmospheric pressure,

$$\left[1 - e^{-\alpha_{co}T_m}(\cos b\alpha_{co}T_m + \frac{1}{b}\sin b\alpha_{co}T_m)\right]$$

$$= (30d + 1.35)\frac{K_c}{U_m} \tag{10.26}$$

Obviously, the product $b\,\alpha_{co}$ in effect represents the atmospheric continuum admitivity during lightning, whereby inserting $\alpha_{co}\,b = (\hat{Y})$ in Equation (10.26) will lead to the solution for $(\alpha + j\omega E)$, where $\hat{Y} = \sigma + j\omega\epsilon$ (isotropic continuum). Therefore,

$$e^{-f_3T_m}\left(\cos f_3T_m + \frac{1}{b}\sin f_3T_m\right)$$

$$= \left[1 - \frac{K_c}{U_m}(30d + 1.35)\right] \tag{10.27}$$

b, as explained earlier, is equal to $\pi/\ln(k_c - 1)$.

The solution for $(\hat{Y})$ from Equation (10.27) can be obtained to some degree of approximation by expanding the exponential and the trigonometric factors in terms of power series. Such a process is left as a problem for students to solve. Since $\hat{Y} = \sigma + j\omega\epsilon$, the flashing frequency ω can be secured.

D. CRITICALLY DAMPED FRONT

For the critically damped front of lightning voltage surge that is given by Equation (2.113A), let

$$\alpha_{ck} = \hat{Y} = \sigma + j\omega\epsilon \tag{10.28}$$

Therefore, at the moment of flashing $t = T_m$ and at atmospheric pressure, we can proceed as:

$$U_m[1 - (1 + \hat{y}T_m)e^{-\hat{y}T_m}]$$

$$\approx 30d + 1.35$$

or

$$e^{-\hat{y}T_m}(1 + \hat{y}T_m) = 1 - \frac{30d + 1.35}{U_m} \qquad (10.29)$$

The solution for $\hat{Y}$ from Equation (10.29) is left as a problem for students to solve.

E. OVERDAMPED FRONT

For an overdamped front of lightning voltage surge as given by Equation (2.114), the same procedure for calculating the flashing frequency W is used as in Sections C and D. Let

$$b\alpha_{cp} = \hat{y} \qquad (10.30)$$

Therefore, at the moment of flashing in air and at atmospheric pressure, we can proceed by letting $t = T_o$:

$$U_m\left[1 - e^{\hat{y}T_o}(\cos h\hat{y}T_o + \frac{1}{b} \sin h\hat{y}T_o) \right]$$

$$30d + 1.35 \qquad (10.31)$$

where, as before

$$\hat{y} = \sigma + j\omega\epsilon$$

Therefore,

$$e^{\omega T_o}\left(\cos h\hat{y}T_o + \frac{1}{b} \sin h\hat{y}T_o \right) = 1 - \frac{30d + 1.35}{U_m} \qquad (10.32)$$

The solution for $\hat{y}$ from Equation (10.31) is left as a problem to solve, after which the flashing frequency ω will be established.

F. FREQUENCY OF FLASHING BASED ON MOMENTS METHOD

In Chapter 3, the solution for the inducing voltage $U_{ic,v}$ caused by lightning conductive and convective strokes was derived and expressed by Equation (3.42).

That solution was established by the process of summation of an infinite number of magnetic moments.

$$U_{ic,v} = -j2\mu_o Qv_o(t - t_o) \sum_{\substack{m_c=1 \\ m_v=1}}^{\infty} \sum_{\substack{n_c=1 \\ n_v=1}}^{\infty} \frac{Y_{n,m}Z_{n,m}}{X_{n,m}} e^{-jX_{n,m}} \qquad (3.42)$$

At atmospheric pressure, the flashing or sparking voltage $U_s \approx 30d + 1.35$. Now let

$$X_{n,m} = C_x t \qquad (10.33)$$

where C_x is the velocity of light along X direction

$$= \omega\sqrt{Y_{n,m}^2 + Z_{n,m}^2}\,t \qquad (10.34)$$

where ω = frequency of flashing in rads/s. Therefore,

$$U_{ic,v} = -j2\mu_o Qv_o(t - t_o) \sum_{m=1}^{\infty} \sum_{n=1}^{\infty} \frac{Y_{n,m}Z_{n,m}e^{-j\omega t\sqrt{Y_{n,m}^2 + Z_{n,m}^2}}}{\beta t\sqrt{Y_{n,m}^2 + Z_{n,m}^2}}$$

$$\approx 30d + 1.35 \text{ at atmospheric pressure} \qquad (10.35)$$

d in Equation (10.35) could be expressed by:

$$d = \sqrt{X_{n,m}^2 + Y_{n,m}^2 + Z_{n,m}^2} \qquad (10.36)$$

From Equations (10.35) and (10.36), we can observe that at any time t and at any point in space, the frequency of flashing could be implied theoretically. The solution for W is left as a problem for students to solve.

10.VI SOLVED EXAMPLES

A. With reference to Figure 10.13, derive a matching function that may fit the probability function $P(r)$ with respect to r/R_g, for CIGRE counter.

Solution
First try the well-known probability function given by:

$$f(\mu) = Ae^{-\alpha u^2} = p(u) \qquad (A1)$$

where $u = r/R_g$ and A, a are constants. Examine several points (≈ 7) along the curve as shown below:

r/R_g	$P(r)$	r/R_g	$P(r)$
0.0	1.0	1.0	0.35
0.2	0.94	1.2	0.25
0.4	0.84	1.4	0.15
0.6	0.74	1.6	0.08
0.8	0.55	2.2	0

Of course, $A = 1$.

$r/R_g > 1$ has a great influence on $P(r)$

$$P(r) = e^{\alpha\left(\frac{r}{R_g}\right)^2} \tag{A2}$$

To ensure acceptable representation of $P(r)$ by Equation (A2), substitute respective values for (r/R_g) and $P(r)$ such that the constant (a) remains within a small deviation. Therefore:

$$0.96 = e^{a(0.2)^2} = e^{0.04a}, \quad a = 1$$

$$0.84 = e^{a(0.4)^2} = e^{0.16a}, \quad a = -0.93$$

$$0.72 = e^{a(0.6)^2} = e^{0.36a}, \quad a = -0.96$$

$$0.52 = e^{a(0.8)^2} = e^{0.64a}, \quad a = -1.02$$

$$0.38 = e^a = e^a, \quad a = -0.97$$

$$0.23 = e^{a(1.2)^2} = e^{1.44a}, \quad a = -0.98$$

$$0.15 = e^{a(1.4)^2} = e^{1.96a}, \quad a = -1.02$$

$$0.08 = e^{a(1.6)^2} = e^{2.56a}, \quad a = -1.02 \tag{A3}$$

To minimize error in securing a close, accurate number for the exponential parameter (a), an averaging process is conducted on all values of (a) calculated resulting in a mean value of -0.985. Taking into account personal errors involved in reading the points for $P(r)$ and the respective r/R_g, this author feels that -0.985 is very close to unity. Therefore,

$$P(r) = e^{\frac{r^2}{R_g^2}} \tag{A4}$$

B. Using the expression of $P(r)$ for the CIGRE 10 kHz counter obtained in Example A, establish a numerical solution for the effective range R_g.

Solution

From Equation (10.4):

$$R_g = \left[2 \int_0^\infty P(r)r\,dr \right]^{1/2} \text{ km} \tag{B1}$$

and from Example A

$$P(r) = e^{\frac{r^2}{R_g^2}} \tag{B2}$$

Therefore, from Equations (B1) and (B2), R_g in p.u. becomes:

$$R_g = \left[2 \int_0^\infty e^{-\left(\frac{r}{R_g}\right)^2} \frac{r}{R_g} \, d\left(\frac{r}{R_g}\right) \right]^{1/2} \text{ p.u.} \tag{B3}$$

Since

$$\int_0^\infty x e^{-x2} \, dx = \frac{1}{2} \tag{B4}$$

therefore, from Equation (B3): $R_g = 1$ p.u. or $= 36.8$ km, which is the effective range given in Table 10.2.

C. The simulation model for measurements in high voltage system involves the element of voltage divider which contains two concentrated elements and one distribution element. The distributed element is represented by the stray impedance Z_{ln}, where

$$Z_{ln} = \frac{L_{s,n}}{C_{s,n}} \tag{C1}$$

Show representation for Z_{ln}, if the surrounding medium is simple and then if it is linear.

Solution

For the simple medium

$$Z_{ln} = \sqrt{\frac{\mu_r}{\epsilon_r}} \frac{L_{s,n}}{C_{s,n}} \tag{C2}$$

where μ_r, ϵ_r is the relative magnetic permittivity and electric permittivity, respectively.

For the linear medium, where the magnetic induction vector $\overline{B}$,

$$\overline{B} = \mu\overline{H} + \mu_1 \frac{\partial \overline{H}}{\partial t} + \mu_2 \frac{\partial^2 \overline{H}}{\partial t^2} + \ldots \tag{C3}$$

and $\overline{D}$ is the electric displacement vector:

$$\overline{D} = \epsilon\overline{E} + \epsilon_1 \frac{\partial \overline{E}}{\partial t} + \epsilon_2 \frac{\partial^2 \overline{E}}{dt^2} + \ldots \tag{C4}$$

Therefore,

$$\overline{B} = \overline{H}\left[\mu + \mu_1 \frac{\partial}{\partial t} + \mu_2 \frac{\partial^2}{\partial t^2} + \ldots + \mu_n \frac{\partial^n}{\partial t^n} \right] \tag{C5}$$

and

$$\overline{D} = \overline{E}\left[\epsilon + \epsilon_1 \frac{\partial}{\partial t} + \epsilon_2 \frac{\partial^2}{\partial t^2} + \ldots + \epsilon_n \frac{\partial^n}{\partial t^n} \right] \tag{C6}$$

Hence, Z_{ln} is given by:

$$= \frac{L_{s,n}}{C_{s,n}} \sqrt{\frac{\mu_o}{\epsilon_o}} \; [\mu + \mu_1 P + \mu_2 P^2 + \ldots + \mu_n P^n]/$$

$$[\epsilon + \epsilon_2 P + \epsilon P^2 + \ldots + \epsilon_n P^n] \tag{C7}$$

where $L_{s,n}$ and $C_{s,n}$ are the stray inductance and stray capacitance in normal atmosphere, and

$$P = \frac{\partial}{\partial t}$$

$$P^2 = \frac{\partial^2}{\partial t^2}$$

$$\vdots$$

$$P^n = \frac{\partial^n}{\partial t^n} \tag{C8}$$

D. Given the induced electric field across the distributed element of voltage divider in the simulation model for H. V., the measurement shown in Figure 10.2 $= \hat{a}_x\, A\, \delta(t)/(1\, +\, y^2)$. Obtain expression for the energy density stored magnetically and electrically.

Solution

$$\overline{E} = \hat{a}_x A\delta(t)/(1 + y^2) \tag{D1}$$

Then, since

$$\overline{\nabla} \times \overline{E} = -\frac{\partial B}{\partial t} \tag{D2}$$

$$\overline{\nabla} \times \overline{E} = \begin{bmatrix} \hat{a}_x & \hat{a}_y & \hat{a}_z \\ (\partial/\partial x) & (\partial/\partial y) & (\partial/\partial z) \\ E_x & 0 & 0 \end{bmatrix} = -\frac{\partial \overline{B}}{\partial t}$$

$$= -\hat{a}_y\left[-\frac{\partial \overline{E}_x}{\partial Z}\right] + \hat{a}_z\left[-\frac{\partial \overline{E}}{\partial y}\right]$$

$$= -\hat{a}_z A\delta(t)\left(\frac{-2y}{(1 + y^2)^2}\right)$$

$$= \frac{\partial \overline{B}}{\partial t} \tag{D3}$$

Therefore,

$$\hat{a}_z\overline{B} = -\int A\delta(t)\,\frac{2y}{(1 + y^2)^2}\,dt$$

$$= -AU_{-1}(t)\,\frac{2y}{(1 + y^2)^2}\ \text{tesla} \tag{D4}$$

where $U_{-1}(t)$ is a step function and $\partial(t)$ is a unit impulse. Since the system is simple,

$$\hat{a}_x\overline{D} = \hat{a}_x\epsilon\overline{E}$$

$$= \epsilon\left[\frac{A\delta(t)}{(1 + y^2)}\right] \tag{D5}$$

Therefore, energy density stored magnetically W_m:

$$W_m = \frac{B^2}{2\mu} = \frac{1}{2\mu}\left[AU_{-1}(t)\,\frac{2y}{(1+y^2)^2}\right]^2 \text{ J/m}^3 \qquad\qquad \text{(D6)}$$

and energy density stored electrically W_e:

$$W_c = \frac{1}{2}\,\epsilon D^2 = \frac{\epsilon}{2}\left[A\delta(t)\,\frac{1}{(1+y^2)^2}\right]^2 \text{ J/m}^3 \qquad\qquad \text{(D7)}$$

10.VII PROBLEMS

1. Refer to the high voltage arm of the divider shown in Figure 10.2. Replace the concentrated capacitive element with an inductive concentrated component having a hard core ferromagnet. Hard magnetic material is characterized as having its inductance linearly proportional to frequency, and at an impulse current surge the inductance acts as a short circuit. Magnetic permeability of such material is on the order of 2 to 3 μm. Through an appropriate mathematical model of differential equations, and if the induced voltage lightning surge wave form front has a sharp linear rise toward its peak, followed by a slow linearly decaying tail, obtain expression for voltage distribution to be divided on each arm of the newly developed voltage divider.

2. Repeat problem 1 if the induced voltage lightning surge wave form is an exponentially decaying sinusoid represented by $A\,e^{-\alpha_1 t}\sin\beta t$.

3. Repeat problem 1 if the induced voltage lightning surge wave form is a single exponential pulse expressed by $A\,e^{-\alpha t}$.

4. Refer to Figure 10.1, which represents a block diagram for typical impulse voltage test arrangement. Transform this diagram into circuit and network elements as well as any feasible active component.

5. Repeat problem 4, replacing the front capacitor with an ideal inductor, obeying the criterion of constant flux linkage theorem, i.e., $di/dt = 0$.

6. Using Figures 10.9A and 10.9B, plot the ratio of relative damping for an active filter with respect to CIGRE against rise times. Obtain an empirical mathematical equation for the developed curve.

7. With reference to Figure 10.8, plot a curve for $(V_a - V_m)$ with respect to T_m, and derive a feasible empirical mathematical equation for the developed curve.

8. Show that $h(t)$ given by Equation (10.4) is the inverse Laplace transform of $G(s)$ expressed by Equation (10.3).

9. Using Equation (10.4), obtain expression for the probability function $P(r)$ with respect to an LFC effective range R_g, and with respect to $1/R_g$. Compare the developed relationship with respect to the curve shown in Figure 10.13 for the RSA-5 counter.

10. Derive an empirical relationship for $P(r)$ in terms of the ratio of distance at flash counter effective range (r/R_g) for the RSA-10 kHz counter shown in Figure 10.13. Then, obtain expression for R_g.

11. Repeat problem 10 for the RSA-5 counter shown in Figure 10.13.

12. Using Equation (10.6) for the density of ground flashes in terms of Y_g and R_g, obtain expression for N_g in terms of probability function $P(r)$. The role of Y_g can be extracted from data given in Table 10.3.

13. Using Equation (10.6), obtain expression for N_g in terms of R_g representing only the performance of RSA-5 counter. Use the data in Table 10.3 to account for the role of Y_g.

14. Using Equation (10.9), develop expression for K_1/K_2 for RSA-10 counter with respect to CIGRE counter; may apply data from Figure 10.13 and Table 10.2 in the process as well as any other pertinent resource.

15. With reference to Figure 10.14, obtain plots for K_s/K_T with respect to Y_g for the following values of $d/R_g = 0.5, 0.7, 0.9, 1.1, 1.3, 1.5, 1.7$, and 1.9. d is the separation station and R_g is the effective counter range.

16. Solve Example C (Section 10.VI) if the atmospheric medium at lightning medium will render all electromagnetic field components tensors. This will also associate tensorial character to μ and ϵ as being of rank two.

17. Obtain approximate solution for the flashing frequency W when the lightning front of voltage surge is associated with superimposed oscillations.

18. Obtain approximate solution for the flashing frequency W when the lightning front of voltage surge is critically damped.

19. Obtain approximate solution for the flashing frequency W when the lightning front of voltage surge is overdamped.

20. Obtain solution for the flashing frequency W from Equation (10.35) at a point 200 mi from the direct place of lightning stroke.

21. Repeat problem 4 to obtain solution for the flashing frequency omega by using Equation (10.35) at a point 100 mi and then 1000 mi from the direct place of lightning stroke.

22. Express the current density vector in Equations (10.11 to 10.13) in the principle system of coordinates and solve for the frequencies of flashing.

23. The set of Equations (10.16) represents the flashing frequency omega when $J_\rho \to \infty$. Revise Equations (10.16) for $\overline{\overline{E}}$ to be represented in the principle coordinate system.

24. Repeat problem 6 for $\overline{J}_\alpha \to \infty$.

25. Repeat problem 6 for $\overline{J}_\tau \to \infty$.

26. Equation (10.24) represents the sparking voltage at lightning breakdown at a separation of (d). If $d = P(\rho, \alpha, \tau)$, a point in space with respect to the place of stroke initiation $P(0, 0, 0)$. Obtain solution for the potential gradient vector.

27. Using the solution of the potential gradient obtained in problem 9, express the developed current density vector with the medium electrical conductivity as a rank two tensor in the principle coordinate system.

REFERENCES

1. **Anderson, R. B., Van Niederk, H. R., Kroninger, H., and Meal, D. V.,** Development and field evaluation of a lightning earth-flash counter, *IEEE Proc.,* 131(Part A, No. 2), 118, 1984.

2. **Anderson, R. B., Van Niekerk, H. R., and Gertenbach, J. J.,** Improved lightning earth-flash counters, *Electron. Lett.,* 9, 394, 1973.

3. **Anderson, R. B., Van Niekerk, H. R., Prentice, S. A., and Mackerras, D.,** Improved lightning flash counters, *Electra,* 66, 85, 1979.

4. **Anderson, R. B. and Eriksson, A. J.,** Lightning parameters for engineering application, *Electra,* 69, 65, 1980.

5. **Anderson, R. B.,** The Lightning Discharge, Ph.D. thesis, University of Cape Town, Cape Town, South Africa, CSIRO Spec. Rep. ELEK 12, Part 1, Committee for Scientific and Industrial Research Organization, Pretoria, 1971, 48.

6. **Barham, R. A. and Mackerras, D.,** Vertical-aerial CIGRÉ-type lightning-flash counter, *Ibid.,* Part 8, 1972, 480.

7. **Barham, R. A.,** A transistorised lightning flash counter, *Electron. Eng.,* 38, 164, 1966.

7a. **Prentice, S. A. and Mackerras, D.,** Recording range of a lightning flash counter, *Proc. IEEE,* 116, 294, 1969.

8. **Deveniza, M. and Uman, M. A.,** Research into Lightning Protection of Distribution Systems II. Results from Florida Field Work 1978 and 1979, paper 83 SM 454-6, No. 855 presented at the IEEE Summer Power Meeting, July 1983.

9. **Golde, R. H.,** Lightning Flash Counter, Electrical Research Association of Great Britain Report, July 1957.

9a. **Horner, F.,** The design and use of instruments for counting local lightning flashes, *Proc. IEEE,* 107(Part B), 321, 1960.

10. **Kroninger, H.,** Newsletter — National Lightning Recording Scheme No. 3, CSIRO Spec. Rep. ELEK 181, September 1979; South African Weather Bureau, Department of Transport, Climate of South Africa, Part 8, General Survey W.B. 28, 1965, p. 298, Fig. 154.

11. **Krider, E. P. and Noggle, R. C.,** A Detection System for Lightning, U.S. Patent 4115732, September 19, 1978.

12. **Nikolopoulos, P. N. and Topalis, F. V.,** Accurate method of representation of high voltage measuring systems and its application in high impulse voltage measurements, *IEEE Proc.,* 136(Part A, No. 2), 66, 1989.

13. **Oladiran, E. O., Pisler, E., and Israelsson, S.,** New lightning flash counter and calibration circuit with improved discrimination of cloud and ground discharges, *IEEE Proc.,* 135(Part A, No. 1), 22, 1988.

14. **Pierce, E. T.,** The influence of individual variations in the field changes due to lightning discharges upon the design and performance of lightning flash counters, *Arch. Met. Geophys. Bioklim. Ser. A.,* 1, 78, 1956.

15. **Prentice, S. A.,** CIGRE lightning flash counters, *Electra,* 22, 179, 1972.

15a. **Malan, D. J.,** Lightning counter for flashes to ground, in *Proc. Int. Conf. Gas Discharges,* Butterworths, London, 1962.

16. **Van Niekerk, H. R.,** Calibration of Lightning Flash Counters, CSIRO Spec. Rep. ELEK 49, Committee for Scientific and Industrial Research Organization, Pretoria, July 1974.

17. **Cobine, J. D.,** *Gaseous Conductors,* Dover Publications, Inc., New York.

18. **Harrington, R. F.,** *Time-Harmonic Electromagnetic Fields,* McGraw-Hill, New York, 1961. Reprinted with permission.

Chapter 11

PRINCIPLE OF PROTECTION FROM H.V. SURGES

11.I GAPLESS ARRESTERS

A conventional arrester made of SCI alone is called "varistor" for variable resistor and the sudden change in resistance is characterized as nonlinear. Gapped arresters offer significant protection, but with certain limitations, especially that their response to many lightning surges is not fast enough. Sometimes even during the first microseconds of lightning stroke, the system to be protected may suffer potentially damaging over-voltage.

Another problem resulting from the use of gapped arrester is the erosion with time of the electrodes due to melting produced by the huge flow of electrons by lightning or switching impulse. The electrodes eventually become unreliable. Also, moisture in arrester housing produces deterioration in the structure of the electrodes, which adds another element to unreliable operation of the gapped arrester.

Because of this problem that may be encountered during the protective function of the gapped arrester, a change has been made to solid state gapless arresters over the last 25 years. During the mid 1970s, manufacturers initiated the utilization of metal oxide devices to protect sensitive equipment such as televisions, stereos, and other electronic devices. Electric utilities turned their attention to the use of the metal oxides varistor (MOV) in the design of gapless arresters after research indicated their reliability, adequate sensitivity, and quick response to over-voltages. However, metal oxides have not yet been designed in the form of large blocks for utility applications.

To proceed toward possible adoption of MOV in power system protection, the Electric Power Research Institute (EPRI)[7] selected two power equipment manufacturers, McGraw-Edison and Westinghouse Electric to conduct a research project targeting the operational design of the largest MOV in power system installations.

Currently almost all electric utilities use gapless arresters in their systems, and in fact, gapped arresters are no longer installed for high voltage system >69 kV; however, gapped arresters are still being used for distribution systems to retain the aspect of compatibility with existing power installations in smooth coordination.

Solid state gapless arresters made use of zinc-oxide (ZnO) which is ground up with small amounts of bismuth, antimony, chromium, nickel, and aluminum combination.

Cylindrical molds of MOV materials are subjected to kiln firing at 1200°C, followed by cooling. The metal oxides transform into millions of tiny grains, primarily ZnO surrounded by insulating layers from other metals. The insulating layers respond to the flow of surge current through the ZnO grains just above the threshold limit.

In 1976, General Electric introduced a new series of gapped arresters in which the varistor element was ZnO instead of SIC. In 1981, EPRI sponsored the research that developed gapless arresters using ZnO grain and ceramic as insulating layers in the MOV. Bonneville Power Administration and the Tennessee Valley Authority carried out evaluations of MOV gapless arresters for 1200 and 550 kV system with very satisfactory and encouraging results.

Acceptance of gapless surge arresters by the utilities and various industries continues to grow, and concomitantly with the increase in production, their cost is more competitive; they are also smaller in size and of lighter weight. They react more quickly to surges than gapped arresters, especially with no sparkover. The insulating layers in the MOV ''sense'' the rise in the surge wavefront immediately, and more rapidly alter the conductive state at a lower voltage level than conventional arresters.

MOV arresters start conduction at about 1.25 to 1.30 times normal voltage, while gapped arresters begin conduction at 1.50 to 2.00 times normal voltage. Consequently, protected equipment from over-voltage will be subjected in less time duration in MOV arresters than in the case of conventional arresters. Therefore, the basic insulation level (BIL) for equipment and systems to be protected will be relatively less by using MOV arresters, and this is also less costly.

Freeman[7] characterized the MOV arresters with reliable nonlinearity, where the leakage current is negligible at normal voltage, while current flow surges to a multi-kiloampere level at a predetermined voltage threshold. Their response to over-voltages is always consistent with every discharge. Also the nature of MOV material and design feature of arresters eliminates the problem of erosion and degradation that usually exists in gapped arresters. In addition, gapless arresters do not depend on spark; only an extreme amount of water leakage can disrupt their response to discharge.

In underground substations, MOV arresters provide significant economic benefits via their epoxy insulation and carbon-painted exteriors safely installed directly to the terminals of the protected device. Gapped arresters, on the other hand, require a separate pad and enclosure for isolation and shielding.

Electric utilities are beginning to employ metal oxide semiconductors in other applications such as series capacitor protection equipment, vacuum interrupter switch gear, and capacitor voltage transformers. Vasu Tahiliani, EPRIs project manager for metal oxide technology, predicts future large-scale production of transient energy absorption devices as well as MOV arresters using metal oxide semiconductors.

11.II INCIDENCE OF H.V. SURGES ON FERROMAGNETS

The problem of normal and effective operation of electronic and electro-mechanical control devices and local circuits in every power system is a major concern. The penetration of surge voltages and high current pulses into those regions leads to damaging effects in most cases unless an adequate and secure protective scheme is provided. Current and voltage surges are random and destructive, usually of short time rise and with high velocity of propagation. Surges enter relaying systems and other control devices through capacitive coupling, inductive coupling of cables with high power equipment, instrument transformers, difference of potential at various points along the transmission system, direct strikes of lightning stroke, and switching operations of lines and cables.

Present protective schemes, although numerous, do not completely provide optimum security against current surges, especially in transmission systems. These schemes include shielding of cables, diverse routing of lines, selective location of control lines and cables, insertion of series inductance and shunt capacitance to flatten the strength of voltage and current surges, and occasionally the connection of linear and nonlinear resistive elements in the power system.

A major point in high voltage system protection is in connection with the primary current transformer, in which the circuit phase angle approaches $90°$ and the moment of maximum current corresponds to zero voltage. In the case of failure of the insulation impulse level the occurrence of the surge current may be destructive to the local and neighboring circuits. Also, switching operation in the primary circuit of the current transformer may occur at the moment of zero voltage leading to the generation of high current pulses. High current surges may enter a neighboring line with different surge impedance, resulting in a series of reflections that may amplify the strength of the current pulse severalfold.

This section presents calculations of surge fields and the positive role a new protective element may serve in protecting against high current surges in high voltage transmission systems. This protective element is called the saturable resistor, which is of the same construction as the closed core reactor, but with a hard magnetic material core. The saturable-resistor is characterized by the following properties:

1. The ohmic values rise sharply as the current passes the pick-up value up to saturation. The maximum current will be more than twice the pick-up value.
2. The resistive component is due to eddy current plus hysteresis losses and is a function of the time varying current flowing in the device and frequency variations.

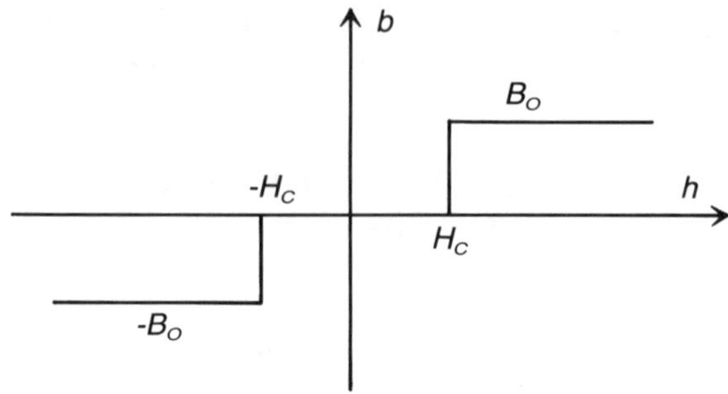

FIGURE 11.1   Ideal displaced magnetization curve of saturable-resistor.

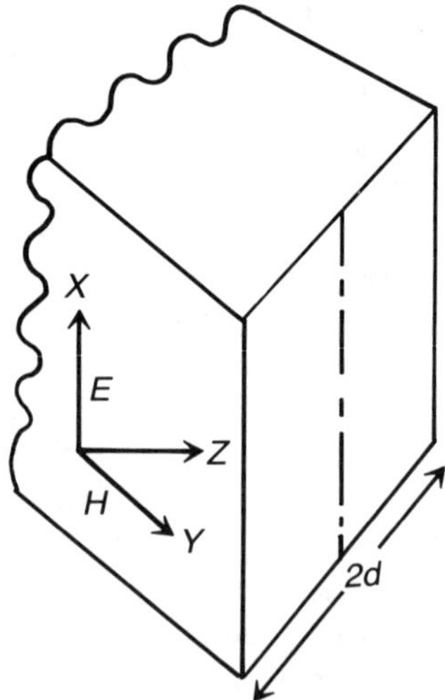

FIGURE 11.2   Incidence of electromagnetic field at the surface of an ALNICO block.

3.   It has an almost constant power factor.
4.   It has a displaced B-H curve.
5.   Its permeability is only a few times that of nonmagnetic material, matching
      the expected value offered by most magnetic materials.

Figure 11.1 illustrates a displaced magnetization curve, while Figure 11.2 shows the incidence of field components at the surface of the core and its propagation toward the center. The intention of the author is to present solutions for:

1.   The induced surge electric field
2.   The surface surge impedance

3. The surge propagating separating plane movement
4. The surging Poynting vector

A. SOLUTION OF THE PENETRATING PLANE AND THE ELECTRIC FIELD

Let σ = electrical conductivity of the core, e = electric field polarized in the x direction, and h, b = magnetic field intensity and induction surges in the y direction. Maxwell's integral equations are expressed below:

$$\int_{line} \bar{e} \cdot d\bar{s} = -\frac{\partial}{\partial t} \int_{surface} \bar{b} \cdot d\bar{a} \tag{11.1}$$

$$\int_{line} \bar{h} \cdot d\bar{s} = \sigma \int \bar{e} \cdot da \tag{11.2}$$

Equations (11.1) and (11.2) become:

$$\int \hat{a}_y h \cdot ds = \sigma \int e \cdot da \tag{11.3}$$

$$\int \hat{a}_x e \cdot ds = -\frac{\partial}{\partial t} \int b \cdot da = -\frac{\partial \phi}{\partial t} \tag{11.4}$$

where ϕ = magnetic flux = $b(t)$ [Area] and $\hat{a}_x$, $\hat{a}_y$ = unit vectors in the x and y directions

$$h(t) = H_m \delta(t) \hat{a}_y \tag{11.5}$$

$$b(t) = B_o \delta_o(t - t_c) \hat{a}_y \tag{11.6}$$

The rate of change of flux/unit length in the x direction is given by:

$$\frac{d\phi}{dt} = 2vb(t) \tag{11.7}$$

where v = velocity of propagation of the separating plane toward the interior of the core.

Refer to Figure 11.3 and consider the loop $|_{nkm'}$. Since no change of magnetic flux occurs there, no induced field exists beyond the separating plane γ. Also consider loop *mkcd* enclosing the plane γ:

$$e_{mk} + e_{cd} = 2vB_o = 2b(t)\frac{d\gamma}{dt} \tag{11.8}$$

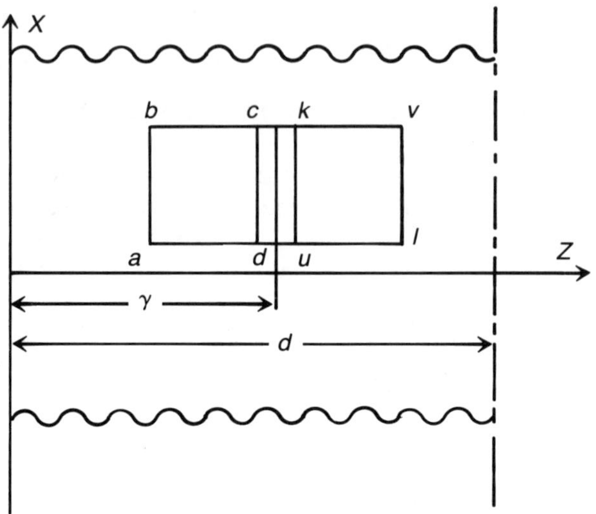

FIGURE 11.3.   (*X-Z*) section of Alnico block.

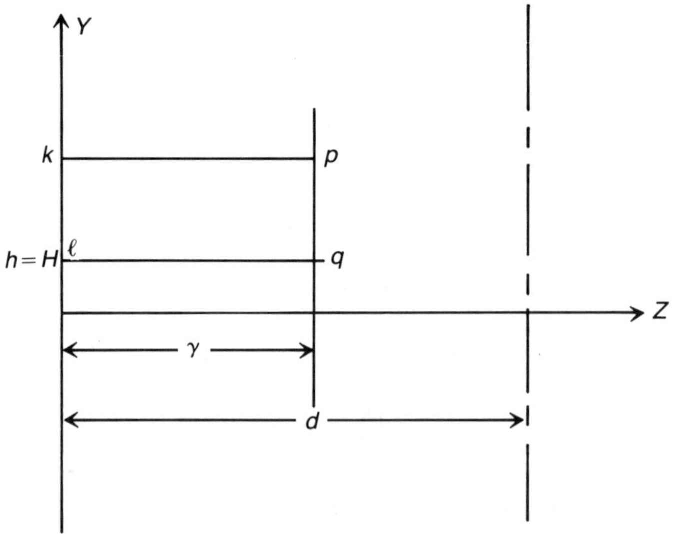

FIGURE 11.4.   (*Y-Z*) section of Alnico block.

Since $e_{mk} = 0$,

$$e_{cd} = 2b(t)\,\frac{d\gamma}{dt} = E \tag{11.9}$$

γ represents the advance of the separating plane in the z direction.

Now, referring to Figure 11.4 to calculate the value of the integral in Equation (11.3), take the loop *nmkl*: $h = H_c$ at the start of the separating plane

$$\oint_{nmkl} U_y h \cdot ds = H - H_c = \sigma \gamma E \tag{11.10}$$

or

$$\frac{\partial h}{\partial z} = \frac{H - H_c}{\gamma} = \sigma E \tag{11.11}$$

where $H =$ the value of the field h at $z = 0$. Therefore, the field h linearly drops from its initial value at $z = 0$ to a point inward where h will again regain the value of H_c at which the separating plane terminates with $r = 0$ at $t = t_c$. The solution for r is

$$\gamma = \sqrt{\frac{Hm}{\sigma B_o}} \left[\frac{v_1(t) - 1}{\delta(t - t_c)} + v(t - t_c)\delta(t_c) \right]^{1/2} \tag{11.12}$$

Then, from Equation (11.9), the electric field becomes

$$e(t) = 2b(t) \frac{d\gamma}{dt}$$

$$= 2B_o\delta(t - t_c)\sqrt{\frac{Hm}{\sigma B_o}} \left[\frac{\delta(t - t_c)\delta(t) - \delta(t - t_c)[v_1(t) - 1]}{\delta^2(t - t_c)} \right.$$

$$\left. + \delta(t_c)\delta(t - t_c) \right] \left[\frac{v_1(t) - 1}{\delta(t - t_c)} + v(t - t_c)\delta(t_c) \right]^{-1/2} \tag{11.13}$$

One can also write that at $t = t_c$

$$e(t_c) = 2B_o\sqrt{\frac{Hm}{\sigma B_o}} [\delta(t_c) - v_1(t_c) + 1][v_1(t_c) - 1 + \delta(t_c)]^{-1/2} \tag{11.14}$$

B. SURFACE SURGE IMPEDANCE, DEPTH OF PENETRATION, AND POYNTING VECTOR

In general, the complex form for the surge impedance Z_s can be expressed by:

$$Z_s = e(t)/h(t)\Omega/m^2 \tag{11.15}$$

where

$$h(t) = H_m \delta(t) \tag{11.16}$$

and $e(t)$ is given by Equation (11.13). However, at $t = t_c$,

$$Z_s(t_c) = 2\sqrt{\frac{B_o}{\sigma H m}}\left[1 - \frac{v_1(t)}{\delta(t_c)} + \frac{1}{\delta(t_c)}\right]\left[\frac{v_1(t_c) - 1 + \delta(t_c)}{\delta^2(t_c)}\right]^{-1/2} \tag{11.17}$$

The depth of penetration (s) can be secured from Equation (11.12) by setting $t = t_c$, where

$$S = \sqrt{\frac{Hm}{\sigma B_o}}\,[v(t_c) - 1 + \delta(t_c)]^{1/2} \tag{11.18}$$

The Poynting vector $\overline{W}$ is expressed by:

$$\overline{W} = \frac{1}{2}\,\overline{e}\overline{h}^x \text{ watts/m}^2 \tag{11.19}$$

where h^x is the conjugate of the magnetic field surge vector.
From Equations (11.13) and (11.5)

$$\overline{W} = \frac{1}{2}\left[2b(t)\frac{dy}{dt}\right][Hm\delta(t)]$$

$$= B_o H_m \delta(t - t_c)\frac{d\gamma}{dt}\delta(t) \tag{11.20}$$

C. CONCLUSIONS

The ohmic values of the saturable resistor increase automatically with respect to the magnitude of surge current and harmonic content, and offer an excellent mechanism of protection against in rush high current pulses in power systems, and especially in high voltage transmission systems subjected to surge incidence. For example, upon connection of the saturable resistor in series with the primary circuit of current transformers where the phase angle is on the order of 90°, a moment of zero voltage will result in a maximum current; however, this current surge will damp substantially with the help of the saturable resistor. Switching operations that coincide with a moment of zero voltage will also lead to a high current surge and the saturable resistor is able to substantially reduce such a surge.

The results presented in this section outline one area of calculation concerning this damping device connected with the induced electric field, propagation of the advancing plane, the induced impulsive field, and the surface surge impedance. However, more analytical and experimental work is needed to generate additional theorems of this surge damper and its transient behavior in a variety of high voltage and current surges.

11.III SOLVED EXAMPLES

A. It is mentioned in the literature that the sparking voltage is almost a linear function of gap separation. Find a relationship for the sparking voltage $(V_s)v_{ys}$ (d), and then with respect to (pd) and implicit relationship v_s P (pressure) for $T_c = 1.2$ μs, which is the time lag element for the peak in lightning surge, based on the specified information below.

Solution

Considering a gap arrester with $T_c = 1.2$ μs and given:

$$V_s = 10 \ (50) \qquad = 500 \ \text{kV}$$

$$V_s = 10 \ (100) \qquad = 1000 \ \text{kV}$$

$$V_s = 10 \ (150) \qquad = 1500 \ \text{kV}$$

$$V_s = 11.25 \ (200) = 2250 \ \text{kV}$$

where (0.5), (1), (1.5), and (2) are the total gap measures in meters for every case. The linear relationship for (V_s) against (d) is generally expressed by:

$$V_s = a(d) + b \qquad \text{(A1)}$$

$$500 = 50a + b \qquad \text{(A2)}$$

$$1000 = 100a + b \qquad \text{(A3)}$$

Therefore, from (A2) and (A3), $500 = 50a$, $a = 10$, and $b = 0$. Next,

$$1000 = 100a + b \qquad \text{(A3)}$$

$$1500 = 150a + b \qquad \text{(A4)}$$

Also, from (A3) and (A4) one can obtain $a = 10$, $b = 0$, then

$$1500 = 150a + b \qquad \text{(A4)}$$

$$2250 = 200a + b \qquad \text{(A5)}$$

Therefore, $750 = 50a$; $a = 15$, $b = 2250 - 3000 = -750$.

The trend as indicated is that (V_s) is directly linear with respect to gap separation (d) up to $d = 50$ cm, after which the model of linear changes into a negative slope graph. Therefore, up to $d = 150$ cm,

$$V_s = 10d \tag{A6}$$

And $d \geq 200$ cm,

$$V_s = (15d - 750) \tag{A7}$$

It is the author's guess that V_s is in a transition mode between Equations (A6) and (A7) as d changes from 150 to 200 cm.

Now, derive empirically a relationship for (V_c) with respect to $[Pd]$, or actually, $K[Pd] = T_c$ for gap arrester.

The following function describes a curve (for the sparking voltage with respect to time lag) to a fair degree of approximation.

$$V_s \approx \frac{100T_c}{1 + 2T_c + 4T_c^2} \tag{A8}$$

However, as mentioned above,

$$T_c = K(Pd)$$

Therefore, Equation (A8) becomes:

$$V_s \approx \frac{100kPd}{1 + 2KPd + 4kP^2d^2} \tag{A9}$$

The next desired result is to derive an implicit empirical relationship between the sparking over-voltage V_s and the prevailing pressure P. From Equation (A7),

$$d = \frac{V_s}{15} + 50$$

Therefore, substituting into

$$V_s \approx \frac{100P\left[\dfrac{V_s}{15} + 50\right]K}{1 + 2KP\left[\dfrac{V_s}{15} + 50\right] + 4KP^2\left[\dfrac{V_s}{15} + 50\right]^2} \tag{A10}$$

Equation (A10) represents an implicit relationship between the sparking potential V_s and the prevailing pressure (P).

11.IV PROBLEMS

1. From Equation (A9), secure condition on (Pd) and then on the pressure (P) for optimum expectation for V_s.

2. Obtain solution for the induced electric field in a hard ferromagnet characterized with a displaced B-H curve if the incident magnetic field intensity surge $h(t) = \hat{a}_y \, H_m \, \delta_o(t) \, e^{-\alpha t}$.

3. From solution of the induced electric field obtained in problem 2, secure solution for the rate of movement of the plane of magnetization r and the surge surface impedance; also obtain solution for the depth of penetration.

4. Repeat solution of problem 10 if the incident magnetic field intensity surge $h(t) = \hat{a}_y \, \delta_o(t) \, [e^{-\alpha_1 t} - e^{-\alpha_2 t}]$.

5. From solution obtained in problem 4, obtain solution for the rate of movement for the plane of magnetization r, the surface surge impedance, and the depth of penetration through a hard ferromagnetic thin material.

6. From data and solutions obtained in problems 2 to 5, obtain solutions for the corresponding Poynting vectors in every case.

7. Repeat solution of problem 2 if the incident magnetic field intensity surge $h(t) = \hat{a}_y \, \delta_o(t) \, \sin wt \, e^{-\alpha t}$.

8. From solution of $e(t)$ secured in problem 7, obtain solutions for the rate of movement for the plane of magnetization and the surface surge impedance.

9. Supplement solutions obtained in problems 7 and 8 to obtain solutions for the Poynting vector and the depth of penetration of the plane of magnetization through a hard ferromagnetic material.

10. With reference to the statement of problem 2, obtain solution for the induced electric field if the material for incidence is a soft ferromagnet characterized by undisplaced B-H curve as shown below.

11. From solutions and base information secured in problem 10, obtain solution for the rate of movement for the plane of magnetization, the surface surge impedance, and the Poynting vector.

12. Repeat problem 10 if the incident magnetic field surge intensity $h(t) = \hat{a}_y \, \delta_o(t) \, [e^{-\alpha_1 t} - e^{-\alpha_2 t}]$.

13. From solutions obtained in problem 12 with relevant base information, secure solutions for the rate of movement for the plane of magnetization, the surface surge impedance, and the Poynting vector.

PRINCIPAL LIST OF SYMBOLS

V_s	Sparkover voltage
$\overline{h},\overline{b}$	Magnetic field intensity and magnetic induction vectors, respectively
ϕ	Magnetic flux in Webers
γ	Advance of the separating plane in meters
H_m	Amplitude of magnetic field in A-T/meters
S	Depth of penetration of magnetizing
$\overline{W}$	Separating plane in meters
B_o	Poynting vector in W/m² amplitude to applied magnetic inducting

REFERENCES

1. **Agarwal, P. D.,** Eddy current losses in solid laminated iron, *AIEE Trans. Power Apparatus Syst.,* 1959.
2. **Alger, P. L.,** *Induction Machines,* Gordon & Breach, New York, 1970.
3. **Alger, P. L., Angst, G., and Schweder, W.,** Saturistors and low starting current induction motors, *AIEE Trans. Power Apparatus Syst.,* 1963.
4. **Hale, J. and Richardson, F.,** Mathematical description of core losses, *AIEE Trans.,* September 1953.
5. **Ramo, S., Whinnery, J., and Duzer, V.,** *Fields and Waves in Modern Radio,* John Wiley & Sons, New York, 1968.
6. **Warrington, W. R.,** *Protective Relays: Their Theory and Practice,* Vol. 2, John Wiley & Sons, New York, 1969.
7. **Freeman, C.,** *EPRI J.,* May 1985.
8. **Denno, K.,** Eddy, current theory in hard, thick ferromagnetic materials, IEEE Conference paper C75-005-4, presented at the IEEE-PES Winter Meeting, New York, January 26-31, 1975.
9. **Cobine, J. D.,** *Gaseous Conductors,* Dover Publ., New York, 1958.

Chapter 12

ENERGY EXTRACTION AND STORAGE FROM LIGHTNING

12.I INTRODUCTION

The process of sustained ionization in the atmosphere preceding the occurrence of lightning breakdown involves the release of ion clusters of positive and negative polarities and the unleashing of an ionization avalanche at the moment of breakdown. The voltage of potential energy among multiple layers of charged clouds is at a multimegavolt level, and the released ionic avalanche carries a multikiloampere electric current. Surface charge distribution, which accumulates at the top and bottom of clouds also maintains huge loads of electric charge on the order of kilo- or even megacoulombs. It is obvious that a vast amount of energy is involved before, during and after lightning in the atmosphere. The question is whether it is feasible or even possible to bring down such tremendous amounts of energy for useful applicability on Earth. The generation of ions is characterized as time varying, amounting to a flow of local electric current before breakdown and to the release of lightning strokes upon breakdown. Before breakdown, the potential difference between clouds and builds up, depending upon atmospheric storm conditions, and hence energy density distribution in the ionizing atmosphere is the electrostatic or the potential density that can be expressed by $\rho V/2$ J/m^3, where ρ is the volume charge density in coulombs/m^3 and V is the potential at any point. The other form of energy is the magnetic or kinetic energy expressed by $JA/2$ J/m^3, where 3 is the inducing current density in amperes/m^2 and A is the magnetic vector potential at any point. Therefore,

$$W_e = \frac{1}{2}\,\rho V \text{ j/m}^3 \tag{12.1}$$

and

$$W_m = \frac{1}{2}\,JA \text{ J/m}^3 \tag{12.2}$$

351

where W_e is the potential energy density in J/m^3 and W_m is the kinetic energy density in J/m^3.

12.II SCALAR ELECTRIC POTENTIAL AND VECTOR MAGNETIC POTENTIAL

A. SCALAR ELECTRIC POTENTIAL

The following is a summary presentation for the basic laws concerning the scalar electric potential V:

1.   Due to a linear charge

$$V = K \int_L \frac{\rho_L \, d\rho}{R} \tag{12.3}$$

where R is a distance from any point on L to the point at which the potential is required; ρ_L is the linear charge density; K is the dielectric constant $= 1/4 \, \pi\epsilon$; and ϵ is the medium permissivity.

2.   Due to a surface charge distribution ρ_s

$$V = K \int_S \frac{\rho_s \, dS}{R} \tag{12.4}$$

where ρ_s is the surface charge density.

3.   Due to a volume charge distribution

$$V = K \int_v \frac{\rho_v \, dv}{R} \tag{12.5}$$

where ρ_v is the volume charge density.

B. MAGNETIC VECTOR POTENTIAL

The following is a summary of relevant laws concerning the vector magnetic potential $\overline{A}$.

1.   Due to a line current distribution

$$\overline{A} = \mu \int_L \frac{I' \, dt'}{4\pi R} \text{ Weber/m} \tag{12.6}$$

where I' is the linear current flow and dt' is the linear differential current element.

2. Due to a surface current distribution

$$\overline{A} = \mu \int_s \frac{J' \, ds'}{4\pi R} \tag{12.7}$$

where ds' is the surface differential area and J'_s is the surface current density vector.

3. Due to volume current distribution

$$\overline{A} = \mu \int_v \frac{J' \, dv'}{4\pi R} \tag{12.8}$$

where J'_v is the volume current density and dv' is the differential volume current.

An important and relevant relationship in our analysis for the electromagnetic phenomenon involved in lightning is

$$\overline{E} = \left[-\overline{\nabla} V - \mu \, \frac{\partial \overline{A}}{\partial t} \right] \text{ V/m} \tag{12.9}$$

where $\overline{E}$ is the total induced electric field vector produced by potential and kinetic field effects.

12.III ENERGY STORED IN CLOUD-CLOUD SYSTEM

This representation is the same as cloud-earth, where in both cases it is modeling or simulation of a condenser system. Before breakdown, to initiate the occurrence of lightning, surface charge accumulation builds at the inner surfaces of the upper and lower cloud layers or between a cloud and the ground.

The dielectric medium between the lightning condenser system is to characterize it as an anisotropic rank two tensor, where

$$\overline{\overline{\epsilon}} = \epsilon_{ij}, \quad i \cdot j = 1,2,3$$

$$= \overline{D}_i/\overline{E}_j \tag{12.10}$$

$\overline{D}_i$ is the electric displacement vector or numerically, the surface charge density in coulombs/m^2 residing on the opposing surfaces.

Potential energy density stored in the two cloud layers' condenser, W_e, is expressed by:

$$W_e = \frac{1}{2} \epsilon E^2 \ \text{J/m}^3$$

$$= \frac{1}{2} \begin{bmatrix} \epsilon_{11} & \epsilon_{12} & \epsilon_{13} \\ \epsilon_{21} & \epsilon_{22} & \epsilon_{23} \\ \epsilon_{31} & \epsilon_{32} & \epsilon_{33} \end{bmatrix} \begin{bmatrix} \epsilon_1 \\ \epsilon_2 \\ \epsilon_3 \end{bmatrix} \ \text{J/m}^3 \qquad (12.11)$$

Equation (12.11) may be expressed in a simpler form by identifying the tensor ϵ_{ij} into the principle axes system as shown below

$$W_e = \frac{1}{2} \begin{bmatrix} \epsilon'_{11} & 0 & 0 \\ 0 & \epsilon'_{22} & 0 \\ 0 & 0 & \epsilon'_{33} \end{bmatrix} \begin{bmatrix} \epsilon_1 \\ \epsilon_2 \\ \epsilon_3 \end{bmatrix} \ \text{J/m}^3 \qquad (12.12)$$

where ϵ'_{11}, ϵ'_{22}, and ϵ'_{33} are elements of λ which can be obtained by the process of diagonalization of the following determinant:

$$\begin{bmatrix} \epsilon_{11} - \lambda & \epsilon_{12} & \epsilon_{13} \\ \epsilon_{21} & \epsilon_{22} - \lambda & \epsilon_{23} \\ \epsilon_{31} & \epsilon_{32} & \epsilon_{33} - \lambda \end{bmatrix} = 0 \qquad (12.13)$$

Upon the threshold limit of field breakdown, the ionic avalanche will bridge the two layer clouds accompanied by the surge of the return lightning conductive stroke and the time varying current due to bound charges in the lower cloud or earth surface. However, in terms of electric potential V and charge density, W_e becomes:

$$W_e = \frac{1}{2} \rho_s V \ \text{J/m}^3 \qquad (12.14)$$

Upon field breakdown and the onset of lightning, the return stroke and its companion, the convective stroke, establish the magnetic or kinetic energy system represented by the flow of (J_c), the cylindrical column conductive current, and J_v, the convective current generated by $\partial D/\partial t$ in the lower cloud or Earth layer. With W_m as the magnetic or kinetic energy density stored by the system self-inductance and current flows, we now write:

$$W_m = \frac{|\bar{J}\ Abar|}{2} \text{ J/m}^3$$

$$= \frac{1}{2}[\bar{J}_c\bar{A}_c + \bar{J}_v\bar{A}_v] \tag{12.15}$$

where $\bar{J}_c$, $\bar{A}_c$ are the conductive current density vector and its associative magnetic vector potential, due to the conductive stroke, and $\bar{J}_v$,$\bar{A}_v$ are the convective current density vector produced by the time varying bound charges and its associative magnetic vector potential, respectively.

The configuration showing J_c and J_v is given by Figure 2.1. In Chapter 3, analysis of the calculation of inducing voltage was carried out by the method of summation of magnetic moments. Equation (3.1) represents the expression of the magnetic vector potential $\overline{AQ}$, while Equation (3.9) gives representation for the $\bar{A}_y$ component of the magnetic potential. Equation (3.12) identifies the total magnetic vector potential for the entire conductive stroke and for the total convective stroke. Then, from relation (3.31), wherein the gradient of the inducing electric potential is expressed in terms of time variation of the magnetic vector potential, the solution for the inducing electric potential is given by Equation (3.38).

W_m may now be expressed as:

$$W_m = \frac{1}{2}\bar{J}_{c,v} \cdot \bar{A}_{c,v} \text{ J/m}^3$$

$$= \left[|J_{c,v}| \sum_{n=1}^{\infty} \sum_{m=1}^{\infty} d^{-j\sqrt{X_{n,m}^2 + Y_{n,m}^2}} \right] |J_{c,v}|$$

$$= |J_{c,v}|^2 \sum_{n=1}^{\infty} \sum_{m=1}^{\infty} e^{-j\sqrt{X_{n,m}^2 + Y_{n,m}^2}} \text{ in J/m}^3 \tag{12.16}$$

where $J_{c,v}$ represents either the conductive or convective current density vector.

The solutions that emerged in Chapter 3 for $\bar{A}_c$ and $\bar{A}_v$ identified $\bar{J}_c$ and $\bar{J}_v$ as having a large amplitude step function.

Let us now consider the potential energy stored in a multicondenser system before and after breakdown. Let M be the number of condensers extending from a certain $\rightarrow$ elevation within the atmospheric belt and close to the Earth's surface as shown in Figure 12.1. Each condenser has a separation between the two cloud layers equal to S and the area is equal to A.

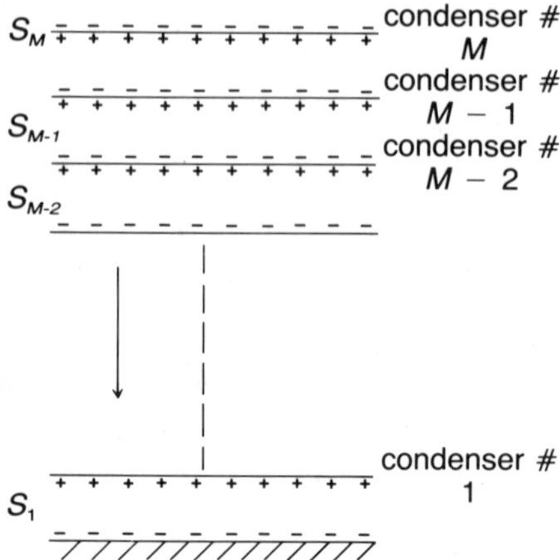

FIGURE 12.1   Multi-condenser system of lightning phenomenon.

Total potential energy density stored in the multicondenser system shown can be expressed by

$$W_{e-total} = M\left[\frac{1}{2}\frac{Q^2}{C}\right] \text{ Joules} \tag{12.17}$$

or the potential energy density is given by:

$$= M\left[\frac{1}{2}\rho_s V_m\right] \text{ J/m}^3 \tag{12.18}$$

where V_m is the electric potential at the mth condenser, Q is the total charge residing on a surface cloud layer in coulombs per unit length,

$$Q = A|\overline{D}| \text{ coulombs/meter} \tag{12.19}$$

and C is the capacitance per condenser

$$= \overline{\epsilon}_{ij}\frac{A}{S}$$

$$= \frac{A}{S}\begin{bmatrix} \epsilon'_{11} & & 0 \\ & \epsilon'_{22} & \\ 0 & & \epsilon'_{33} \end{bmatrix}$$

$$= \frac{A}{S}[\epsilon'_{11} + \epsilon'_{22} + \epsilon'_{33}] \text{ farads} \tag{12.20}$$

In Equation (12.18), V_m is the inducing electric potential developed in Equation (2.51) when lightning current stroke is a large step function, while Equation (2.95) is valid when the lightning current stroke is represented by a sharp rising front and a relatively slow decaying tail. The inducing potential calculated by the method of magnetic moments is given by Equation (3.38), where lightning current stroke was taken as a large step function. The inducing voltage is given by:

$$V_{mc-v} = -j2\mu_o J_{c,v} v_o(t - t_o) \sum_{n=1}^{\infty} \sum_{m=1}^{\infty} Y_{n,m} Z_{n,m} \left(\frac{e^{-jX_{n,m}}}{X_{n,m}} \right) \text{ volts} \quad (12.21)$$

12.IV FEASIBILITY OF ENERGY EXTRACTION

Throughout the discussion of energy storage in the ionized atmosphere during thunderstorms, the author indicated that in the period of time preceding field breakdown, i.e., lightning, the nature of energy storage is potential, while during lightning it is kinetic. Potential energy density is given by ($\frac{1}{2} \rho_s V$) J/m^3 and the kinetic energy density is given by ($\frac{1}{2} \bar{J} \cdot \bar{A}$) J/m^3.

In this section an exploratory approach is presented regarding the feasibility of extracting the huge amount of potential and kinetic energy available in the ionized charged atmosphere before and during the phenomenon of lightning.

Figure 12.2 shows the feasibility diagram for an energy extraction system comprising a set of thin conducting plates, large in area, suspended from a high elevation in the atmospheric belt in a descending pattern with a separation of 5 between each plate. The area of each plate equals (A) m^2. The author visualizes that this multisurface conducting system could be erected by a reliable suspending system that can withstand severe weather conditions. The collecting plates are assumed to be highly conducting, receiving surface charges in prelightning periods, which could be transformed throughout the descending conducting plates system down to the final collecting plate located above ground by an elevation of (h) meters.

During atmospheric discharge, i.e., the occurrence of lightning, conductive as well as convective currents with their associative magnetic vector potential descend through the multiconductive system up to the final collective layer as shown in Figure 12.3.

A. VOLTAGE-CHARGE COLLECTION STATION

In the conceptual process of energy extraction in the period preceding lightning, a system of suspended multiconducting surfaces is expected to descend surface charge accumulation to an elevated charge collection station, which will perform the equivalent or conceptual static voltage generator energy output from the static voltage collector. This could be expressed according to:

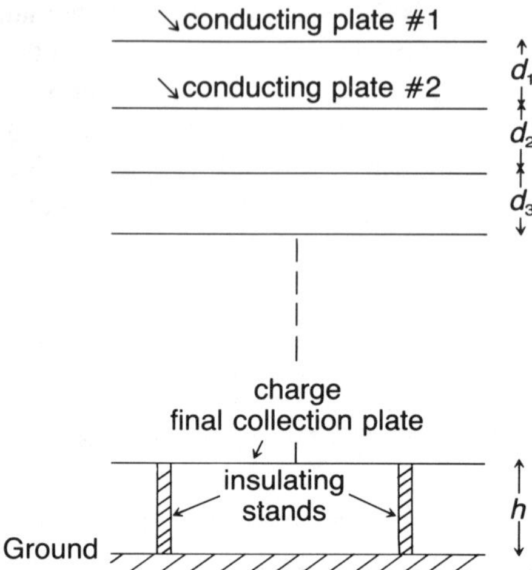

FIGURE 12.2   Energy extraction system preceding lightning.

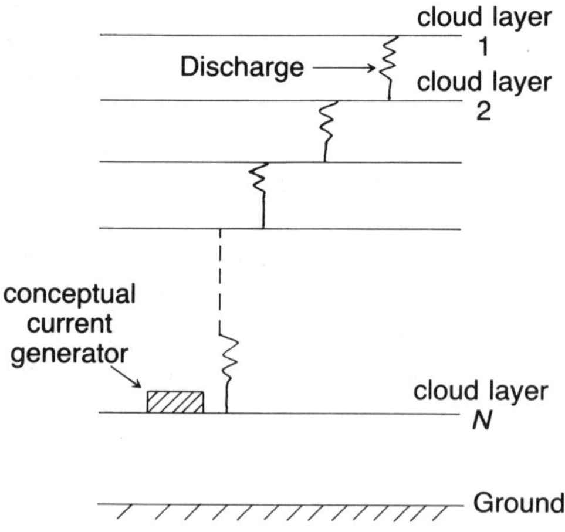

FIGURE 12.3   Energy extraction during lightning.

1.   Inducing voltage derived by conventional means, and with constant current in the return stroke and time varying bound charges:

$$W_{e_1} = M\left[\frac{1}{2}\rho_s V_{ic_1 - v_1}\right] \text{ J/m}^3 \tag{12.22}$$

where $V_{ic_1 v_1}$ is the inducing electric potential per one condenser given by Equation (2.51), and M is the number of condensers for collection and transportation to Earth voltage-charge collector.

2. Inducing voltage derived by conventional means, but where lightning surge is a sharp linear rise in the wave front and a slow decaying tail:

$$W_{e_2} = M\left[\frac{1}{2} \rho_s V_{ic_2 - v_2}\right] \text{ J/m}^3 \tag{12.23}$$

where $V_{ic_2} - v_2$ is the inducing electric potential per one condenser given by Equation (2.81).

3. Inducing voltage derived by the method of moments:

$$W_{e_3} = M\left[\frac{1}{2} \rho_s V_{ic_3 - v_3}\right] \text{ J/m}^3 \tag{12.24}$$

where $V_{ic_3} - v_3$ is the inducing electrical potential per one condenser given by Equation (3.38).

B. MAGNETIC POTENTIAL-CURRENT COLLECTION STATION

Magnetic vector potential $\overline{A}$ has been derived only for the case of a step function for lightning surge and is given by Equation (3.38). It was solved through the method of magnetic moments:

$$W_{e_m} = M\left[\frac{1}{2} \overline{J}_{c,v} \cdot \overline{A}_{c,v}\right] \text{ J/m}^3 \tag{12.25}$$

where $\overline{A}_{c,v}$ is the magnetic vector potential expressed by Equation (3.35), and $\overline{J}_{c,v} = Q$ shown in Equation (3.35) and it represents the amplitude of either the conductor or convective current density vector. The expressions for the vector $\overline{A}$ produced by a sharp, linearly rising wave front and a slow decaying tail are left as problems for students.

The author wishes to emphasize that modes involving the collection of electric charge associated with the buildup of voltage across any condenser, and the other mode involving receipt of electric current with its associative magnetic vector potential, are only based on conceptual, exploratory and feasibility analyses. The other feasible mode of energy extraction from the charged atmosphere mentioned in the literature involves the propagation of electromagnetic energy to a receiving Earth station to be transmitted from a floating collection station.

12.V MHD — MODE FOR ENERGY STORAGE AND EXTRACTION

A. THE MHD PHENOMENON

MHD is the abbreviation for Magnetohydrodynamics, which is based on the interaction of moving conducting fluids and a transverse component of stationary

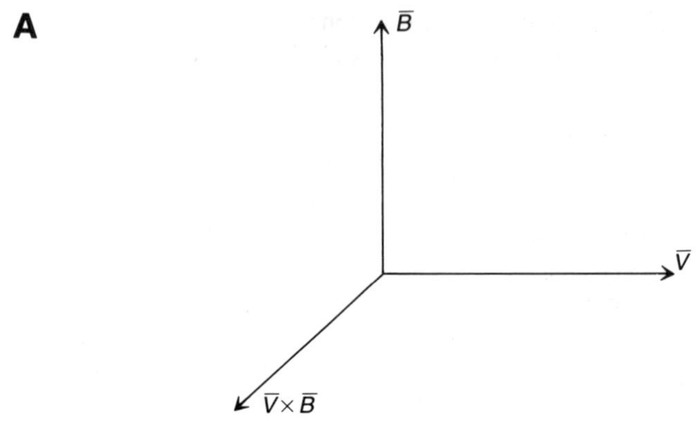

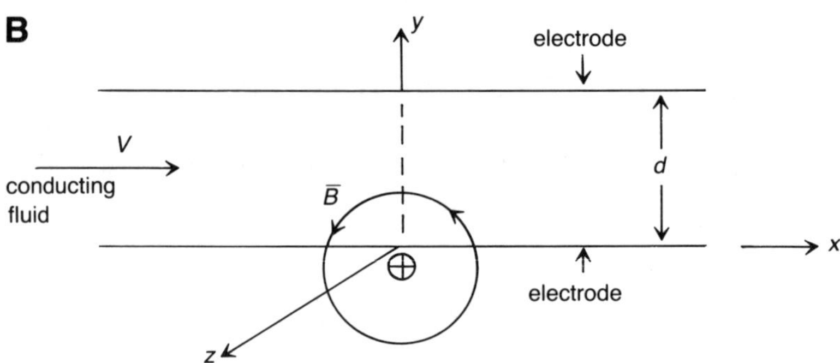

FIGURE 12.4   MHD phenomenon.

or time varying magnetic fields. The cross-interaction will result in having electric field intensity vector E expressed by:

$$\bar{v} x \bar{B} = \bar{E} \text{ Volts/m}$$

$$\text{or} \quad |E| = |\bar{v}| \, |\bar{\beta}| \sin \alpha \tag{12.26}$$

where α is the angle between the $\bar{v}$ and $\bar{\beta}$ vectors, $\bar{v}$ is the velocity of conducting fluid in m/s, and $\bar{\beta}$ is the prevailing magnetic induction vector in tesla.

Figure 12.4a illustrates the MHD interaction process in dimensional terms, while Figure 12.4b shows the interaction between the $\bar{B}$ field from a line conductor and a flowing conducting fluid.

In cases in which the $\bar{\beta}$ and $\bar{v}$ components are uniform, the buildup of electric potential difference (e) across (d):

$$e = |\bar{v}| \, |\beta| \sin \alpha(d) \text{ Volts} \tag{12.27}$$

B. MHD EQUATIONS

1. Flow field equations:

    a. Continuity equation

    $$\frac{D\rho}{Dt} + \rho\overline{\nabla} \cdot \overline{v} = 0 \qquad (12.28)$$

    where

    $$\frac{D}{Dt} = \frac{\partial}{\partial t} + \overline{v} \cdot \overline{\nabla} \qquad (12.29)$$

    b. Momentum equation:

    $$\rho(\overline{v} \cdot \overline{\nabla})\overline{v} = -\overline{\nabla}\rho + \overline{J}x\overline{B} \qquad (12.30)$$

    c. Energy equation:

    $$\rho(\overline{v} \cdot \overline{\nabla})\left[\frac{1}{2}\,|v^2| + \frac{P}{\rho} + C_v T\right] = \overline{E} \cdot \overline{J} \qquad (12.31)$$

where ρ is the mass density in kg/m³, ρ is the hydrodynamic pressure, $\overline{v}$ is the velocity of the conducting fluid, $\overline{E}$ is the existing electric field vector, C_v is the fluid specific heat at constant volume, and T is the fluid temperature in degrees Kelvin. Equations (12.28) to (12.31) are valid for an inviscid and incompressible fluid.

2. Electromagnetic field equations:

    a. Charge continuity equation

    $$\overline{\nabla} \cdot \overline{J} + \frac{\partial \rho_e}{\partial t} = 0 \qquad (12.32)$$

    b. Ampere's law

    $$\overline{\nabla}x\overline{B} = \mu_o\left[\overline{J}_c + \frac{\partial\overline{D}}{\partial t}\right]$$

    $$\overline{\nabla} \cdot \overline{B} = 0 \qquad (12.33)$$

    c. Faraday's equation

    $$\overline{\nabla}x\overline{E} = -\frac{\partial B}{\partial t} \qquad (12.34)$$

d.   Ohm's law

$$\bar{J}_c = \sigma[\bar{E}_a + \bar{v} \times \bar{B}] \qquad (12.35)$$

e.   Energy equation

$$W_{e-m} = \bar{E}_a \cdot \bar{J}_t - \frac{J^2_t}{\sigma} \qquad (12.36)$$

where $\bar{J}_t = \bar{J}_c + \bar{J}_v$ is the total current density vector
$\bar{J}_c$ is the conductive current density vector

$\bar{J}_v$ is the convective current density $= \dfrac{\partial \bar{D}}{\partial t}$

$\bar{D}$ is the electric displacement vector
$\bar{E}_a$ is an applied electric field vector
ρ_e is the electric charge density in coulombs/m^3.

C. MHD PHENOMENON IN CHARGED ATMOSPHERE

MHD phenomenon commences immediately after the initiation of voltage breakdown and the release of current surge. The current surge in the conductive and in the convective strokes generates a very powerful magnetic flux density vector. The interaction of the resulting magnetic field with the flow of conducting atmospheric cloud is indeed an MHD phenomenon leading to the induction of an intensive electric field and a further contribution to voltage buildup among atmospheric charged layers. Consequently, the author sees a significant contribution of MHD phenomenon to the concept of energy storage and possible extraction at earth stations, as indicated by Figure 12.2. Therefore, W_{MHD} as the energy density generated by a single parallel surface-cloud system is expressed by:

$$W_{MHD} = M\left[\frac{1}{2}\bar{J}_{1,2} \cdot \bar{A}_{1,2}\right] \qquad (12.37)$$

where

$$\bar{J}_1 = \bar{\sigma}[\bar{v} \times \bar{B}_c] \qquad (12.38)$$

$$\bar{J}_2 = \bar{\sigma}[\bar{v} \times \bar{B}_v] \qquad (12.39)$$

where $\bar{J}_1$ is the MHD-induced current density produced by the interaction of the moving charged cloud and the $\bar{B}$ field generated by the conductive lightning stroke identified as $\bar{B}_c$; $\bar{J}_2$ is the MHD-induced current density produced by the interaction of the charged moving cloud with the $\bar{B}$ field generated by the convective stroke identified as $\bar{B}_v$; and $A_{1,2}$ represents the magnetic vector potential produced by J_1 or J_2 described above. W_{MHD} expressed by Equation (12.37) is indeed another form of kinetic energy that can be extracted on a conceptual basis only.

12.VI SOLVED EXAMPLES

A. This involves obtaining a solution for the magnetic vector potential resulting from knowledge of the inducing voltage at a point in the atmospheric belt given by Equation (2.51).

Solution

From the following:

$$\overline{\nabla}V_{ic,v} = -\mu \frac{\partial \overline{A}}{\partial t} \tag{A1}$$

or

$$\overline{A} = -\frac{1}{\mu} \int_t \overline{\nabla}_{ic,v} \, dt \tag{A2}$$

In cylindrical systems,

$$\overline{\nabla}V_{ic,v} = \hat{a}_\rho \frac{\partial V}{\partial \rho} + \hat{a} \frac{1}{\rho} \frac{\partial V}{\partial} + \hat{a}_\gamma \frac{\partial V}{\partial \gamma}$$

$$= \hat{a}_\rho E_\rho \tag{A3}$$

$\hat{a}\rho E\rho$ is given by Equation (2.41), where

$$\hat{a}_\rho E_\rho = J_{c,v}v_1(t) \sqrt{\frac{j}{\pi}} \frac{\sqrt{\rho - R}}{R\hat{y}} \sum_{n=0}^{\infty} \frac{j^n e^{-j(\rho - R)}}{n!2^n}$$

$$\left[-\frac{(\rho - R)^n}{R^n} - (\rho - R)^{n-1} - \frac{(\rho - R)^{n-\frac{3}{2}}}{\sqrt{2n}} + \frac{j2^n(n-1)!}{\sqrt{2}} \right] \tag{A4}$$

where $R = R_c$ or $= R_v$. Therefore from Equation (12.26)

$$\overline{A} = -\frac{1}{\mu} J_{c,v}t \text{ (the rest of right-hand-side of Equation (A4))} \tag{A5}$$

However, (t) in Equation (12.26) is limited to the time at which the surge current practically diminishes, usually $t \leq 15$ μs. The duration of 15 μs represents a little more than the time span of the decaying surge tail, usually about 12.5 μs.

B. Consider the conductive current stroke as a line conductor carrying a current $I_c \, \delta_o(t)$. The conducting fluid has a velocity vector in axial direction between parallel cloud layers. Express $\bar{J}_1$ if σ is a scalar.

Solution

The $\bar{B}$ field due to a conducting line current $I_c \, \delta_o(t)$ is given x-y plane by:

$$B_{oy} = \frac{I_c \delta_o(t)\mu}{2\pi} \frac{x}{X^2 + y^2} \tag{B1}$$

$$B_{ox} = \frac{I_c \delta_o(t)\mu}{2\pi} \frac{y}{X^2 + y^2} \tag{B2}$$

Let $\bar{v} = \hat{a}_z \, \mu$ m/s.

$$J_1 = \sigma\hat{a}_z\mu X[\hat{a}_x B_{ox} + \hat{a}_y B_{oy}]$$

$$= \sigma[\hat{a}_y\mu B_{ox} - \hat{a}_x\mu B_{oy}] \tag{B3}$$

12.VII PROBLEMS

1. Obtain solution for the kinetic energy that could be extracted from lightning cycles, considering the breakdown surge as a sharp, linearly rising wave-front to its peak followed by a slow decaying tail. Assume the possibility of building (M) suspended collecting energy condensers.
2. Repeat problem if the surge current breakdown is a decaying sinusoid given by $K \, e^{-\alpha t} \cos$ wt. K and α are constants.
3. Repeat problem if the surge current breakdown is a double exponential pulse given by $K[e^{-\alpha_1 t} - e^{-\alpha_2 t}]$. K, α_1, and α_2 are constants.
4. With reference to Equation (12.22), convert the functional elements ρ_s and $V_{ic_1 - v_1}$ in terms of the electric displacement vector ($\bar{D}$) and the electric field intensity $\bar{E}$. Consider the surrounding medium as totally anisotropic, i.e., ϵ is a rank two tensor.
5. Repeat problem 4 with respect to Equation (12.23).
6. With reference to Equation (12.25), convert the functional elements $\bar{J}_{c,v}$ and $\bar{A}_{c,v}$ and their representation in terms of the corresponding $\bar{B}$ and $\bar{H}$ fields. Consider the surrounding medium as linear material.
7. Repeat problem 6 if the surrounding medium is totally anisotropic.
8. Repeat problem 4 if the surrounding medium is a linear material.
9. With reference to Figure 2.1 showing the configuration for the conductive return stroke and the convective coaxial current, calculate and sketch the

magnetic flux density distribution for J_c and J_v. Express the corresponding magnetic vector potentials associated with J_c and J_v.

10. From solutions for A_c and A_v, present solutions for the stored kinetic energy per unit volume.

11. From solutions for $\overline{B}_c$ and $\overline{B}_v$ secured in problem 9, and given the velocity vector for the charged cloud to be entirely transverse to the plane containing B_c and B_v, solve for the MHD-induced current densities. Assume the medium as anisotropic such that σ is a rank two tensor.

12. From solutions for the MHD-induced $\overline{J}_c$ and $\overline{J}_v$, solve for the associated magnetic vector potentials. Also obtain expressions for the developed energy density per unit volume.

PRINCIPAL LIST OF SYMBOLS

W_e	Electric energy density
W_m	Magnetic energy density
ρ	Electric charge density
$\overline{A}$	Magnetic vector potential
ρ_L	Linear charge density
I'	Linear current flow — amp
J	Surface current density — amp/m^2
MHD	Magnetohydrodynamic

REFERENCES

1. **Denno, K. and Fouad, A. A.,** Effects of the induced magnetic field on the magnetohydrodynamic channel flow, *IEEE Trans. Electr. Devices,* ED-19(No. 3), 322, 1972.
2. **Denno, K.,** Generation aspects of weakened fusion plasma in MHD channel, *J. Appl. Sci. Eng.,* A3, 213, 1978.
3. **Harrington, R. F.,** *Time Harmonic Electromagnetic Fields,* McGraw-Hill, New York, 1961.
4. **Rudenberg, R.,** *Electrical Shock Waves in Power Systems,* Harvard University Press, Cambridge, MA, 1968.

Chapter 13

EFFECTS OF ELECTROMAGNETIC FIELDS ON HEALTH

13.I CONTROVERSY OVER IMPACT OF ELECTROMAGNETIC FIELDS (EMFs) ON HUMAN HEALTH[1]

The article[1] that forms the basis of discussion in this section pointed out the increasing drive to utilize electric power generation and transmission centered on the design of high voltage alternators and transmission systems, coupled with continuous concern about the effects of emitted EMFs on human health.

Temptation for high voltage system is based on economic and engineering motivations. For example, a single 765-kV line can carry as much power as 30 138-kV lines at $1/10$ the construction cost and with only $1/13$ the amount of right-of-way land required per kilowatt. EMFs produced by the 765 kV line are much more intense than those in the 138 kV line, and this begs the question about field effects on public health, such as subtle biologic effects. Some experiments indicated that even low level fields may influence the biochemical activity of some body tissue. Because of increasing concern regarding the effects of EMFs on the public health, ERPI sponsored several research projects with Battelle, Pacific Northwest Laboratories, Enertech Consultants, and Electric Research and Management, Inc. EPRI also cooperated with the Department of Energy (DOE) to secure reliable meaningful information on biological effects that can be used to present quantitative risk assessments.

Assessing field effects on humans requires a consistent pattern of measurements regarding exposure to EMFs over continuous or interrupted duration, close proximity of conducting objects to standing or sitting person, and whether the person is moving or stationary. The Douglas[1] article, for example, reported that a man standing under a 765-kV line will experience a field of 10 kV/m, while another person a few hundred feet away will experience a field of 2 kV/m.

EPRI sponsored research projects with other electric utilities as well as other research centers are for the ultimate goal of securing reliable information for the Electric and Magnetic Fields (EMF).

Previous researchers developed what is known as the activity system model, which focuses on an activity factor, defined as the ratio between the exposure measured during some activity (e.g., riding a tractor) and the exposure that would have been experienced by the same person standing erect on wet grass with wet shoes (thus maximizing field distortion). The activity factors measured were for horseback riding, 82%; working bent over in knee-high grass, 52%; and sitting on an open top tractor surrounded by waist-high (1 m) vegetation, 30%.

Using data obtained from values of activity factors coupled with total exposure to EMF per year, an annual time exposure for that particular activity is established. Information gathered from activities covering 18 farms that were criss-crossed by H.V. transmission lines indicated in one case with exposure to 345 kV that cumulative exposure averaged 60 kV/m/h/year or equivalent to a 10-kV/m field for 6 h/year, or a 1-kV/m field for 60 h/year. However, this report indicated that 90 to 98% of the field exposure in farm activities came from contact with low level fields having the same intensity as those prevailing in residential housing.

Another environment for exposure to EMF is the electric utility works area. A pilot research program was conducted by the utilities in which the workers wore a device developed by the Bonneville Power Administration called Electric field exposure meter (EFEM), which can be placed in a shirt pocket or on a hard hat. Data collected for time exposure using this device indicated that utility workers received much less field exposure than anticipated.

Approximately 15 years ago Soviet workers operating in utility in switch yards complained about adverse reactions due to exposure to high voltage fields. They indicated that they experienced a loss in appetite and energy, and diminished sex drive after a long-term occupational exposure to high voltage fields. The Soviets responded by setting time limits for exposure to electric fields to >5 kV/m. Research work conducted in other European countries, however, disputed the Soviet findings and indicated that other factors that exist in the switch yard may have contributed to complaints such as noise, oil fumes, and exposure to a variety of chemicals by the Soviet utility workers.

A paper published in 1979 in the *American Journal of Epidemiology*, authored by Nancy Wertheimer and Ed Leeper, indicated that the number of children who died from cancer was around 24% higher because they lived in homes subjected to high current configuration, as opposed to the control group. However, when a group of researchers from Brown University attempted a similar study, they found no evidence to support the report by Wertheimer and Leeper. The methodology was analyzed for EPRI by H. Daniel Roth Associates, Inc., which indicated that Wertheimer and Leeper failed to consider other potential causes of cancer such as pollution, genetics, and diet; that the selection of controls may have been biased; the researchers did not consider other environmental variables such as neighborhood and traffic; and no adjustment was made for the effect of family cancer pattern. The Roth analysis also pointed to the drawback

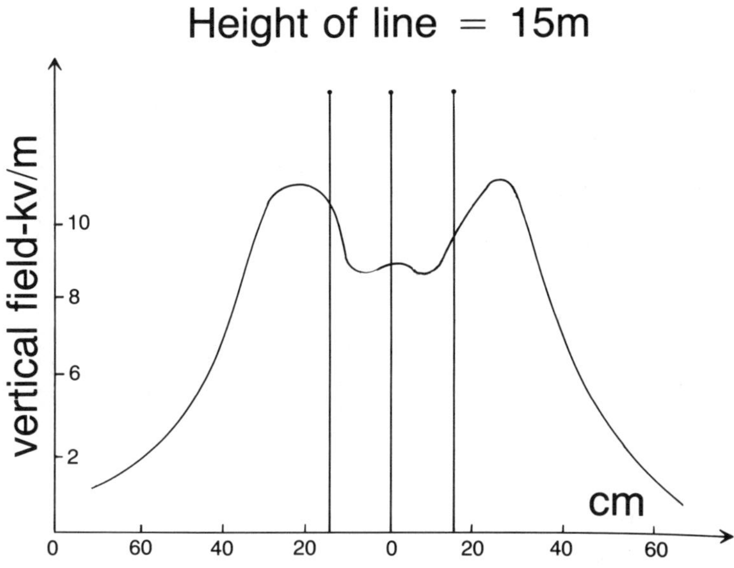

FIGURE 13.1   Field due to 765 kV line.

of the work by researchers at Brown University, in that they failed to consider other potential causes of cancer in their results.

Another study that centered on research in cell biology using tissue cultures exposed to EMFs provided useful information on the effects of EMF on living organisms. The primary interaction between EMF and a biological cell appears to occur mainly at the cell membrane rather than inside the cell. The function of the membrane is the control of ion transport and its interaction with EMF may disrupt that function. Some studies indicated that while noticeable effects on biological cells have been observed, no effect was found for the whole organism. Research was conducted on animals subjected to EMFs over a sustained period of time under a source of 30 kV/m whose surface effect on the head of female swine is the same as 10 kV/m at the head of a human. Currents induced internally were, of course, different — the swine felt the EMF effect while the human could not.

Figure 13.1 shows the electric field distribution across a distance around the centerline under a transmission line, while Figure 13.2a shows the percentage of time spent under an electric field in homes only and for farmers subjected to EMFs from domestic appliances and high voltage power lines. Figure 13.2b shows a trend for the time spent under an electric field for farmworkers and rats with respect to induced current density in A/cm^2 $\times$ 10^{-9}.

Douglas and colleagues[1] discussed the controversy over the economy and the impact of EMFs on HV transmission lines. The article indicated that by the year 2000 the U.S. will require between 108,000 to 209,200 mi of new transmission lines. Disallowing the installation of the 765 kV system in order to avoid

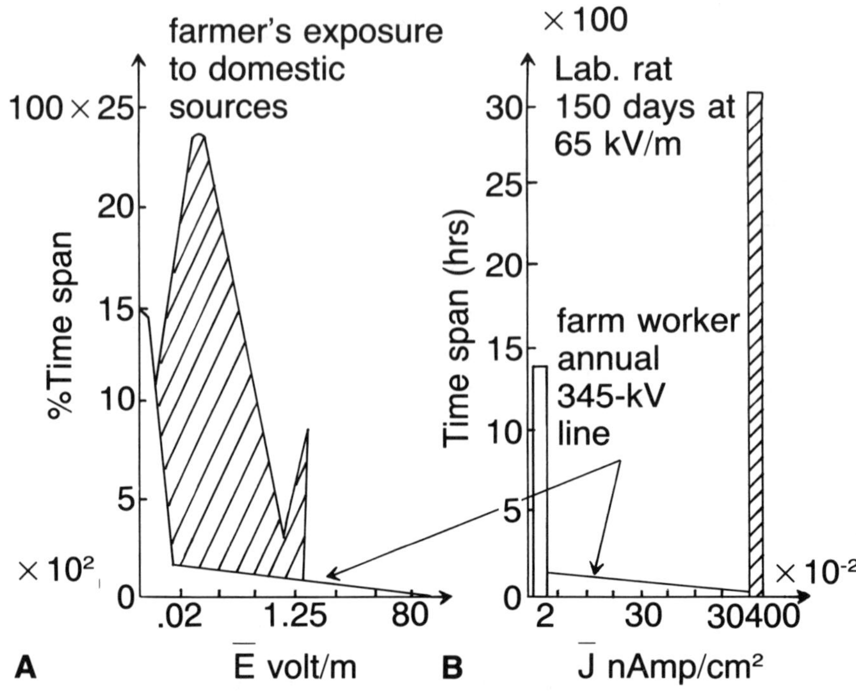

FIGURE 13.2   Electric effects on farm workers and rats.

the public protests and concern over EMF impacts will result in 21% more lines, 22% more right-of-way land, and 16% more peak energy loss at an additional cost of about $3 billion.

During the construction of a ±400 kV DC line between Minnesota and North Dakota in 1977, public opposition was intense, and after the line was completed in 1978, residents complained of health problems (nosebleeds, fatigue, miscarriages, skin irritations, and stress) they believed resulted from the EMFs. In response to public complaints, the Minnesota Environmental Quality Board (MEQB) launched an inquiry into the complaints, and a scientific advisory panel concluded that there was no foundation for any adverse impact resulting from EMFs. They investigated revised records from the Dairy Herd Improvement Association of the University of Minnesota and reported that no adverse effects had been detected from line-EMF on milk production and herd reproduction.

It is clear to the author of this book as well as to the people at EPRI that more extensive research based on definitive, sustained data is needed to arrive at a satisfactory and meaningful conclusion about the problem of EMF impact from HV lines above 765 kV.

13.II CONNECTION BETWEEN ELECTROMAGNETIC FIELDS AND CHILDHOOD CANCER[2]

Subsequent to the Wertheimer-Leeper report, a similar study begun in 1980 affirmed the concept of a very loose connection between EMFs from HV lines and cancer occurrence among children living in close proximity to the system. However, in 1986, the call for a more conclusive connection between childhood cancer and EMF from HV transmission lines led to a study in Denver conducted by Savitz, Wachtel, and Barnes funded by the utility. This study group carried out point measurements of EMFs from HV lines on children living in homes under HV lines or in the vicinity, and concluded that a modest connection exists between EMF effect from HV lines and childhood cancer, but found that the effect is much weaker than from the EMFs of domestic appliances.

Leonard Sagan, manager of EPRIs research into field effects, said that no direct or strong correlation exists between EMF effects from HV lines and children's cancer, but that a stronger correlation does exist between the *pattern* of power lines outside the homes and cancer. Sagan indicated that crowded lines and layout configuration usually results in a densely populated neighborhood, which brings with it many sources of air and water contamination. All add to the cancer potential, especially for children, and that due to the complexity of this problem, Sagan sees more need to continue research in this area.

Figure 13.3 presents a useful picture of the sources of exposure to EMF by measuring v_s magnetic field strength in milligauss. Figure 13.4 shows the distribution of risk ratios for magnetic flux density produced by turning appliances on and off as well as by heavy current configuration.

SOURCES OF EXPOSURE

Exposure to electric and magnetic fields is an inevitable consequence of living in a society that uses electricity. The relative contribution of different sources to overall exposure is not well documented, but it appears that household and workplace appliances and equipment provide at least as much exposure as power lines. EPRI has several projects under way to measure and analyze public and occupational exposure patterns.[2]

It is indicated in the Sagan article that EPRI is currently funding research projects based on epidemiologic sustained data that may identify a reliable conclusion for the linkage between EMF impact on telephone workers, electric utility workers, and laboratory research animals and cancer. EPRIs other projects focus on the mechanisms of interaction between organisms and EMFs, for example, how EMFs affect biological materials at the cellular level, the effects on membrane behavior, and macromolecular activities, all subjects which lead to meaningful findings concerning the possible disease processes in animals.

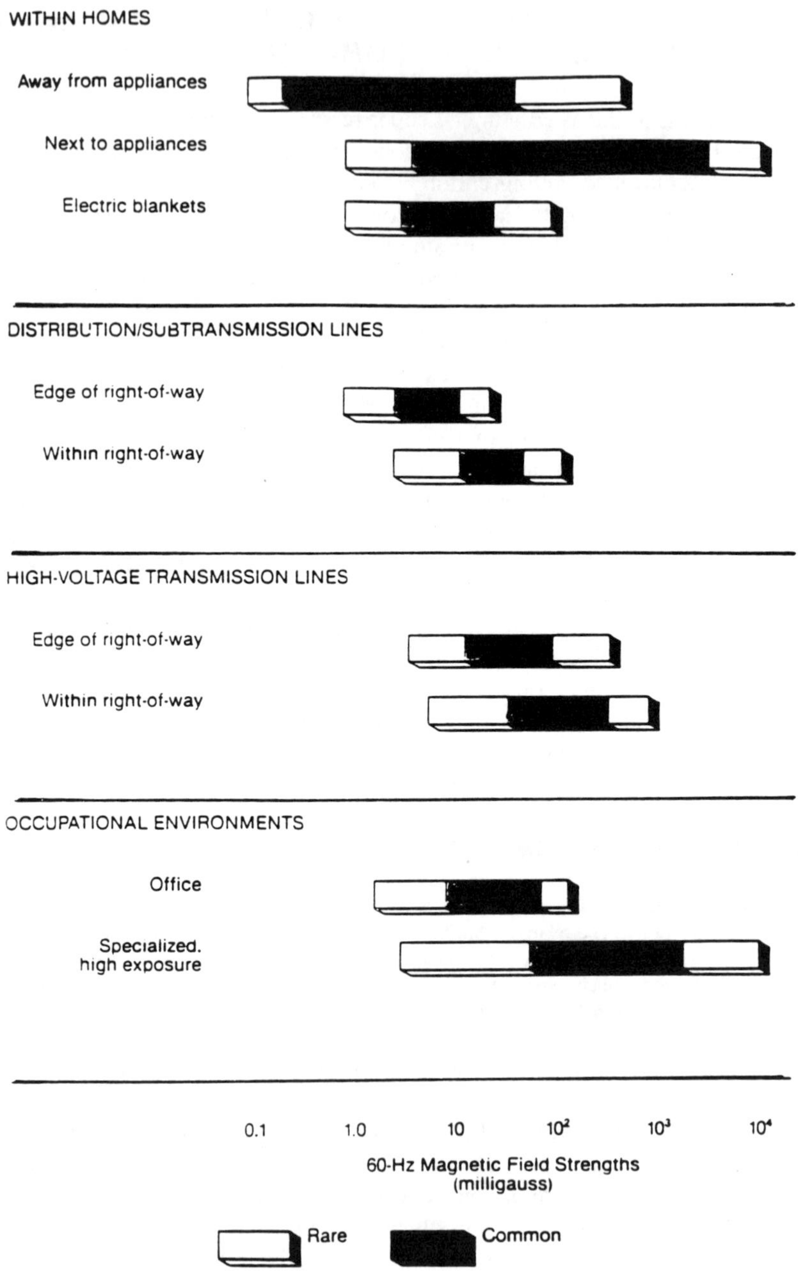

FIGURE 13.3   Sources of exposure.

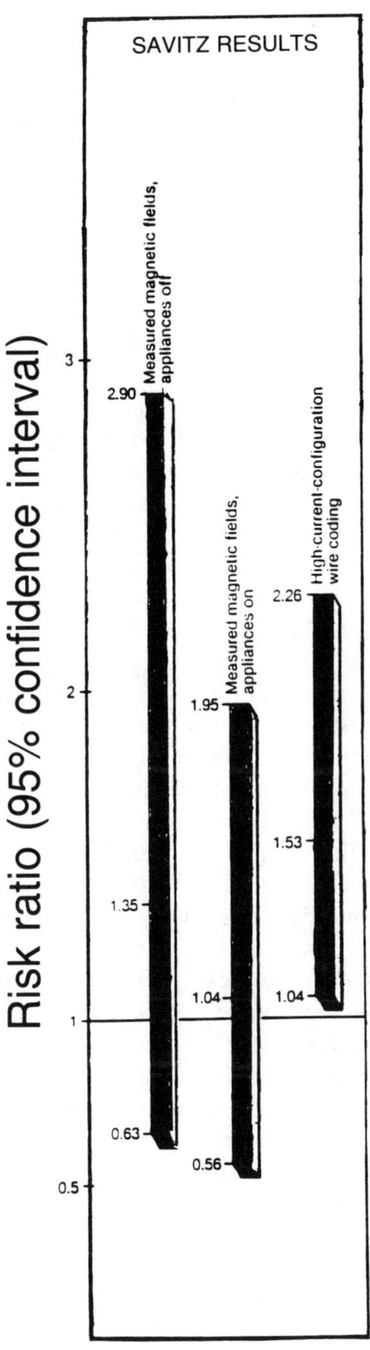

FIGURE 13.4   Distribution of risk ratios.

13.III FURTHER RESEARCH ON ELECTROMAGNETIC FIELDS[3]

Moore et al.[3] indicated that EPRI, with other funding agencies, such as the (U.S.) National Cancer Institute, Sweden's National Institute of Occupational Health, France's Electricité de France, several Canadian utilities, and the World Health Organization, continues to sponsor research projects to obtain conclusive results about EMF adverse impacts on humans as well as animals, and that an estimated $15 million/year is spent on this research in approximately 12 countries.

According to a paper issued in June 1989 by the U.S. Congress' Office of Technology Assessment and prepared by Indira Nair, Granger Morgan, and Keith Florig of the Department of Engineering and Public Policy at Carnegie-Mellon University, recent epidemiologic studies indicate an increasing reflection on the role of risky exposure under EMF and the linkage to serious diseases such as cancer and the alteration of circadian rhythms. EPRIs Leonard Sagan emphasized that research is progressing on a broad scale that encompasses epidemiology, basic science, and exposure assessment.

Figure 13.5 illustrates the distribution of magnetic flux density in milligauss over a spectrum of durations at indoor as well as outdoor locations. Figure 13.6 illustrates the distribution of magnetic flux density in milligauss from appliances, distribution lines, and an HV line of 500 kV. The graph indicates that magnetic flux densities from electrical appliances can exceed those experienced directly under utility power lines; however, the field emanating from appliances is only generated when the devices are turned on, whereas the effects of fields from power lines are constant for a person remaining in that place over an extended period of time.

The discussion in this chapter has thus far centered on informational assessment about the adverse impact of EMFs on humans and animals possibly produced by indoor domestic electrical appliances and outdoor fields generated by low voltage distribution and HV transmission lines. The results of these studies demand more intensive work that may establish a definitive and consistent role for EMFs in producing adverse effects on humans and animals.

In the next section this author proceeds to present quantitative as well as qualitative information for the EMF generated by lightning on people and other species. It is a statistical fact that electrical fields due to lightning that are usually measured in the vicinity of thunderstorms range from 5 to 300 kV/m. It is also known in terms of order of magnitude that lightning discharge usually occurs at atmospheric pressure at a breakdown field of 32 kV/cm. A 5 kV/m field created by lightning is considered to be about the same field that can be produced from a 765.5 kV transmission line, and in the opinion of this author, may create adverse health problems. In this section, quantitative analysis is presented for the impact of EMFs from charged cloud layers shortly after field breakdown or the occurrence of lightning. This author agrees with *EPRI Journal* presentations that more extensive and broad research is required to arrive at more definitive

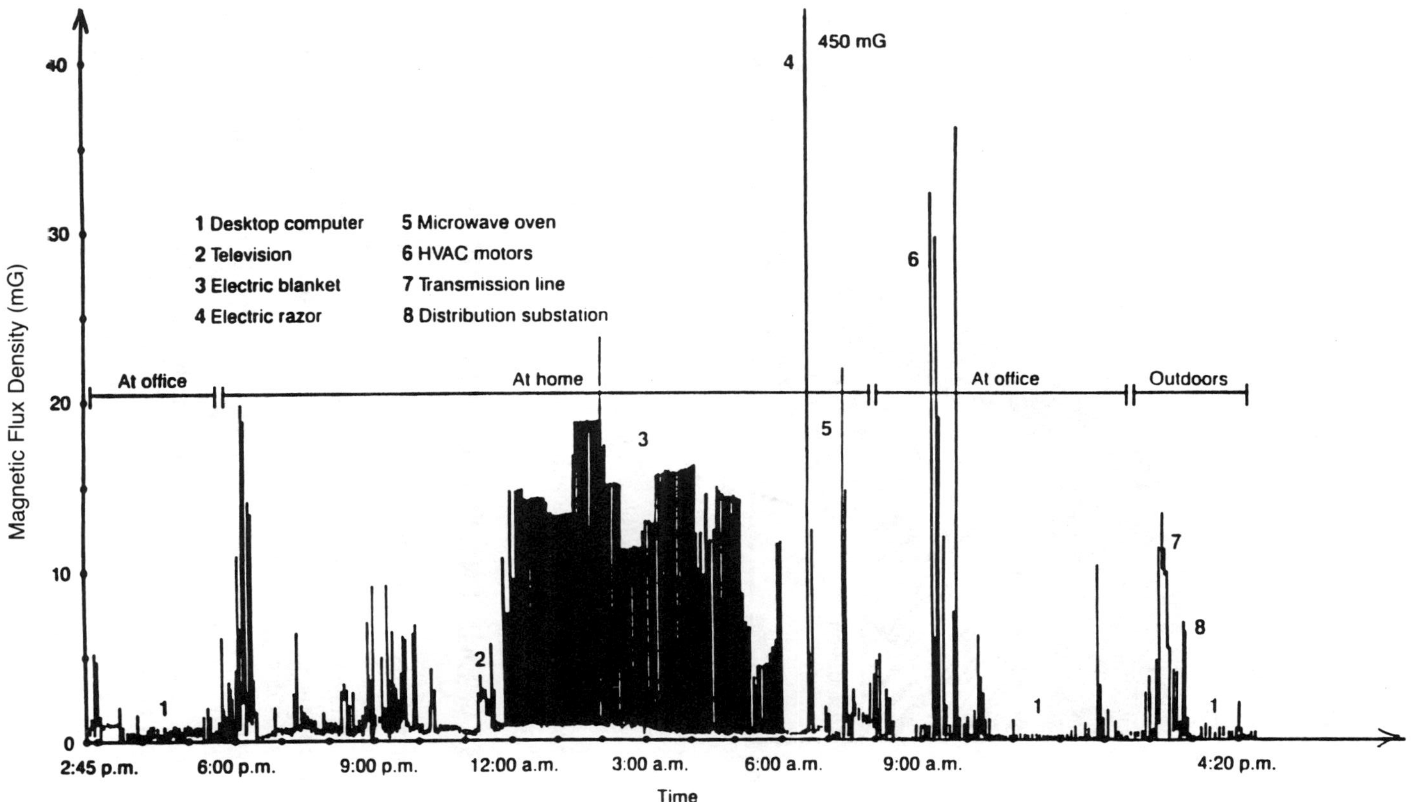

FIGURE 13.5 Distribution of magnetic induction over time.

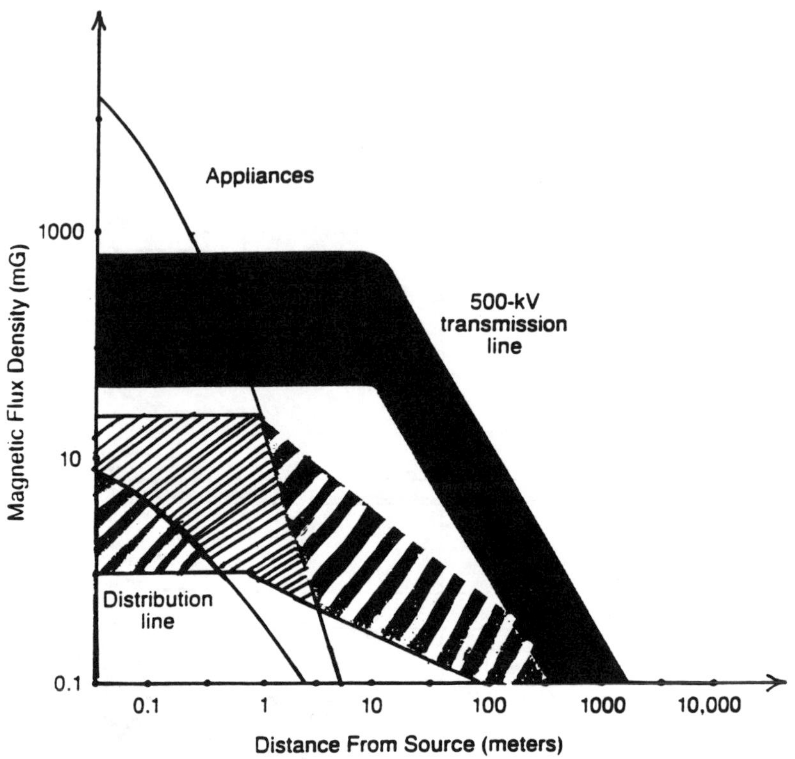

FIGURE 13.6   Distribution of magnetic induction over distance.

data on the impact of EMF on health and ways in which their adverse effects may be reduced.

13.IV EMF FROM LIGHTNING DISCHARGE

In Chapter 2, the author presented solutions for the EMF intensities generated by the conductive and time varying bound charges induced in the earth or in another cloud layer. Solutions obtained were based on a constant magnitude of discharge current and an actual surge waveform represented by a sharp linear rise and a slow decaying linear tail.

However, EMFs produced by the linear but closely representing lightning surge is the set of Equations (2.74). Expressions E_ρ, E_α, and E_τ as well as H_ρ, H_α, and H_τ at any point in space influenced by the inducing EMF are given in the equations below. Exact representation for the inducing EMF developed by an actual but linear waveform of lightning surge is given by:

$$\overline{E}_\rho = \frac{-J_{c,v}(t)}{2\hat{y}} \sum_{n=0}^{\infty} \left[\frac{e^{jn\alpha}}{nR_{c,v}^2} J_n(K|\rho - R_{c,v}|)H_n^{(2)} \right.$$

$$(K|\rho - R_{c,v}|) + \frac{e^{jn\alpha}}{nR_{c,0}} H_n^{(2)} \frac{(\rho - R_{c,v})^{n-1}}{2^{n-1}(n-1)!}$$

$$+ \sqrt{\frac{2}{\pi}} (-j)^{n+\frac{1}{2}} J_n \frac{e^{jn\alpha}}{nR_{c,v}} \frac{e^{-j(\rho - R_{c,v})}}{(\rho - R_{c,v})^2}$$

$$\left. + \frac{e^{jn\alpha}}{nR_{c,v}} j^{\left(n+\frac{3}{2}\right)} \sqrt{\frac{2}{\pi(\rho - R_{c,v})}}\, e^{-j(\rho - R_{c,v})} \right] \tag{13.1}$$

$$\overline{H}_\rho = \frac{j\tau J_{c,v}(t)}{2} \sum e^{jn\alpha} J_n(K|\rho - R_{c,v}|)H_n^{(2)}(K|\rho - R_{c,v}|) \tag{13.2}$$

$$\overline{E}_\alpha = \frac{j J_{c,v}(t)}{2\hat{y}} \sum_{n=0}^{\infty} e^{jn\alpha} J_n(K|\rho - R_{c,v}|)H_n^{(2)}(K|\rho - R_{c,v}|) \tag{13.3}$$

$$\overline{H}_\alpha = \frac{\tau J_{c,v}(t)}{2} \sum_{n=0}^{\infty} \frac{e^{jn\alpha}}{nR_{c,v}^2} J_n(K|\rho - R_{c,v}|)$$

$$H_n^{(2)}(K|\rho - R_{c,v}|) - \frac{e^{jn\alpha}}{nR_{c,v}} H_n^{(2)} \frac{(K|\rho - R_{c,v}|)}{2^{n-1}(n-1)!}$$

$$+ \sqrt{\frac{2}{\pi}} (-j)^{n+\frac{1}{2}} J_n(K|\rho - R_{c,v}|) \frac{e^{-j(\rho - R_{c,v})}}{(\rho - R_{c,v})^2}$$

$$+ j^{\left(n+\frac{3}{2}\right)} \sqrt{\frac{2}{\pi(\text{rho} - R_{c,v})}}\, e^{-j(\text{rho} - R_{c,v})} \frac{e^{jn\alpha}}{nR_{c,v}} \tag{13.4}$$

$$\overline{E}_\tau = \frac{J_{c,v}(t)R_{c,v}\tau K^2}{2\hat{y}} \sum_{n=0}^{\infty} \frac{e^{jn\alpha}}{n}$$

$$J_n(K|\rho - R_{c,v}|)H_n^{(2)}(K|\rho - R_{c,v}|) \tag{13.5}$$

$$\overline{H}_\tau = 0 \tag{13.6}$$

For points relatively close to the location of lightning discharge, where ρ is small, E_ρ, H_ρ, E_α, H_α, E_τ, and H_τ can be simplified by substituting the following limiting processes:

$$X = (K|\rho - R_{c,v}|)$$

$$\lim_{X \to 0} J_n(X) = \frac{1}{n!} \left(\frac{X}{2}\right)^n \tag{13.7}$$

$$\lim_{X \to 0} H_n^{(2)}(X) = \lim_{x \to 0}[J_n(X) - jH_n^{(2)}(X)]$$

$$= \frac{1}{n!} \left(\frac{X}{2}\right)^n + j\frac{(n-1)!}{\pi} \left(\frac{2}{X}\right)^n \tag{13.8}$$

R_c is on the order of centimeters, while R_v may extend to order of kilometers. Also, peak $J_c(t)$ A_c could be on the order of kiloamperes or even megamperes, while $A_v J_v$ is on the order of kiloamperes.

Observations on EMF calculations indicate relatively immense EMFs that are far above the 10 kV/m that may be experienced by a person standing under a 765-kV transmission line. However, EMFs from lightning discharge are of short duration, and aside from the severe damage that could be inflicted by direct stroke, are of little adverse effect.

13.V PROBLEMS

From the set of Equations (13.1) to (13.8), solve the following problems:

1. Simplify solutions of E_ρ, E_α, and E_τ under a relatively small value for ρ.
2. Simplify solutions of H_ρ, H_α, and H_τ under a relatively small value for ρ.
3. From simplified solutions for $\bar{E}$ fields established in problem 1, obtain numerical values for E_ρ, E_α, and E_τ for:

$$\rho \; 2 \text{ km}, \; R_{c,v} = 2 \text{ cm}, \; AJ_c = AJ_v = 1 \text{ Mamp}$$
$$\hat{Z} = j\omega\mu_o, \; \hat{y} = j\omega\epsilon_o$$

A is the discharge column cross-section.

4. From simplified solutions for $\bar{H}$ fields established in problem 2, obtain numerical values for H_ρ, H_α, and H_τ for

$$\rho = 4\text{km}$$

$$R_{c,v} = 1 \text{ cm}$$

$$AJ_c = AJ_v = 1 \text{ Mamp}$$

$$\hat{Z} = jW\mu_o, \quad \hat{Y} = jW\epsilon_o$$

5. From simplified solutions of $\overline{E}$ and $\overline{H}$ fields obtained in problems 1 and 2, obtain numerical values of all possible Poynting vectors.

6. From results secured in problem 5, obtain numerical values of all Poynting vectors for:

$$\rho = 2 \text{ km}$$

$$R_{c,v} = 2 \text{ cm}$$

$$AJ_c = AJ_v = 1 \text{ M-amp}$$

$$\hat{Z} = jW\mu_o, \quad \hat{Y} = jW\epsilon_o$$

REFERENCES

1. **Douglas, J., Malé, R., Kavet, R., Newell, G., Perhac, R., and Sagan, L.,** Electromagnetic fields and human health, *EPRI J.,* p. 14, July/August 1984.
2. **Shepard, M., Sagan, L., Black, R., Sussman, S., and Rafferty, C.,** EMF: the debate on health effects, *EPRI J.,* p. 4, October/November 1987.
3. **Moore, T., Black, R., Kheifets, L., Rafferty, C., Sagan, L., and Sussman, S.,** Pursuing the science of EMF, *EPRI J.,* p. 4, January/February 1990.

INDEX